COMPLEX VARIABLES
Harmonic and Analytic Functions

COMPLEX VARIABLES
Harmonic and Analytic Functions

FRANCIS J. FLANIGAN
San Diego State University, California

Dover Publications, Inc., New York

Published in Canada by General Publishing Company, Ltd., 30 Les-
mill Road, Don Mills, Toronto, Ontario.
Published in the United Kingdom by Constable and Company, Ltd.,
10 Orange Street, London WC2H 7EG.

This Dover edition, first published in 1983, is an unabridged and
corrected republication of the work originally published in 1972 by
Allyn and Bacon, Inc., Boston.

International Standard Book Number: 0-486-61388-7

Manufactured in the United States of America
Dover Publications, Inc., 180 Varick Street, New York, N.Y. 10014

Library of Congress Cataloging in Publication Data

Flanigan, Francis J.
  Complex variables.

  Reprint. Originally published: Boston : Allyn and Bacon, 1972.
  Bibliography: p.
  Includes index.
  1. Functions of complex variables. 2. Harmonic functions. 3.
Analytic functions. I. Title.
QA331.F62 1983   515.9   82-17732

Contents

Chapter 3   Complex Numbers and Complex Functions

SECTION

APPENDIX

Chapter 4   Integrals of Analytic Functions

SECTION

Chapter 5   Analytic Functions and Power Series

SECTION

Chapter 6   Singular Points and Laurent Series

Chapter 7   The Residue Theorem and the Argument Principle

Chapter 8   Analytic Functions as Conformal Mappings

Preface

This book originated at the University of Pennsylvania in the Spring of 1968 as a set of lecture notes addressed to undergraduate math and science majors. It is intended for an introductory one-semester or quarter-and-a-half course with minimal prerequisites; it is neither a reference nor a handbook.

We approach complex analysis via real plane calculus, Green's Theorem and the Green's identities, "determination by boundary values," harmonic functions, and steady-state temperatures. The conscientious student will compute many line integrals and directional derivatives as he works through the early chapters. The beautiful Cauchy theory for complex analytic functions is preceded by its harmonic counterpart.

The young student is likely to assume that an arbitrary differentiable function defined somewhere enjoys the remarkable properties of complex analytic functions. From the beginning we stress that

(i)   the analytic $f(z) = u(x, y) + iv(x, y)$ is much better behaved than the arbitrary function encountered in freshman calculus or the first ε, δ course;

(ii) this is because $u(x, y)$, $v(x, y)$ satisfy certain basic partial differential equations;

(iii) one can obtain much useful information about solutions of such equations without actually solving them.

In developing integration theory, we emphasize the analytic aspects at the expense of the topological or combinatorial. Thus, a complex function $f(z)$ is defined to be analytic at a point if it is continuously

complex differentiable in a neighborhood of that point. The Cauchy
Integral Theorem is thereby an easy consequence of Green's Theorem
and the Cauchy-Riemann equations. Goursat's remarkable deepening
of the Integral Theorem is discussed, but is not proved. On the other
hand, we make much of the standard techniques of representing a
function as an integral and then bounding that integral (the *ML*-
inequality) or differentiating under the integral sign. The integral
representation formulas (Green's Third Identity, the Poisson Integral,
the Cauchy Formula) are the true heroes of these chapters.

The second half of the book (Chapters 5–8) is motivated by two
concerns: the integration of functions which possess singularities,
and the behavior of analytic mappings $w = f(z)$. Power series are
developed first; thence flows the basic factorization

$$f(z) = f(z_0) + (z - z_0)^n g(z);$$

from this comes all the rest. The book concludes with a discussion (no
proof) of the Riemann Mapping Theorem.

The author recalls with pleasure many, many hours spent discussing
complex analysis with Professor Jerry Kazdan at the University of
Pennsylvania and nearby spots. Particular thanks are due Professor
Kazdan and Professor Bob Hall for reading the manuscript and making
many usable suggestions. Finally, the author is happy to record his
gratitude to the staff of Allyn and Bacon for encouragement and prompt
technical assistance over the months and miles.

FRANCIS J. FLANIGAN

COMPLEX VARIABLES
Harmonic and Analytic Functions

1

Calculus in the Plane

Section 1.1 DOMAINS IN THE xy-PLANE

1.1.0 Introduction

Here's what we'll do in the first few chapters:

1. We examine the geography of the xy-plane. Some of this will be familiar from basic calculus (for example, distance between points), some may be new to you (for example, the important notion of "domain"). We must also consider curves in the plane.

2. We consider real-valued functions $u(x, y)$ defined in the plane. We will examine the derivatives (partial derivatives, gradient, directional derivatives) and integrals (line integrals, double integrals) of these functions. Most of (1) above will be necessary for (2). All this happens in this chapter.

3. We next focus attention on a particular kind of real-valued function $u(x, y)$, the so-called harmonic function (Chapter 2). These are very interesting in their own right, have beautiful physical interpretations, and point the way to complex analytic functions.

4. At last (Chapter 3) we consider points (x, y) of the plane as complex numbers $x + iy$ and we begin our study of complex-valued functions of a complex variable. This study occupies the rest of the book.

One disadvantage of this approach is the fact that complex numbers and complex analytic functions (our chief topic) do not appear until the third chapter. Admittedly, it would be possible to move directly from step (1) to step (4), making only brief reference to real-valued functions.

On the other hand, the present route affords us

    (i) a good look at some very worthwhile two-variable real calculus, and

   (ii) an insight into the reasons behind some of the magical properties of complex analytic functions, which (as we will see) flow from (a) the natural properties of real-valued harmonic functions $u(x, y)$ and (b) the fact that we can multiply and divide points in the plane. In the present approach the influences (a) and (b) will be considered separately before being combined.

One effect we hope for: You will learn to appreciate the difference between a complex analytic function (roughly, a complex-valued function $f(z)$ having a complex derivative $f'(z)$) and the real functions $y = f(x)$ which you differentiated in calculus. Don't be deceived by the similarity of the notations $f(z)$, $f(x)$. The complex analytic function $f(z)$ turns out to be much more special, enjoying many beautiful properties not shared by the run-of-the-mill function from ordinary real calculus. The reason (see (a) above) is that $f(x)$ is merely $f(x)$, whereas the complex analytic function $f(z)$ can be written as

$$f(z) = u(x, y) + iv(x, y),$$

where $z = x + iy$ and $u(x, y)$, $v(x, y)$ are each real-valued *harmonic* functions related to each other in a very strong way: the Cauchy–Riemann equations

$$\frac{\partial u}{\partial x} = \frac{\partial v}{\partial y} \qquad \frac{\partial v}{\partial x} = -\frac{\partial u}{\partial y}.$$

In summary, the deceptively simple hypothesis that

$$f'(z) \text{ exists}$$

forces a great deal of structure on $f(z)$; moreover, this structure mirrors the structure of the harmonic $u(x, y)$ and $v(x, y)$, functions of *two* real variables.

All these comments will make more sense after you have read Chapter 4. Let us begin now at the beginning.

1.1.1 The Algebraic Structure in $\mathbb{R}^2$

Throughout these pages $\mathbb{R}$ denotes the set of all real numbers. By $\mathbb{R}^2$ (read "$\mathbb{R}$-two") we mean the set of all ordered pairs (x, y) with both

x and y in $\mathbb{R}$. "Ordered" pair means $(x, y) = (x_1, y_1)$ if and only if $x = x_1$ and $y = y_1$. We call these pairs (x, y) *points*. Some points in $\mathbb{R}^2$ are $(3, -2)$, $(1, 0)$, $(0, 1)$, $(\pi, -1)$, $(0, 0)$. It is customary to denote the typical point (x, y) by z; thus, $z = (x, y)$. We'll also use $z_0 = (x_0, y_0)$ and $\zeta = (\xi, \eta)$. Here, ζ, ξ, η are the lower-case Greek letters "zeta," "xi," "eta," respectively.

We may add and subtract points in $\mathbb{R}^2$. Thus, if $z = (x, y)$ and $\zeta = (\xi, \eta)$, we define

$$z + \zeta = (x + \xi, y + \eta),$$
$$z - \zeta = (x - \xi, y - \eta).$$

For example, if $z = (1, 2)$ and $\zeta = (3, -1)$, then

$$z + \zeta = (1 + 3, 2 + (-1)) = (4, 1),$$
$$z - \zeta = (1 - 3, 2 - (-1)) = (-2, 3).$$

We may also multiply a point of $\mathbb{R}^2$ by a number in $\mathbb{R}$. Thus, if $z = (x, y) \in \mathbb{R}$ and $c \in \mathbb{R}$, we define

$$cz = c(x, y) = (cx, cy).$$

For example, if $z = (1, 2)$ and $c = 5$, then

$$cz = 5(1, 2) = (5, 10).$$

Note how strongly the definitions of addition, subtraction, and multiplication in $\mathbb{R}^2$ depend on the corresponding operations in the real numbers themselves.

Preview

In Chapter 3, we will define the product $z\zeta$ of points z and ζ in $\mathbb{R}^2$ (note that in the product cz above, the factor c was required to be in $\mathbb{R}$). When this new product is defined, the set $\mathbb{R}^2$ will be called the *complex numbers* and thereafter denoted by $\mathbb{C}$. Although it would be easy to define the product $z\zeta$ now, we feel it is more instructive and dramatic to squeeze as much as we can from the familiar real calculus first.

Pictures

The representation of $\mathbb{R}^2$ as the xy-plane should be familiar to you. It is standard to denote the origin $(0, 0)$ by 0. This leads to no confusion, as we will see. Note that if we draw line segments from 0 to z and 0 to ζ, then the sum $z + \zeta$ is the fourth corner of the parallelogram determined by the two segments. You should examine Figure 1.1 (geometric!) and the definition of $z + \zeta$ (algebraic!) until convinced of this.

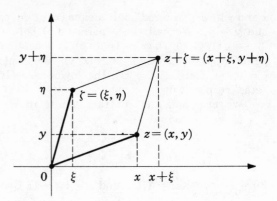

Figure 1.1

Exercises to Paragraph 1.1.1

1. Let $z = (2, -2)$, $\zeta = (-1, 5)$. Compute
   (a) $z + \zeta$,
   (b) $\zeta - z$,
   (c) $2z - \zeta$,
   (d) $z + 4\zeta$.
2. Given z, ζ as in Exercise 1, solve the following for $w = (u, v)$:
   (a) $z + 2\zeta + 3w = 0$,
   (b) $2z + w = -\zeta$.

1.1.2 The Distance Structure in $\mathbb{R}^2$

Let $z = (x, y) \in \mathbb{R}^2$. We define the *norm* (or *length, modulus, absolute value*) of z (denoted $|z|$) by

$$|z| = \sqrt{x^2 + y^2}.$$

For example, if $z = (1, -2)$, then $|z| = \sqrt{1 + 4} = \sqrt{5}$.

Note that the norm is a nonnegative real number (square root!), $|z| \geq 0$, and, in fact, $|z| = 0$ if and only if $z = 0 \ (= (0, 0))$. The definition of $|z|$ agrees with the famous Pythagorean theorem for right triangles,

$$|z|^2 = x^2 + y^2,$$

as Figure 1.2 shows.

Now we use the above to define the *distance* between $z = (x, y)$ and $z_0 = (x_0, y_0)$ as the norm of their difference $z - z_0$; that is, distance from z to $z_0 = |z - z_0|$. See Figure 1.3. Since $z - z_0 = (x - x_0, y - y_0)$, we have the formula

$$|z - z_0| = \sqrt{(x - x_0)^2 + (y - y_0)^2}.$$

Figure 1.2

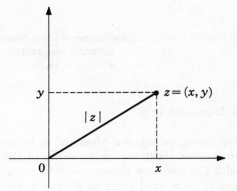

Figure 1.3

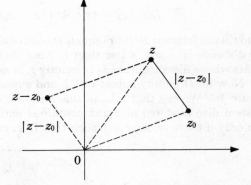

For example, if $z = (1, -2)$ and $z_0 = (2, 5)$, then the distance from z to z_0 is

$$|z - z_0| = \sqrt{(1 - 2)^2 + (-2 - 5)^2} = \sqrt{1 + 49} = \sqrt{50}.$$

Exercises to Paragraph 1.1.2

1. Let $z = (-1, 4)$, $z_0 = (2, 2)$. Compute:
   (a) $|z|$,
   (b) $|z_0|$,
   (c) $|z - z_0|$,
   (d) $|z_0 - z|$.
2. Compute the distance from z_0 to z, with z, z_0 as in Exercise 1.
3. Sketch in the plane the sets of points z determined by each of the following conditions. Here, $z_0 = (1, 1)$.
   (a) $|z| = 1$.
   (b) $|z| < 1$.

(c) $|z - z_0| = 1$.

(d) $|z - z_0| \geq 1$.

4. Establish the following useful inequalities. Sketch!

 (a) $|z + \zeta| \leq |z| + |\zeta|$ (triangle inequality).

 (b) $|x| \leq |z|, |y| \leq |z|$ where $z = (x, y)$.

1.1.3 Domains in $\mathbb{R}^2$

We intend to define a "domain" to be an open connected subset of $\mathbb{R}^2$. Hence, we must first make sense of the words "open" and "connected." The notion of distance will be crucial for this.

Let z_0 be a fixed point in $\mathbb{R}^2$ and $r > 0$ a given positive number. We denote by $D(z_0; r)$ the *disc of radius r centered at z_0*, defined as the set

$$D(z_0; r) = \{z \in \mathbb{R}^2 \mid |z - z_0| < r\}.$$

Read this as follows: $D(z_0; r)$ equals the set of all points z in $\mathbb{R}^2$ such that the distance $|z - z_0|$ is less than r. Note that its "rim," the circle of points whose distance from z_0 is exactly r, is *not* included in $D(z_0; r)$.

Now let Ω be any subset of $\mathbb{R}^2$, and suppose $z \in \Omega$ (z is in Ω). See Figure 1.4. We say that z is an *interior point* of Ω if and only if there exists a disc centered at z and contained entirely inside Ω; that is, if and only if there exists $r > 0$ such that $D(z; r) \subset \Omega$.

Figure   1.4

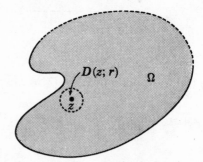

Examples of Interior Points

1. Let $\Omega = \mathbb{R}^2$. Then every point z is an interior point of $\mathbb{R}^2$ because any disc around z will surely be contained in $\mathbb{R}^2$.

2. Let Ω be itself a disc, $\Omega = D(z_0; r)$. If z is any point of $D(z_0; r)$, then we may find a smaller disc $D(z; r_1)$ contained inside $D(z_0; r)$; see Figure 1.5. Thus, every point of $D(z_0; r)$ is an interior point of $D(z_0; r)$.

3. This time let Ω consist of a disc $D(z_0; r)$ together with its "rim," the set of points z satisfying $|z - z_0| = r$. Now not all points of Ω are interior points. More precisely, the points of $D(z_0; r)$ are interior points

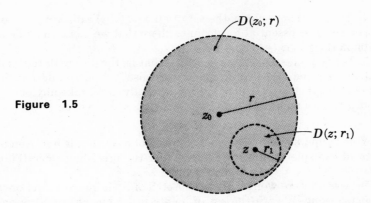

Figure   1.5

of Ω (why?), but no point z on the rim of Ω is an interior point because we cannot surround such a point (on the borderline) with a disc that fits inside Ω. Perfectly reasonable. See Figure 1.6.

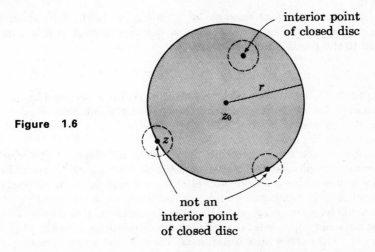

Figure   1.6

At last we may define "open set." A subset Ω of $\mathbb{R}^2$ is *open* if and only if every point of Ω is an interior point of Ω. Thus, every point of an open set is "well inside" the set; none of its points are on its boundary.

Examples of Open Sets

1. The entire plane $\mathbb{R}^2$ is open.
2. The empty set $\varnothing$, the set with no points, is open. Since there are no points, we needn't worry about discs around them.
3. Each disc $D(z_0; r)$ is open. From now on we may refer to these as "open discs," to distinguish them from discs with rims included ("closed discs").

4. The upper half-plane, the set of all $z = (x, y)$ satisfying $y > 0$, is an open set. It is essential for openness here that we do not include any points on the x-axis ($y = 0$).

5. Let Ω be the set $D(z_0; r) - \{z_0\}$, that is, the disc with the center point z_0 removed (called the "punctured disc"). You should convince yourself that this set Ω is open. More generally, if we take any open set S and remove a point z, the new set $S - \{z\}$ is again open.

You should be able to give an example of a set that is not open (see the third example of interior points under the preceding topic (Figure 1.6)).

In case you are wondering, a subset S of $\mathbb{R}^2$ is *closed* if and only if its complement $\mathbb{R}^2 - S$ (the set of points in $\mathbb{R}^2$ but not in S) is open. One example of a closed set is $\mathbb{R}^2$, since its complement $\mathbb{R}^2 - \mathbb{R}^2$ is the empty set $\varnothing$, which is open. Similarly, the empty set is closed (why?). It can be shown that the only subsets of $\mathbb{R}^2$ which are both open and closed are $\mathbb{R}^2$ and the empty set $\varnothing$. Some sets are neither open nor closed.

Another important example of a closed set is the following: Let $\overline{D}(z_0; r)$ denote the set of points whose distance from z_0 is less than or equal to the positive number r; that is,

$$\overline{D}(z_0; r) = \{z \in \mathbb{R}^2 \mid |z - z_0| \leq r\}.$$

This is the "closed" disc (rim included) of radius r centered at z_0. We discussed this set in Example 3 under interior points. See Figure 1.6. You should convince yourself that it *is* closed in the sense that its complement $\mathbb{R}^2 - \overline{D}(z_0; r)$ is open.

Now we continue our definition of "domain." We must define what we mean for an open set to be connected. Actually, it is possible to define connectedness for any subset of $\mathbb{R}^2$, not just the open subsets, but this takes more work and would be superfluous at the moment.

Let Ω be an open subset of $\mathbb{R}^2$; Ω is *disconnected* if and only if there exist nonempty open sets Ω_1, Ω_2 which are disjoint (no points in common, $\Omega_1 \cap \Omega_2 = \varnothing$) and whose union is Ω. Thus, a disconnected open set Ω may be decomposed into two "smaller," nonoverlapping, nonempty, open sets. An open set Ω is *connected* if and only if it is not disconnected. See Figure 1.7.

Challenge

You might try to prove that an open set Ω is connected if and only if any two points in Ω may be linked by a path made of a finite number of straight-line segments lying entirely in Ω.

Thus, there are two equivalent notions of connectedness for an open set Ω: (1) it can't be broken into disjoint open pieces, or (2) any pair of points may be linked by a path in Ω. Both are reasonable.

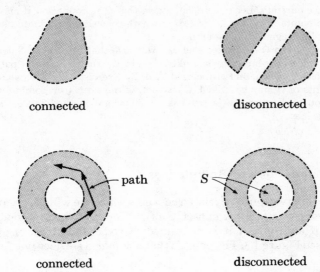

connected   disconnected

connected   disconnected

Figure 1.7

At last we make our basic definition. A subset Ω of $\mathbb{R}^2$ is a *domain* if and only if it is open and connected. The domains we will encounter most frequently are $\mathbb{R}^2$ itself, the open discs $D(z_0; r)$, the punctured discs, and the upper half-plane (see the fourth example under the preceding topic). Not all domains, of course, are quite so symmetric as these; this is fortunate or unfortunate, depending on your point of view.

Exercises to Paragraph 1.1.3

1. (a) Sketch the set S of points $z = (x, y)$ satisfying $x \geq 0$.
   (b) Verify that the subset of interior points of S is determined by the condition $x > 0$.
   (c) Is the subset in (b) a domain?

2. (a) Sketch the "annulus" $\Omega = \{z \mid 1 < |z| < 2\}$.
   (b) There is a hole in Ω. Is Ω connected?
   (c) Verify that Ω is a domain.

3. Rather than designate one of the standard domains by a capital letter, we often speak of "the unit disc $|z| < 1$," "the punctured disc $0 < |z| < 1$," and so on. Sketch the sets determined by each of the following conditions and decide which are domains. Here, z_0 is an arbitrary but fixed point.
   (a) $|z| > 1$.
   (b) $1 \leq |z| \leq 2$.
   (c) $|z - z_0| < 1$.
   (d) $|z - z_0| \leq 2$.

4. Let Ω be a domain and let S be a nonempty subset of Ω satisfying (i) S is open, (ii) its complement $\Omega - S$ is open (sometimes stated as S is *closed in* Ω). Prove $S = \Omega$; that is, $\Omega - S$ is empty.

5. *Challenge question.* Let Ω be open. Prove that Ω is connected if and only if any two points in Ω may be linked by a path consisting of a finite number of straight-line segments lying entirely in Ω.

   Hint: Given Ω connected and $z_0 \in \Omega$, prove that the subset S consisting of all points of Ω which may be linked to z_0 by the specified type of path is not empty (clear!), open, and also closed in Ω. By Exercise 4, $S = \Omega$, whence any two points of Ω may be linked. Constructing the set S (the points for which what you wish to prove is true) is an important method in dealing with connectedness.

6. Is a disc a circle?

1.1.4 Boundaries and Boundedness

These are two quite unrelated concepts which we will use frequently.

Boundedness first. A subset S of $\mathbb{R}^2$ (open or not) is said to be *bounded* if and only if it is contained in some disc $D(z_0; r)$ of finite radius r; see Figure 1.8. The point is that a bounded set does not "escape"

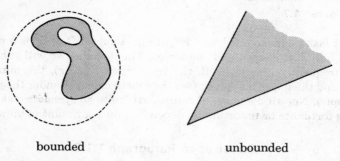

bounded   unbounded

Figure   1.8

to infinity. Examples of bounded sets are any open disc, any closed disc, a single point, any finite set of points, and (of course) the empty set $\varnothing$. The plane $\mathbb{R}^2$, the upper half-plane, the x-axis are each unbounded.

The bounded domains (the open disc again) are an important sub-family of the family of all domains.

Now let us discuss the notion of the boundary of a set. Let S be any subset of $\mathbb{R}^2$. A point z of $\mathbb{R}^2$ is a *boundary point* of S if and only if every open disc $D(z; r)$ centered at z contains some points in S and also some points not in S. Note that we do not require a boundary point of S to be an element of S. In fact, if S is itself an open disc, then its boundary points are precisely those on the rim of the disc, and none of these is a member of the disc.

It is worthwhile noting that, given a set S, a point z is a boundary point of S if and only if it is a boundary point of $\mathbb{R}^2 - S$, the complement of S. A brief meditation should convince you that this is a reasonable property for a boundary point to possess.

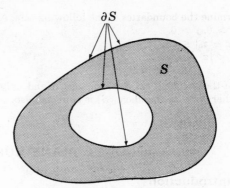

Figure   1.9

Finally, we define the *boundary* (or *frontier*) of a set S to be the collections of all boundary points of S. The boundary of S is denoted ∂S. See Figure 1.9.

Examples of Boundaries

1. Let $S = D(z_0; r)$, an open disc as usual. Then ∂S is the circle of points z satisfying $|z - z_0| = r$.
2. If $\overline{S} = \overline{D}(z_0; r)$, the closed disc, then $\partial \overline{S}$ is the same set as ∂S in Example 1, namely, the circle of radius r centered at z_0.
3. If Ω is the punctured disc $D(z_0; r) - \{z_0\}$, then $\partial \Omega$ consists of the circle $|z - z_0| = r$ together with the point z_0.
4. The boundary of $\mathbb{R}^2$ is empty.
5. The boundary of the upper half-plane $(y > 0)$ is the x-axis $(y = 0)$. This is an example of an unbounded set with a nonempty boundary. Don't confuse the two notions.

Preview

Our model of a nice bounded domain is the open disc $D(z_0; r)$. Its boundary is a very nice curve, a circle. In Section 1.2 we continue our study of domains by studying curves in the plane. In our applications, these curves will almost always arise as the boundaries of certain domains. Once we have completed our study of these boundary curves, we will begin at last to discuss the functions that live on our domains.

Exercises to Paragraph 1.1.4

1. Which of the following sets are bounded?
    (a) $|z| \geq 1$.
    (b) A subset of a bounded set.
    (c) $0 < |z - z_0| < 1$.
    (d) The graph of $y = \sin x$.

2. Determine the boundaries of the following sets. As usual, $z = (x, y)$.
    (a) $x > 0, y > 0$.
    (b) $|z - z_0| \le 2$.
    (c) $0 < |z - z_0| < 2$.
    (d) $0 < x < 1, y$ arbitrary.
3. Prove that a plane set S is bounded if and only if its closure $\bar{S}$ (that is, S together with its boundary ∂S) is bounded also.

Section 1.2 PLANE CURVES

1.2.0 Introduction

Curves—we know them when we see them, and yet to get an adequate terminology is surprisingly troublesome. First, therefore, let us look at the most important example, the circle, in some detail. This should make us more willing to accept the technical definitions to follow.

We let $0 = (0, 0)$ denote the origin of $\mathbb{R}^2$ as usual, and $C = C(0; r)$ be the circle of radius $r > 0$ centered at the origin, that is, the set of points z satisfying $|z| = r$. So far, C is a "static" set of points. Now we "parametrize" C as follows:

Let $[0, 2\pi]$ denote the interval of real numbers t satisfying $0 \le t \le 2\pi$. Let $\alpha = \alpha(t)$ be the function that assigns to each t in $[0, 2\pi]$ the point $\alpha(t)$ of $\mathbb{R}^2$, given by

$$\alpha(t) = (x, y) = (r \cos t, r \sin t).$$

We note first that $|\alpha(t)| = r$ (recall $\sin^2 t + \cos^2 t = 1$) so that each point $\alpha(t)$ does in fact lie on the circle $C(0; r)$. We indicate this last statement briefly by writing

$$\alpha: [0, 2\pi] \to C(0; r).$$

We can say even more. As the real number t increases from $t = 0$ to $t = 2\pi$, the point $\alpha(t)$ travels once around the circle in a counterclockwise direction. See Figure 1.10. Note for instance that $\alpha(0) = (r, 0)$, the starting point, and then

$$\alpha\left(\frac{\pi}{2}\right) = (0, r), \qquad \alpha(\pi) = (-r, 0),$$

$$\alpha\left(\frac{3\pi}{2}\right) = (0, -r), \qquad \alpha(2\pi) = (r, 0),$$

and we're back where we started. Note also that the number t is the angle (in radians) between $\alpha(t)$ and the x-axis.

Now things are no longer static. The function α has imposed a direction of travel (counterclockwise) around the circle. We emphasize this by differentiating α with respect to its variable t. Let us write

$$\alpha(t) = (\alpha_1(t), \alpha_2(t))$$

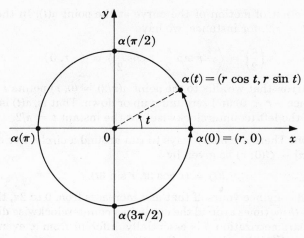

Figure 1.10

so that $\alpha_1(t) = r \cos t$, $\alpha_2(t) = r \sin t$ (the "coordinate functions" for α). We differentiate $\alpha(t)$ by differentiating its coordinates:

$$\alpha'(t) = \langle \alpha'_1(t), \alpha'_2(t) \rangle;$$

that is,

$$\alpha'(t) = \langle -r \sin t, r \cos t \rangle.$$

The pointed brackets here remind us that $\alpha'(t)$ is to be regarded as a vector (arrow), the "velocity vector" of α. This is usually depicted with tail end at the point $\alpha(t)$; see Figure 1.11. The velocity vector points

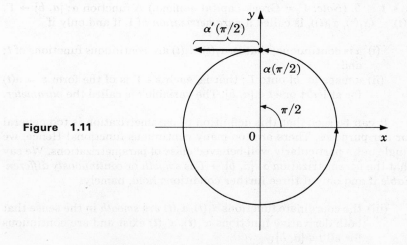

Figure 1.11

in the direction of motion of the curve at the point $\alpha(t)$. In the particular case $t = \pi/2$, for instance, we have

$$\alpha'\left(\frac{\pi}{2}\right) = \left\langle -r \sin \frac{\pi}{2}, r \cos \frac{\pi}{2} \right\rangle = \langle -r, 0 \rangle.$$

Thus the arrow that we affix to the point $\alpha(\pi/2) = (0, r)$ points r units to the *left* (since $-r < 0$) and zero units up or down. That is, $\alpha(t)$ is moving directly to the left (counterclockwise!) at the instant $t = \pi/2$.

Caution: There are many ways to run around a circle. For example, let $\beta: [0, 2\pi] \to C(0; r)$ be given by

$$\beta(t) = (r \cos 3t, r \sin 3t).$$

You should convince yourself that as t increases from 0 to 2π, the point $\beta(t)$ travels *three* times around the circle in a counterclockwise direction. This new parametrization β is essentially different from α, even though the point set $C(0; r)$ is the same in both cases.

Here is a *clockwise* parametrization of $C(0; r)$. Let $\gamma: [0, 1] \to C(0; r)$ be given by

$$\gamma(t) = (r \sin 2\pi t, r \cos 2\pi t).$$

Then $\gamma(0) = (0, r)$ is the "starting point" in this parametrization. By locating the points $\gamma(\frac{1}{4})$, $\gamma(\frac{1}{2})$, $\gamma(\frac{3}{4})$, $\gamma(1)$, you should convince yourself that $\gamma(t)$ traverses $C(0; r)$ once in a *clockwise* manner. Note also that γ was defined on the interval $[0, 1]$, not on $[0, 2\pi]$.

1.2.1 Piecewise-smooth Curves

Now we make our definitions in the spirit of the preceding examples.
Let Γ be a subset of $\mathbb{R}^2$ and $[a, b]$ an interval of real numbers t, $a \le t \le b$. (*Note:* Γ = Greek capital gamma.) A function $\alpha: [a, b] \to \Gamma$, $\alpha(t) = (\alpha_1(t), \alpha_2(t))$, is called a *parametrization* of Γ if and only if

(i) α is continuous; that is, $\alpha_1(t)$, $\alpha_2(t)$ are continuous functions of t; and
(ii) α maps $[a, b]$ onto Γ; that is, each $z \in \Gamma$ is of the form $z = \alpha(t)$ for at least one $t \in [a, b]$. The variable t is called the *parameter*.

It can be seen that this definition of parametrization is too general for our purposes. There are too many continuous functions! Hence, we single out a particularly well-behaved class of parametrizations. We say that the parametrization $\alpha: [a, b] \to \Gamma$ is *smooth* or *continuously differentiable* if and only if three further conditions hold, namely:

(iii) the coordinate functions $\alpha_1(t)$, $\alpha_2(t)$ are *smooth* in the sense that both derivative functions $\alpha'_1(t)$, $\alpha'_2(t)$ exist and are continuous for all $t \in [a, b]$;

(iv) for each $t \in [a, b]$, the *velocity vector* $\alpha'(t)$, defined as $\langle \alpha'_1(t),$ $\alpha'_2(t) \rangle$, is different from the zero vector $\langle 0, 0 \rangle$;

(v) if, moreover, $\alpha(a) = \alpha(b)$ (the curve is a closed loop), then $\alpha'(a) = \alpha'(b)$ as well.

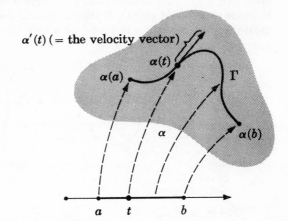

Figure   1.12

Let us discuss this definition; see Figure 1.12. Condition (iii) assures us that the velocity vector exists and depends continuously on the parameter t. We remark also that, by the derivatives $\alpha'_1(a)$, $\alpha'_2(a)$, $\alpha'_1(b)$, $\alpha'_2(b)$ at the end points $t = a$ and $t = b$, we mean one-sided derivatives only. For instance,

$$\alpha_1(a) = \lim_{t \to a} \frac{\alpha_1(t) - \alpha_1(a)}{t - a} \qquad (t > a),$$

$$\alpha'_1(b) = \lim_{t \to b} \frac{\alpha_1(t) - \alpha_1(b)}{t - b} \qquad (t < b).$$

Condition (iv) may be interpreted as follows: If we regard $\alpha(t)$ as a point moving along Γ, then its instantaneous direction is pointed out by the velocity (or tangent) vector $\alpha'(t)$. The vanishing of this vector—say, $\alpha'(t_1) = \langle 0, 0 \rangle$—would mean that the moving point $\alpha(t)$ stops when $t = t_1$. It simplifies things greatly if we rule out this possibility.

We saw above three examples of smooth parametrizations of the circle. Note that each of these examples satisfies condition (iii) because sines and cosines can be differentiated again and again. How would you prove that the parametrization $\alpha(t) = (r \cos t, r \sin t)$ given in Paragraph 1.2.0 satisfies condition (iv)? *Hint:* Use the Pythagorean theorem $\sin^2 t + \cos^2 t = 1$ to derive a smashing contradiction from the assumption $\alpha'(t) = \langle 0, 0 \rangle$.

The most important plane curves for our purposes are the circle and straight line. You will find these treated at some length in the exercises.

Here are some more useful notions. The parametrization α is *simple* if and only if the function α restricted to the "open" interval (a, b)—that is, for t satisfying $a < t < b$—is one-to-one. In other words, if t_1 and t_2 are strictly between a and b and if $\alpha(t_1) = \alpha(t_2)$, then $t_1 = t_2$. Geometrically, this means that the curve doesn't cross itself, except perhaps at the end points. If the end points are equal (that is, if $\alpha(a) = \alpha(b)$) then we say that α is *closed*, or a *loop*. See Figure 1.13.

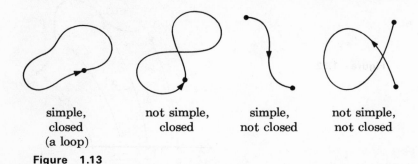

| simple,
closed
(a loop) | not simple,
closed | simple,
not closed | not simple,
not closed |

Figure   1.13

The parametrization $\alpha(t) = (r \cos t, r \sin t)$ for $0 \le t \le 2\pi$ is a simple closed smooth parametrization of the circle $C(0; r)$. On the other hand, $\beta(t) = (r \cos 3t, r \sin 3t)$ for $0 \le t \le 2\pi$ is not simple because each point on the circle corresponds to three values of t (except for the starting point $(r, 0)$ which corresponds to four values).

Actually, smooth parametrizations are not quite general enough. We wish to allow curves with a finite number of "corners" such as triangles and rectangles. At a corner, of course, we would not expect a *unique* direction or velocity vector. Again let $\alpha\colon [a, b] \to \Gamma$ be a parametrization, $\alpha(t) = (\alpha_1(t), \alpha_2(t))$. We say that α is a *piecewise-smooth parametrization* of Γ if and only if there exists a finite set of values $a = a_0 < a_1 < a_2 < \cdots < a_n = b$ such that the function α restricted to the intervals $[a_0, a_1], [a_1, a_2], \ldots, [a_{n-1}, a_n]$ gives in each case a smooth parametrization of the subsets $\Gamma_0, \Gamma_1, \ldots, \Gamma_{n-1}$ of Γ defined by $\Gamma_k = \{\alpha(t) \mid t \in [a_k, a_{k+1}]\}$.

Thus, a piecewise-smooth parametrization is one built up from smooth parametrizations joined end to end. In particular, a smooth parametrization is piecewise-smooth (let $n = 1$ in the definition above).

Example

We will parametrize the right triangle Γ with vertices at the points $(0, 0)$, $(0, 1)$, $(1, 1)$; see Figure 1.14. It is simply a matter of building α in three parts. Thus, let $\alpha\colon [0, 3] \to \Gamma$ as follows:

$$\alpha(t) = \begin{cases} (t, 0) & \text{for } t \in [0, 1] \\ (1, t - 1) & \text{for } t \in [1, 2] \\ (3 - t, 3 - t) & \text{for } t \in [2, 3] \end{cases}$$

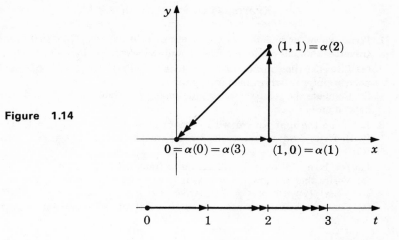

Figure 1.14

You should let t increase from 0 to 3 and check that $\alpha(t)$ travels once around the triangle in a counterclockwise direction. Now we examine a corner point, say, $\alpha(2) = (1, 1)$. By differentiating $\alpha(t) = (1, t - 1)$, we see that the velocity vector $\alpha'(2^-)$ in which t approaches 2 from the left $(t < 2)$ is given by $\alpha'(2^-) = \langle 0, 1 \rangle$. This points upward; the point $\alpha(t)$ is climbing in the y-direction as t increases from $t = 1$ to $t = 2$. Likewise, by differentiating $\alpha(t) = (3 - t, 3 - t)$, we compute $\alpha'(2^+) = \langle -1, -1 \rangle$. This vector, with tail end fixed at the corner $\alpha(2) = (1, 1)$, points toward the origin $(0, 0)$ as expected.

Comments

1. We have not yet defined "curve." Let's attend to this now. Suppose $\alpha: [a, b] \rightarrow \Gamma$ is a piecewise-smooth parametrization of the set Γ. Then the pair (Γ, α) is termed a *piecewise-smooth curve*. However, it is common practice to speak of "the curve Γ," omitting mention of α, or "the curve $z = \alpha(t)$." We will use both terms.

2. We are claiming that the set Γ is a curve if it has a piecewise-smooth parametrization (roughly, a velocity vector). This corresponds to our intuition. Things are more delicate than you may imagine, however. It is possible to find a continuous $\alpha: [a, b] \rightarrow \Gamma$, $\alpha(t) = (\alpha_1(t), \alpha_2(t))$, where Γ is the two-dimensional unit square! A space-filling curve! An example was given by the mathematician Peano in the nineteenth century. In Peano's example, however, the coordinates $\alpha_1(t)$, $\alpha_2(t)$ were not differentiable functions. Of course most of the familiar functions of calculus are infinitely differentiable. (Can you think of any that are not?)

3. Some authors omit mention of the function $\alpha(t)$ and write simply "Let $x = \alpha_1(t)$, $y = \alpha_2(t)$ give a curve...."

Exercises to Paragraph 1.2.1

1. Write down a smooth parametrization $\alpha: [0, 2\pi] \to C(0, 1)$ such that $\alpha(t)$ traverses the unit circle five times counterclockwise as t varies from 0 to 2π.

2. (a) Describe the journey of $\beta(t) = (\cos kt, \sin kt)$ with $0 \le t \le 2\pi$ and k a nonzero integer (which may be negative).
   (b) Compute the velocity vector $\beta'(t)$ and observe how its length and direction depend on k.

3. Let Γ be the unit square with corners $z_0 = 0$, $z_1 = (1, 0)$, $z_2 = (1, 1)$, $z_3 = (0, 1)$. Write down a counterclockwise parametrization $\gamma: [0, 4] \to \Gamma$ such that $\gamma(0) = z_0$.

4. Let $f(x)$ be a continuously differentiable function, $a \le x \le b$.
   (a) Verify that the graph $y = f(x)$ is parametrized smoothly by the mapping $\alpha(x) = (x, f(x))$, $a \le x \le b$. Compare $x = t$.
   (b) Compute the velocity vector $\alpha'(x)$. How is it related to the slope of the graph $y = f(x)$?

1.2.2 Length of a Curve

Let $\alpha: [a, b] \to \Gamma$ be a piecewise-smooth parametrization of the curve Γ. As t varies from a to b, what distance does the point $\alpha(t)$ cover? The following definition offers a quick answer to this question—and one that is reasonable as well. We define

$$\text{length } (\alpha) = \int_a^b |\alpha'(t)| \, dt.$$

We remark first that since the coordinate functions $\alpha_1(t)$ and $\alpha_2(t)$ are piecewise-smooth, the integrand

$$|\alpha'(t)| = \{(\alpha'_1(t))^2 + (\alpha'_2(t))^2\}^{1/2}$$

is bounded and piecewise-continuous, and so the integral exists as a finite positive number; this is a standard result of calculus.

Why does this integral deserve to be called a length? In answer, we note first that the integral definition agrees with the usual distance formula in the case of parametrized line segments. For example, if Γ is the line segment from z_0 to z_1 in $\mathbb{R}^2$, then the function $\alpha: [0, 1] \to \Gamma$ given by

$$\alpha(t) = z_0 + t(z_1 - z_0)$$

parametrizes a journey from $z_0 = \alpha(0)$ to $z_1 = \alpha(1)$. It is easy to see that $\alpha'(t) = z_1 - z_0$ so that

$$\text{length } (\alpha) = \int_0^1 |z_1 - z_0| \, dt = |z_1 - z_0|$$

as desired. See Figure 1.15.

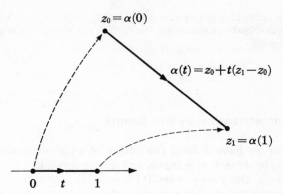

Figure 1.15

A second justification of our definition of length is this: The quantity $|\alpha'(t)|$, being the length of the velocity vector, gives the instantaneous speed of the moving point $\alpha(t)$, provided we interpret the parameter t as time. Thus, $|\alpha'(t)| \, dt$ is the product of instantaneous speed by the differential of time and (we recall happily) distance = speed × time.

It is also possible to define length (α) using polygonal approximations and taking limits. Our present integral definition becomes a theorem in such an approach. The approximation method makes better geometric sense, perhaps, but requires more work.

Example

Let $\alpha: [0, 2\pi] \to C(0; r)$, $\alpha(t) = (r \cos t, r \sin t)$, be the standard parametrization of the circle of radius r. It is easy to see that $|\alpha'(t)| = r$ (Pythagoras!), whence

$$\text{length } (\alpha) = r \int_0^{2\pi} dt = 2\pi r,$$

as expected.

If, instead, we parametrize $C(0; r)$ by the function $\beta(t) = (r \cos 3t, r \sin 3t)$, we get $\beta'(t) = \langle -3r \sin 3t, 3r \cos 3t \rangle$, so that $|\beta'(t)| = 3r$ and length $(\beta) = 6\pi r$. This underscores the fact that the moving point $\beta(t)$ travels around $C(0; r)$ three times. That is, our definition of length via the integral gives actual distance traveled (compare the mileage gauge in an automobile).

Exercises to Paragraph 1.2.2

1. Compute length (α), where $\alpha: [0, 3\pi/2] \to C(0; R)$ is given by $\alpha(t) = (R \cos 2t, R \sin 2t)$.

2. Compute the length of the graph of $y = f(x) = x^2, 0 \le x \le 1$.

3. Suppose a curve is given by $x = x(t)$, $y = y(t)$, with $a \leq t \leq b$. Write $\dot{x}$, $\dot{y}$ for dx/dt, dy/dt (Newton's notation). Verify that the length of the curve is given by the formula

$$\int_a^b \sqrt{\dot{x}(t)^2 + \dot{y}(t)^2} \, dt.$$

1.2.3 Parametrization by Arc Length

Thus far we have defined the notion of a piecewise-smooth parametrization $\alpha : [a, b] \to \Gamma$ of a curve, and have a formula for total distance traveled along the curve, namely, length $(\alpha) = \int_a^b |\alpha'(t)| \, dt$. We refine this slightly as follows: If τ is any value in $[a, b]$, then the distance traveled by the point $\alpha(t)$ as t varies from $t = a$ to $t = \tau$ is given by

$$\int_a^\tau |\alpha'(t)| \, dt.$$

Thus we get the distance traveled by integrating with respect to the parameter (variable) t.

Now we are going to improve this situation. We will construct a new parametrization of Γ which is "equivalent" to the original parametrization (we will make this precise in a moment), but which is more natural in that it takes into account the geometry of Γ. More precisely, this new parametrization (we call it $\sigma = \sigma(s)$) will have the property that the distance traveled by the moving point $\sigma(s)$ between two points $\sigma(s_1)$ and $\sigma(s_2)$ is equal to $s_2 - s_1$. That is, the function σ, considered as a mapping from an interval of the s-axis onto Γ, preserves distances or "arc length." See Figure 1.16.

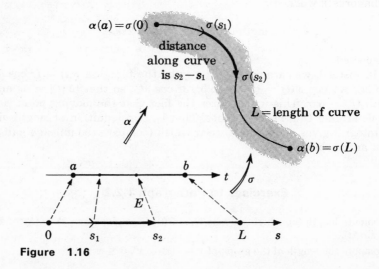

Figure 1.16

First some terminology. Let $[a_1, b_1]$ and $[a_2, b_2]$ be intervals on the s- and t-axes, respectively. A function $E\colon [a_1, b_1] \to [a_2, b_2]$ is an *equivalence* if and only if it is one-to-one, onto, increasing, continuous, and piecewise differentiable with an everywhere positive first derivative. This means that $E(a_1) = a_2$, $E(b_1) = b_2$, and as s increases from a_1 to b_1, $t = E(s)$ increases from a_2 to b_2.

Now we obtain a result to be used continually.

THEOREM 1

Let $\alpha\colon [a, b] \to \Gamma$ be a piecewise-smooth parametrization of the curve, and let length $(\alpha) = L > 0$. Then there exists a unique piecewise-smooth parametrization $\sigma\colon [0, L] \to \Gamma$ equivalent to α (in the sense that $\sigma(s) = \alpha(E(s))$ where $E\colon [0, L] \to [a, b]$ is an equivalence) which satisfies

(i) *the tangent (velocity) vector $\sigma'(s)$ has unit length, $|\sigma'(s)| = 1$;*
(ii) *the distance traveled along Γ from $\sigma(0)$ to $\sigma(s)$ is exactly s.*

Note: Here the parameter s is called *arc length*. The unit tangent vector $\sigma'(s)$ will be frequently denoted $T(s)$.

Proof: (a) It is clear that if σ exists, then it is unique.

(b) Also, if a function σ exists satisfying (i), then it is not hard to see that it satisfies (ii) as well, for the distance from $\sigma(0)$ to $\sigma(u)$ is given by

$$\int_0^u |\sigma'(s)|\, ds = \int_0^u ds = u,$$

since $|\sigma'(s)| = 1$. Thus, it suffices to construct σ satisfying (i).

(c) It is sufficient to prove the result for *smooth* parametrizations α, for if we obtain an arc length parametrization on each smooth section of the curve Γ, then it is not hard to see that we may fit these together "end to end" to obtain a *piecewise*-smooth parametrization of the full curve Γ. You should convince yourself of this.

Thus, we assume that α is smooth for the remainder of this proof.

(d) Now we get $E(s)$. Define $F(\tau) = \int_a^\tau |\alpha'(t)|\, dt$ for $\tau \in [a, b]$. We have $F(a) = 0$, and as τ increases from a to b, $F(\tau)$ increases from 0 to $F(b) = L$. By the Fundamental Theorem of Calculus, $F'(\tau) = |\alpha'(\tau)|$ and hence is positive and continuous. Thus, $F\colon [a, b] \to [0, L]$ is an equivalence (though it goes in the wrong direction!).

Let $E\colon [0, L] \to [a, b]$ be the inverse function of F. That is, $E(F(\tau)) = \tau$, $F(E(s)) = s$. You should convince yourself that $E'(s) = 1/F'(E(s))$ (just differentiate the relation $F(E(s)) = s$, using the Chain Rule). We may now conclude that $E'(s) > 0$, and hence E is an equivalence.

(e) As mentioned above, we next define $\sigma(s) = \alpha(E(s))$; that is, $\sigma(s) = (\alpha_1(E(s)), \alpha_2(E(s)))$. It is straightforward to check that σ is also a piecewise-smooth parametrization.

(f) We have $\sigma'(s) = \langle \alpha'_1(E(s))E'(s), \alpha'_2(E(s))E'(s) \rangle$ by the Chain Rule. This implies $|\sigma'(s)| = |E'(s)| \, |\alpha'(E(s))|$. But

$$E'(s) = \frac{1}{|\alpha'(t)|}$$

(see above) and $E(s) = t$. Hence, $|\sigma'(s)| = 1$. The theorem is proved.

Comments

1. Note how often we used $|\alpha'(t)| > 0$ in the proof.
2. This result is absolutely basic, both in function theory and in the differential geometry of curves. The reason is that arc length s is a "natural" or "intrinsic" parameter, coming from the geometry of the curve itself. Hence, the remarkable fact that the tangent vector $\sigma'(s)$ always has unit length.

Parametrizing the Circle by Arc Length

Let $\alpha: [0, 2\pi] \to C(0; r)$, $\alpha(t) = (r \cos t, r \sin t)$ as usual. We have seen that $|\alpha'(t)| = r$ and $L = \text{length }(\alpha) = 2\pi r$. As in the proof above, we define

$$s = F(\tau) = \int_0^\tau |\alpha'(t)| \, dt = \int_0^\tau r \, dt = r\tau.$$

Thus, the inverse E of F is given by $E(s) = \tau = s/r$, and finally,

$$\sigma(s) = \alpha(E(s)) = \alpha\left(\frac{s}{r}\right) = \left(r \cos \frac{s}{r}, r \sin \frac{s}{r}\right).$$

By differentiating $\sigma(s)$ with respect to s, you may verify that $|\sigma'(s)| = 1$.

Moral

We may assume that our curves are parametrized by arc length s and have *unit* tangent vectors $T(s)$. This is convenient for theoretical work. Admittedly, for a particular curve (the circle, say), other parametrizations may prove even more useful than arc length.

Exercises to Paragraph 1.2.3

1. (a) Parametrize the semicircle $|z| = R$, $y \geq 0$ by the usual counterclockwise angle θ measured (in radians) from the positive x-axis.
   (b) Reparametrize this semicircle by arc length in an equivalent way. Write down the equivalence function E.
   (c) Compute the velocity vectors in each case and compare their lengths.
2. Given a parametrization $\alpha: [a, b] \to \Gamma$ such that $|\alpha'(t)| = k$ is constant for $a \leq t \leq b$.
   (a) Compute $L = \text{length }(\alpha)$ in terms of the given information.

(b) Write down an equivalence $E: [0, L] \to [a, b]$ such that $\sigma(s) = \alpha(E(s))$ is the arc length parametrization.

3. Parametrize the straight-line segment Γ from z_0 to z_1 by arc length, as follows (here, $z_0 \neq z_1$):

   (a) Verify that $\zeta = (z_1 - z_0)/|z_1 - z_0|$ has unit length.

   (b) Verify that $\sigma: [0, L] \to \Gamma$ with $L = |z_1 - z_0|$, $\sigma(s) = z_0 + s\zeta$, is the desired parametrization.

4. Suppose $\alpha: [a, b] \to \Gamma$ is a smooth parametrization with the property that $|\alpha'(t)| = 1$ for $a \leq t \leq b$. Verify the following:

   (a) Length $(\alpha) = b - a$.

   (b) If we write $L = b - a$, then the map $\sigma: [0, L] \to \Gamma$ defined by $\sigma(s) = \alpha(a + s)$ is the arc length parametrization of Γ equivalent to α. The point here is that α is essentially (up to a rigid change of parameter) the arc length parametrization because $|\alpha'(t)| = 1$.

5. *Reversing the parametrization.* We will want to "travel backwards" along a given parametrized curve without worrying about details. The example with which we begin should provide the necessary intuition.

   (a) Let Γ be the unit semicircle $|z| = 1$, $y \geq 0$, parametrized as usual by $\alpha: [0, \pi] \to \Gamma$, $\alpha(t) = (\cos t, \sin t)$. Verify that the point $\alpha(t)$ travels from $(1, 0)$ to $(-1, 0)$ as t increases from $t = 0$ to $t = \pi$.

   (b) Now write $t = \pi - \tau$. Check that, as τ *increases* from $\tau = 0$ to $\tau = \pi$, the value of t *decreases* from $t = \pi$ to $t = 0$.

   (c) Define $\gamma(\tau) = (\cos(\pi - \tau), \sin(\pi - \tau))$. Check that $\gamma(\tau)$ traverses the semicircle Γ starting at $(-1, 0)$ and ending at $(1, 0)$. Thus $\gamma(\tau)$ is the reverse of $\alpha(t)$. (Picture!)

   (d) This procedure generalizes to any curve $(\Gamma, \alpha(t))$; that is, $\alpha: [a, b] \to \Gamma$. First verify that if we put $t = b - (\tau - a)$, then increasing τ from a to b causes t to decrease from b to a. Can you derive this relation between t and τ?

   (e) Verify that $\gamma: [a, b] \to \Gamma$, $\gamma(\tau) = \alpha(b - (\tau - a))$, defines a parametrization of the set Γ that reverses the journey taken by $\alpha(t)$.

   (f) Explain why the parametrizations $\alpha(t)$ and $\gamma(t)$ of the curve Γ are not equivalent. Why is this geometrically obvious?

Notation: Given a curve Γ with parametrization $\alpha(t)$ understood, it is standard to denote by $-\Gamma$ the curve with the reverse parametrization $\gamma(\tau)$. Thus, the notation $\Gamma + (-\Gamma)$ indicates a journey over the curve Γ from start to end, followed by the reverse journey back to the original starting point.

1.2.4 Jordan Curves and Jordan Domains

Now we put Sections 1.1 and 1.2 together with an eye toward the future. We wish now to define "Jordan domain." In later sections our functions will frequently live on such domains.

A *Jordan curve* is a simple, closed curve (piecewise-smooth, of course). Thus, a Jordan curve looks like a loop (closed) that does not cross itself (simple). Examples: circles, triangles, rectangles.

Here is an intuitively plausible theorem about Jordan curves.

JORDAN CURVE THEOREM

If Γ *is a Jordan curve, then its complement* $\mathbb{R}^2 - \Gamma$ *consists of two disjoint domains, one bounded (the "inside") and one not (the "outside"), each domain having the curve* Γ *as its boundary. If a point inside* Γ *be joined by a path to a point outside* Γ, *then the path must cross* Γ.

This is one of those geometrically obvious theorems that are quite difficult to prove. It takes a fair amount of rigorous topology to give a complete proof (more than the French mathematician Jordan had at his disposal in the 1890s when he pointed out that the obvious fact required a proof!). We will use this theorem without proof. You might enjoy the challenge of proving it in the special case where Γ is a circle $C(0; r)$. See Figure 1.17.

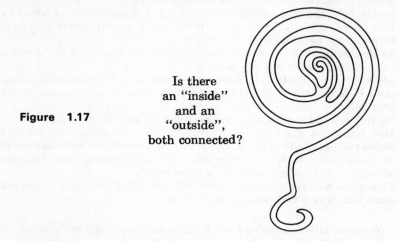

Figure   1.17

Is there
an "inside"
and an
"outside",
both connected?

One theme of the chapters to come is the relation of the behavior of a function defined on some domain to its behavior on the boundary of that domain. To accomplish this gracefully, we wish to rule out domains with unpleasant or pathological boundaries. Therefore, we single out the class of Jordan domains for special emphasis. A *Jordan domain* is a bounded (contained inside some disc) domain Ω whose boundary $\partial\Omega$ is the union of a finite number of disjoint Jordan curves, with each curve parametrized so that, as a point moves along the curve in the direction of parametrization, the domain Ω is always lying to its left. In this case we say that the boundary curves are *positively oriented*.

Examples of Jordan Domains

1. The open disc $\Omega = D(0; r)$ with the standard counterclockwise parametrization of the circular boundary.

2. The annulus

$$\Omega = \{z \in \mathbb{R}^2 \mid 0 < r_0 < |z| < r_1\}.$$

Note parametrization of boundary circles. Ω is to the left of a point moving counterclockwise on the outside circle or clockwise on the inside circle.

3. Figure 1.18 shows a Jordan domain with several holes. Note orientations.

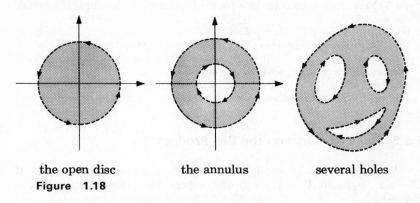

    the open disc   the annulus   several holes

Figure   1.18

Warning: Some authors define a Jordan domain to be a bounded domain whose boundary consists of *precisely one* positively oriented Jordan curve, and not finitely many such curves as we have done. To these people the annulus is *not* a Jordan domain.

k-Connected Jordan Domains

The Jordan domain Ω is said to be *k-connected*, where k is an integer ≥ 1 if its boundary consists of k distinct Jordan curves. This is equivalent to its complement $\mathbb{C} - \Omega$ consisting of k disjoint connected components. Thus, the disc is 1-connected ("simply connected," no holes inside); the annulus is 2-connected.

We will use the phrase "k-connected Jordan domain" in the statements of theorems to remind the reader that our Jordan domains may have holes.

Exercises to Paragraph 1.2.4

1. Which of the following are Jordan curves?
   (a) The circle $|z| = 1$.
   (b) The x-axis.
   (c) A figure-eight.
   (d) The boundary of the annulus $1 < |z| < 2$.
   (e) A straight-line segment.
   (f) The graph of $y = \sin x$.

2. Prove the Jordan Curve Theorem in the special case $\Gamma = C(0; r)$.

   Hint: In this case, the two disjoint domains (components) forming $\mathbb{R}^2 - \Gamma$ may be characterized by simple inequalities. Some elementary real analysis will help in proving that a path from inside to outside must intersect Γ.

3. Given a Jordan curve Γ, how would you define its "interior" (the set of points "inside" Γ) in a mathematically effective way? Don't appeal to the Jordan Curve Theorem! (For one approach, see Section 7.2.)

4. True or false?

   (a) The xy-plane is a Jordan domain.

   (b) The set of points enclosed by a positively oriented Jordan curve is a Jordan domain.

   (c) A Jordan domain is bounded.

   (d) A Jordan curve separates the plane into two Jordan domains.

   (e) The annulus $1 < |z| < 2$, with $|z| = 1$ and $|z| = 2$ oriented counterclockwise and clockwise, respectively, is a Jordan domain.

   (f) The domain $|z| > 1$, consisting of all points *outside* the unit disc, is a Jordan domain, provided we orient the circle $|z| = 1$ in a clockwise fashion.

1.2.5 Some Remarks on the Dot Product

We wish to make explicit a result that is frequently useful. If $U = \langle u_1, u_2 \rangle$ and $V = \langle v_1, v_2 \rangle$ are vectors, then their dot product is defined as

$$U \cdot V = u_1 v_1 + u_2 v_2.$$

Thus, if $U = \langle 2, 3 \rangle$ and $V = \langle 1, -4 \rangle$ then $U \cdot V = -10$.

We note that $|U| = \sqrt{U \cdot U}$.

The dot product is easy to compute, but its geometric meaning is less clear. But consider

THEOREM 2

$U \cdot V = |U| \, |V| \cos \theta$, *where θ is the angle between U and V in the plane $\mathbb{R}^2$.*

Note: Since $\cos \theta = \cos (-\theta)$, the direction of θ is immaterial.

Proof: We have, from trigonometry,

$$u_1 = |U| \cos \omega, \qquad u_2 = |U| \sin \omega.$$
$$v_1 = |V| \cos \varphi, \qquad v_2 = |V| \sin \varphi.$$

(See Figure 1.19.)

Thus,

$$\begin{aligned} U \cdot V &= u_1 v_1 + u_2 v_2 \\ &= |U| \, |V| (\cos \omega \cos \varphi + \sin \omega \sin \varphi) \\ &= |U| \, |V| \cos (\omega - \varphi). \end{aligned}$$

But $\theta = \omega - \varphi$. Done.

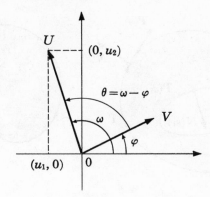

Figure   1.19

This result enables us to deal effectively with some geometric notions. We say that U and V are *perpendicular (orthogonal, normal)* if and only if $U \cdot V = 0$, for $U \cdot V = 0$ if and only if one of the vectors is zero or $\cos \theta = 0$, in which case θ is a right angle, $\theta = \pm(\pi/2)$.

Exercises to Paragraph 1.2.5

1. Let $U = \langle 3, 7 \rangle$, $V = \langle -3, 8 \rangle$.
   (a) Compute $U \cdot V$.
   (b) Compute $|U|$, $|V|$.
   (c) Compute $\cos \theta$, where θ is the angle between U and V.
   (d) How would you get θ from $\cos \theta$?
2. Let $U = \langle 3, 7 \rangle$.
   (a) Construct a nonzero vector V such that $U \perp V$.
   (b) Construct a vector with unit length in the same direction as V above.
3. *The Cauchy–Schwarz Inequality.* Prove $|U \cdot V| \geq |U| \, |V|$ for all vectors U, V.
   Hint: Cosine.
4. *The Triangle Inequality.* Prove $|U + V| \leq |U| + |V|$ for all vectors U, V.
   Hint: Show $|U + V|^2 \leq (|U| + |V|)^2$ via Cauchy–Schwarz.

1.2.6 The Outward Normal Vector *N(s)*

This will be essential in our study of the behavior of a function on the boundary $\partial\Omega$. Suppose Ω is a Jordan domain. Let Γ be one of the Jordan curves that form part of $\partial\Omega$. Let $\sigma: [0, L] \to \Gamma$ be the parametrization of Γ by arc length. We will denote the unit tangent vector $\sigma'(s)$ by the customary $T(s)$.

Now, to each point $\sigma(s)$ of Γ (consider s fixed), we wish to affix a vector $N(s)$ that satisfies

   (i) $|N(s)| = 1$ (unit vector),
   (ii) $N(s) \perp T(s)$; ($N(s)$ is "normal" or perpendicular to the curve),

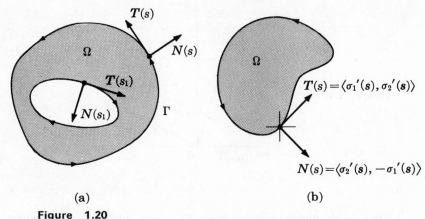

(a)   (b)

Figure   1.20

(iii) $N(s)$ points outward, away from the domain Ω. Such a vector
$N(s)$ is called the *outward normal vector*. See Figure 1.20(a).

We construct $N(s)$ with our bare hands as follows: Given $\sigma(s) =
(\sigma_1(s), \sigma_2(s))$, we may compute $T(s) = \sigma'(s) = \langle \sigma'_1(s), \sigma'_2(s) \rangle$. We now
define

$$N(s) = \langle \sigma'_2(s), -\sigma'_1(s) \rangle.$$

Now let's verify that this $N(s)$ has properties (i), (ii), (iii) listed
above. From $|T(s)| = 1$ follows immediately that $|N(s)| = 1$, as required
by (i) above. To verify that the vector $N(s)$ defined here satisfies (ii), you
may check immediately that the "dot product" $N(s) \cdot T(s) = 0$ and recall
that the vanishing of the dot product of two vectors is equivalent to
their perpendicularity. (Recall Theorem 2 above.)

To verify requirement (iii), "outwardness," note that when both
$\sigma'_1(s) > 0$ and $\sigma'_2(s) > 0$, so that the vector $T(s)$ points upward and to
the right (see illustration) with Ω on its left, then $N(s)$ points downward
(since its y-component $-\sigma'_1(s)$ is negative) and to the right (since
$\sigma'_2(s) > 0$) whence $N(s)$ points away from Ω. See Figure 1.20(b).

Sample Construction of N(s)

Let $\sigma: [0, 2\pi r] \to C(0; r)$ be the arc length parametrization of the
circle of radius r. We saw above that $\sigma(s) = (r \cos (s/r), r \sin (s/r))$.
Thus, the unit tangent vector is

$$T(s) = \left\langle -\sin \frac{s}{r}, \cos \frac{s}{r} \right\rangle.$$

That is, $\sigma'_1(s) = -\sin (s/r)$, $\sigma'_2(s) = \cos (s/r)$. In accord with the recipe
above, we define

$$N(s) = \langle \sigma'_2(s), -\sigma'_1(s) \rangle = \left\langle \cos \frac{s}{r}, \sin \frac{s}{r} \right\rangle.$$

Let's verify that $N(s)$ points outward at the point one-eighth of the distance around the circle from the starting point $(r, 0)$. This point is given by $\sigma(s)$ when $s = L/8 = 2\pi r/8 = \pi r/4$, that is, the point is

$$\sigma\left(\frac{\pi r}{4}\right) = \left(\frac{r\sqrt{2}}{2}, \frac{r\sqrt{2}}{2}\right).$$

We compute also, using the formula for $N(s)$ above, that

$$N\left(\frac{\pi r}{4}\right) = \left\langle \frac{\sqrt{2}}{2}, \frac{\sqrt{2}}{2} \right\rangle.$$

Since both coordinates here are positive, this vector points to the right and upward, and therefore away from the interior of the circle, as expected. See Figure 1.21.

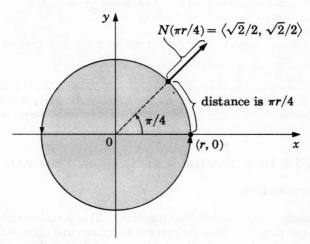

Figure 1.21

Exercises to Paragraph 1.2.6

1. Let Γ be the boundary of the unit square with corners at $z_0 = 0$, $z_1 = (1, 0)$, $z_2 = (1, 1)$, $z_3 = (0, 1)$. We write $\Gamma = \Gamma_0 + \Gamma_1 + \Gamma_2 + \Gamma_3$ (a classical notation), where Γ_0 is the segment from z_0 to z_1, Γ_1 from z_1 to z_2, etc. Verify that if Γ is given a positive (counterclockwise) orientation, then the outward normal vector at any point of Γ_0 is $\langle 0, -1\rangle$, of Γ_1 is $\langle 1, 0\rangle$, of Γ_2 is $\langle 0, 1\rangle$, of Γ_3 is $\langle -1, 0\rangle$. Sketch the square and some normal vectors.

2. Let Γ be a positively oriented Jordan curve and let Ω be the Jordan domain consisting of all points interior to Γ. Let $z_0 \in \Gamma$ with $T = T(z_0)$, $N = N(z_0)$ the unit tangent and outward normal vectors to Γ at the point z_0 (no need to mention parametrization). Let V be any vector with tail end at z_0.

(a) Verify that $V = aT + bN$, where $a = V \cdot T$, $b = V \cdot N$. This decomposition relates V and Ω.

(b) Convince yourself (sketch!) that the vector $aT = (V \cdot T)T$ (the *tangential component* of V) measures the tendency of V to point in the tangential direction, whereas $bN = (V \cdot N)N$ (the *normal component* of V) measures the tendency of V to leave Ω by crossing Γ at right angles. In particular, if $V \cdot N < 0$ then V is pointing *into* Ω from z_0 rather than out of Ω.

(c) Interpret $V \cdot T > 0$, $V \cdot N = 0$. Sketch!

(d) Interpret $V \cdot T < 0$, $V \cdot N = 0$.

(e) Interpret $V \cdot T = 0$, $V \cdot N = |V| > 0$.

3. *A summary in simplified notation.* Let Ω be the Jordan domain interior to a simple closed curve $\Gamma = \partial\Omega$, which is positively oriented with respect to Ω. We parametrize Γ by arc length, denoting the typical point $z = z(s) = (x(s), y(s))$. There is no need to use σ (or α or β, etc.) when there is only one parametrization being discussed. Verify the following assertions: (a) The tangent vector $T(s) = z'(s) = \langle x'(s), y'(s) \rangle$. Here

$$z'(s) \text{ means } \frac{dz}{ds}(s), \qquad x'(s) \text{ means } \frac{dx}{ds}(s), \qquad \text{etc.}$$

(b) $(x'(s))^2 + (y'(s))^2 = 1$.

(c) The outward normal vector $N(s)$ at the point $z(s)$ on Γ is equal to

$$\langle y'(s), -x'(s) \rangle.$$

We remark that most of our applications of $T(s)$ and $N(s)$ are simple uses of assertions (a), (b), and (c). Also, these formulas still apply if Ω is a k-connected Jordan domain, $k > 1$, and $\partial\Omega$ consists of k simple closed curves.

Section 1.3 DIFFERENTIAL CALCULUS IN TWO VARIABLES

1.3.0 Introduction

Our study of curves involved functions $\alpha(t)$ of a real variable t with values in the plane $\mathbb{R}^2$. Now we reverse directions and discuss functions with domain of definition in $\mathbb{R}^2$ which assume real numbers as values.

We will be particularly interested in how a function defined in the plane varies in the vicinity of a fixed point z; hence, differentiation. This is a somewhat more complicated notion for functions of $z = (x, y)$ than for functions of x alone because, roughly speaking, there is much more room to vary. We will be led to defining two different sorts of derivative, the directional derivative and the gradient, and showing how they relate.

Here are some basics. Let Ω be a domain (open connected subset of $\mathbb{R}^2$). A *function* u from Ω into $\mathbb{R}$ is a rule that assigns to each $z \in \Omega$ a real number $u(z)$. If $z = (x, y)$, we may also write $u = u(z) = u(x, y)$. The familiar expression $u: \Omega \to \mathbb{R}$ will also be used.

Here is a handy notation. If S is any subset of $\mathbb{R}$, then

$$u^{-1}(S) = \{z \in \mathbb{R}^2 \mid u(z) \in S\}.$$

We emphasize that $u^{-1}(S)$ is simply the notation for a set; we do *not* claim that the function u has an inverse.

An important definition: The function $u: \Omega \to \mathbb{R}$ is *continuous in* Ω if and only if whenever S is an open interval (end points not included) in $\mathbb{R}$, then $u^{-1}(S)$ is an open subset of $\mathbb{R}^2$. This definition of continuity may be translated into an equivalent limit statement: u is continuous in Ω if and only if, for each fixed $z_0 \in \Omega$, $\lim_{z \to z_0} u(z) = u(z_0)$. The translation from the "open set" definition to the "limit" statement is possible because both open sets and limits are defined in terms of distance $|z - z_0|$. We leave it to you to make your own private peace with this.

We remark that in this book we almost never mention the graph of the function u (which is the surface consisting of all points $(x, y, u(x, y))$ in three-dimensional space).

1.3.1 Continuously Differentiable Functions

We assume you have some acquaintance with the two partial derivatives $u_x(x, y)$ and $u_y(x, y)$ of a function $u(x, y)$. We say that $u(x, y)$ is *continuously differentiable in the domain* Ω if and only if both its partial derivatives exist and are continuous at each point (x, y) of Ω. Most of the functions we study will be of this type.

Here is some useful notation. We let $\mathscr{C}(\Omega)$ denote the set of all real-valued functions $u(x, y)$ which are continuous throughout Ω. Also, $\mathscr{C}^1(\Omega)$ denotes the set of all functions continuously differentiable in Ω, as defined just above. Note that $u \in \mathscr{C}^1(\Omega)$ implies $u_x, u_y \in \mathscr{C}(\Omega)$. One basic thing we wish to prove is

$$\mathscr{C}^1(\Omega) \subset \mathscr{C}(\Omega);$$

that is, a continuously differentiable function is continuous (Corollary 4, p. 33). This is the analog of the calculus theorem that states that if $f'(x)$ exists, then $f(x)$ is continuous.

Most properties of continuously differentiable functions flow from the following basic analysis.

Theorem 3

Let $u \in \mathscr{C}^1(\Omega)$ *with* $z_0 = (x_0, y_0) \in \Omega$. *For each point* $z = (x, y)$ *in* Ω *we may write*

$$u(z) = u(z_0) + u_x(z_0) \cdot (x - x_0) + u_y(z_0) \cdot (y - y_0)$$
$$+ \varepsilon_1 \cdot (x - x_0) + \varepsilon_2 \cdot (y - y_0),$$

where $\varepsilon_1, \varepsilon_2$ *approach zero as* z *approaches* z_0 *in* Ω.

Proof: We will use the Mean-Value Theorem of ordinary calculus in each variable. For (x, y) close enough to (x_0, y_0), we have the standard trick

$$u(x, y) - u(x_0, y_0) = \{u(x, y) - u(x_0, y)\} + \{u(x_0, y) - u(x_0, y_0)\}.$$

The Mean-Value Theorem may now be applied to the x-variable in the
first bracket and to y in the second bracket. This yields equality of the
right-hand side of the last equation with

$$u_x(x_1, y) \cdot (x - x_0) + u_y(x_0, y_1) \cdot (y - y_0)$$

for x_1 between x_0 and x and y_1 between y_0 and y. See Exercise 4. The
continuity of the partial derivatives allows us to conclude that $u_x(x_1, y)$,
$u_y(x_0, y_1)$ approach $u_x(x_0, y_0)$, $u_y(x_0, y_0)$, respectively, as (x, y) approaches
(x_0, y_0). Hence, we may write

$$u_x(x_1, y) = u_x(x_0, y_0) + \varepsilon_1$$
$$u_y(x_0, y_1) = u_y(x_0, y_0) + \varepsilon_2$$

where the ε's tend to zero as (x, y) approaches (x_0, y_0). Substituting these
into the expression for $u(x, y) - u(x_0, y_0)$ above completes the proof of
the theorem.

Interpretation of the Theorem

This is classic differential calculus. To study the function u near the
point z_0, we break u into two parts,

$$u(z) = L(z; z_0) + E(z; z_0)$$

where

$$L(z; z_0) = u(z_0) + u_x(z_0) \cdot (x - x_0) + u_y(z_0) \cdot (y - y_0)$$
$$E(z; z_0) = \varepsilon_1 \cdot (x - x_0) + \varepsilon_2 \cdot (y - y_0)$$

are the "linear approximation" and "error term," respectively. Note
that z_0 is held fixed, so L and E are treated as functions of z alone.
Moreover, we see

(i) $L(z; z_0)$ contains the value $u(z_0)$ and the "first derivative"
information of u;

(ii) $L(z; z_0)$ is a rather simple function of x, y;

(iii) $E(z; z_0)$ carries the complicated information about u in $\varepsilon_1, \varepsilon_2$;

(iv) However, $E(z; z_0)$ vanishes rapidly as z approaches z_0. In fact,
$E(z; z_0)$ gets small fast enough that

$$\lim_{z \to z_0} \frac{|E(z; z_0)|}{|z - z_0|} = 0,$$

as you may check easily, using $|x - x_0|, |y - y_0| \le |z - z_0|$;

(v) Hence, the error $u(z) - L(z; z_0)$ is very small for z close to z_0.
That is, *$u(z)$ is closely approximated by the simpler linear function
$L(z; z_0)$ for z near z_0.*

All this depends on the hypothesis that u is continuously
differentiable.

Example

Given $u(z) = u(x, y) = x^2 y$, $z_0 = (x_0, y_0) = (2, 3)$. The theorem assures us that

$$u(z) = 12 + 12(x - 2) + 4(y - 3) + \varepsilon_1(x - 2) + \varepsilon_2(y - 3)$$

where $\varepsilon_1, \varepsilon_2$ approach zero as (x, y) approaches $(2, 3)$.

Because $u(x, y)$ is so nice (a polynomial in x, y), we may actually compute $\varepsilon_1, \varepsilon_2$ in terms of x, y. In fact, if we use the trick $x = (x - 2) + 2$, $y = (y - 3) + 3$, we see that

$$
\begin{aligned}
u(x, y) &= ((x - 2) + 2)^2((y - 3) + 3) \\
&= 12 + 12(x - 2) + 4(y - 3) + 3(x - 2)^2 \\
&\quad + 4(x - 2)(y - 3) + (x - 2)^2(y - 3),
\end{aligned}
$$

so we might take

$$\varepsilon_1 = 3(x - 2) + 4(y - 3), \qquad \varepsilon_2 = (x - 2)^2.$$

Further Comments

1. Theorem 3 is a first attempt at some ultimate theorem that would assure us that if u were nice enough, then it could be expanded in a convergent power series

$$u(z) = u(x, y) = \sum_{m,n=0}^{\infty} c_{mn}(x - x_0)^m(y - y_0)^n$$

where the coefficients c_{mn} are related to the higher-order partial derivatives of u at z_0. Compare the example just above.

2. Now we use Theorem 3 twice: to prove Corollary 4 and the Chain Rule. (It will also be applied in our discussion of the Cauchy–Riemann equations in Chapter 3.) The Chain Rule will then yield the main results on the directional derivative.

COROLLARY 4

Let $u \in \mathscr{C}^1(\Omega)$ and let $z_0 \in \Omega$. Then u is continuous at z_0. In other words,

$$\mathscr{C}^1(\Omega) \subset \mathscr{C}(\Omega).$$

Proof: We leave it to you to show that $u(x, y)$ approaches $u(x_0, y_0)$ as (x, y) approaches (x_0, y_0) in Ω. Use Theorem 3.

A Version of the Chain Rule

Suppose $u \in \mathscr{C}^1(\Omega)$. If Γ is a curve lying inside Ω parametrized by $\alpha: [a, b] \to \Gamma$, then restricting the function u to the curve Γ leads to a composite function of t, namely, $u(\alpha(t))$. This is a function of a single real variable as in ordinary calculus. Hence, we ask for the standard derivative,

$$\frac{d}{dt} u(\alpha(t)) = ?$$

Theorem 3 leads to the following answer.

CHAIN RULE

Let u and $\alpha(t) = (\alpha_1(t), \alpha_2(t))$ be as above. Then

$$\frac{d}{dt} u(\alpha(t)) = u_x(\alpha(t))\, \alpha'_1(t) + u_y(\alpha(t))\alpha'_2(t).$$

Proof: We form the standard difference quotient for the derivative at t_0 and then do what comes naturally. Thus,

$$\frac{u(\alpha(t)) - u(\alpha(t_0))}{t - t_0} = \frac{u_x(\alpha(t)) \cdot (\alpha_1(t) - \alpha_1(t_0)) + u_y(\alpha(t)) \cdot (\alpha_2(t) - \alpha_2(t_0))}{t - t_0}$$

$$+ \frac{\varepsilon_1 \cdot (\alpha_1(t) - \alpha_1(t_0)) + \varepsilon_2 \cdot (\alpha_2(t) - \alpha_2(t_0))}{t - t_0}$$

by Theorem 3, since $x = \alpha_1(t)$, $y = \alpha_2(t)$. Now let t approach t_0 so that $\alpha(t)$ approaches $\alpha(t_0)$, and check that everything works out. Done.

We mentioned earlier that a curve is often introduced by a phrase such as "Let $x = x(t)$, $y = y(t)$ define a curve...," with no reference to the mapping α. In this case the Chain Rule becomes

$$\frac{d}{dt} u(x(t), y(t)) = u_x(x(t), y(t))x'(t) + u_y(x(t), y(t))y'(t).$$

Comment

Compare this with ordinary calculus:

$$\frac{d}{dt} f(x(t)) = f'(x(t))x'(t).$$

Exercises to Paragraph 1.3.1

1. Let $u(z) = u(x, y) = x - x^2 + 3y^2$. Compute the linear approximation $L(z; z_0) = u(z_0) + u_x(z_0)\,(x - x_0) + u_y(z_0)\,(y - y_0)$ to u at $z_0 = (1, 2)$ in two ways as follows:
   (a) by calculating the partial derivatives at z_0, and
   (b) by writing $x = (x - 1) + 1$, $y = (y - 2) + 2$ in the expressions for $u(x, y)$ and truncating the resultant expression after the first-degree term.
   (c) Verify that $|E(z; z_0)|/|z - z_0| \to 0$ as $z \to z_0$ (see the discussion following Theorem 3).

2. Let u be as in Exercise 1, and $\alpha(t) = (\cos t, \sin t)$ the standard parametrization of the unit circle. Compute $(d/dt)\, u(\alpha(t))$ as a function of t in two ways:
   (a) by first computing $u_x(x, y)$, $u_y(x, y)$ and then letting $x = \cos t$, $y = \sin t$ in the Chain Rule,
   (b) by differentiating $u(\cos t, \sin t)$ with respect to t.

3. *The argument function.* Let $z = (x, y)$ not lie on the nonpositive x-axis. Then z may be assigned polar coordinates (r, θ), where $r = |z|$, $\theta = \arctan(y/x)$, and $-\pi < \theta < \pi$.
   (a) Verify that, once these agreements have been made, the argument function $\arg z = \theta$ is continuous in the domain Ω consisting of $\mathbb{R}^2$ with all points $(x, 0)$, $x \leq 0$, deleted. (Ω = the slit plane.)
   (b) Verify that $\arg z$ cannot be extended to the domain $\mathbb{R}^2 - \{0\}$ so as to be continuous at $(x, 0)$, $x < 0$.
   (c) Observe that the convention $\pi < \theta < 3\pi$ leads to a different version ("branch") of the argument function, which differs from the "principal branch" in (a) at each z by a constant term 2π. Neither version is more natural than the other.
   (d) Verify that, for any branch of $\theta = \arg z$,

$$\theta_x = (\arg z)_x = -\frac{y}{r^2}, \qquad \theta_y = (\arg z)_y = \frac{x}{r^2}$$

   where $r^2 = x^2 + y^2$.
   (e) Verify that $\theta_{xx} + \theta_{yy} = 0$ identically. That is, $\theta = \arg z$ is a solution to the "Laplace equation" and is thereby termed *harmonic in* Ω (= the slit plane defined in (a)).

4. Recall the Mean-Value Theorem of ordinary calculus and verify its use in the proof of Theorem 3. See any calculus book!

5. Supply details in the proof of Corollary 4.

1.3.2 The Directional Derivative

The partial derivative $u_x(x, y)$ is defined as

$$u_x(x, y) = \lim_{s \to 0} \frac{u(x + s, y) - u(x, y)}{s}.$$

Given the point (x, y), the number $u_x(x, y)$ is the rate of change of the value $u(x, y)$ with respect to change in x, the value of y being held fixed. There is a similar interpretation of $u_y(x, y)$, of course.

Now we ask about change in an arbitrary direction, not just the direction of the x-axis or y-axis. To be specific, suppose u is a function defined in a domain Ω containing the point $z = (x, y)$ and suppose that V is a unit vector, $|V| = 1$, with tail fixed at z. The vector V points out a particular direction of travel through the fixed point z. We ask: "What is the rate of change (with respect to distance in the plane) of the value of the function u at the point z, in the direction of the vector V?" See Figure 1.22. Note the three mathematical data: function, point, direction.

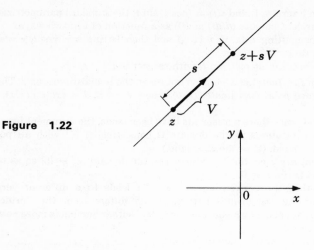

Figure   1.22

Let's translate this into ordinary calculus. The set of all points $z + sV$ in $\mathbb{R}^2$, where s is a real number, is a straight line through z (which is given by $s = 0$). This line has the same direction as V in the sense that as s increases from $-\infty$ to ∞, the point $z + sV$ travels along the line through z in the direction indicated by V. In fact, s is the arc length parameter along this straight line because $|V| = 1$. (Think about this!) Thus the question at the end of the last paragraph becomes, "What is the rate of change of the value of u at the point z when restricted to the parametrized straight line $z + sV$?"

This calls for a definition and a theorem. We define the *directional derivative* of u at z in *the direction* V (denoted $u'(z; V)$) as the ordinary derivative

$$u'(z; V) = \frac{d}{ds} u(z + sV) \bigg|_{s=0} .$$

Note here that z and V are given, so that $u(z + sV)$ is a function of s alone, just as in ordinary calculus.

It is immediate (since z is given by $s = 0$) that

$$u'(z; V) = \lim_{s \to 0} \frac{u(z + sV) - u(z)}{s} ,$$

an equivalent form of the definition.

At this point we find ourselves in the familiar position of having a nice theoretical definition at hand, but without a generally useful method of computing. How can we compute $u'(z; V)$ in terms of known (that is, partial) derivatives of u? The answer utilizes the Chain Rule obtained above. Here it is.

THEOREM 5

Let $u \in \mathscr{C}^1(\Omega)$ and $z \in \Omega$. If $V = \langle k_1, k_2 \rangle$ is a unit vector, then the directional derivative is given by

$$u'(z; V) = u_x(z)k_1 + u_y(z)k_2.$$

Proof: We write

$$\sigma(s) = z + sV = (x + sk_1, y + sk_2).$$

Thus $\sigma'_1(s) = k_1$, $\sigma'_2(s) = k_2$. Now apply the Chain Rule with $s = 0$. Done.

Remark: If V points in the usual direction of the x-axis (that is, if $V = \langle 1, 0 \rangle$), then Theorem 5 reduces to

$$u'(z; V) = u_x(z),$$

the familiar partial derivative, as expected.

Example of a Directional Derivative

What is $u'(z_0; V)$, where $u(z) = u(x, y) = x^2 y$, $z_0 = (1, 1)$ and V is the unit vector $\langle -2/\sqrt{13}, 3/\sqrt{13} \rangle$? We offer two methods for computing this:

1. *Short method:* Apply the theorem. We have $u_x(x, y) = 2xy$, $u_y(x, y) = x^2$ so that $u_x(z_0) = 2$, $u_y(z_0) = 1$. By the theorem, then,

$$u'(z_0; V) = 2(-2/\sqrt{13}) + 1(3/\sqrt{13}) = -1/\sqrt{13}.$$

2. *Long method:* Direct computation. Since $(x, y) = z_0 + sV = (1 - 2s/\sqrt{13}, 1 + 3s/\sqrt{13})$, we see that $u(x, y)$, when restricted to the line through z_0, yields a function of s alone, namely,

$$u(z_0 + sV) = \left(1 - \frac{2s}{\sqrt{13}}\right)^2 \left(1 + \frac{3s}{\sqrt{13}}\right)$$

$$= 1 - \frac{s}{\sqrt{13}} - \frac{8s^2}{13} + \frac{12s^3}{13\sqrt{13}}.$$

The derivative of this function of s at $s = 0$ is immediately seen to be $-1/\sqrt{13}$, in agreement with the answer obtained by the first method.

We conclude that $u(x, y) = x^2 y$ is decreasing as (x, y) varies through the point $(1, 1)$ in the direction pointed out by the given V. This is because $u'(z_0; V) < 0$.

Exercises to Paragraph 1.3.2

1. Compute the directional derivative $u'(z_0; V)$, where $u(x, y) = x - x^2 + 3y^2$, and z_0 and V are as follows (note $|V| = 1$):
   (a) $z_0 = (0, 0)$, $V = \langle 1, 0 \rangle$;
   (b) $z_0 = (0, 0)$, $V = \langle 1/\sqrt{2}, 1/\sqrt{2} \rangle$;
   (c) $z_0 = (1, 2)$, $V = \langle 0, -1 \rangle$;
   (d) $z_0 = (1, 2)$, $V = \langle 0, 1 \rangle$;
   (e) $z_0 = (1, 2)$, $V = \langle 1/\sqrt{2}, 1/\sqrt{2} \rangle$;
   (f) $z_0 = (1, 2)$, $V = \langle 1/\sqrt{2}, -1\sqrt{2} \rangle$.

2. (a) Compute a unit vector U that points in the same direction as $\langle 3, -4 \rangle$.
   (b) Compute the directional derivative of $u(x, y) = e^x \cos y$ at $z_0 = (1, \pi/2)$ in the direction pointed out by $\langle 3, -4 \rangle$.
   Hint: Use (a).

3. (a) Let $u \in \mathscr{C}^1(\mathbb{R}^2)$, $\alpha(\theta) = (\cos \theta, \sin \theta)$ (the usual parametrization of the unit circle). Verify that $(du/d\theta)(\alpha(\theta)) = u'(\alpha(\theta); T(\theta))$, where $T(\theta)$ is the unit tangent vector to the circle at the point $\alpha(\theta)$. (The *tangential derivative* of u.)
   (b) Conjecture the values of the tangential derivatives of the functions $r = |z| = (x^2 + y^2)^{1/2}$ and $\theta = \arg z$ (where defined). *Hint:* How does r change as you move around the circle?

4. Suppose the unit vector V makes a counterclockwise angle ω with the vector $\langle 1, 0 \rangle$ (horizontal).
   (a) Verify $V = \langle \cos \omega, \sin \omega \rangle$.
   (b) Verify $u'(z_0; V) = u_x(z_0) \cos \omega + u_y(z_0) \sin \omega$. This formula underscores the role of the direction angle ω.

1.3.3 The Gradient Vector

Now we define another sort of derivative for $u(x, y)$. It is a vector in the plane and not a number (as is the directional derivative). We will see that it has a close relation with the directional derivative, however.

As usual, Ω is a domain in the plane. Let $u \in \mathscr{C}^1(\Omega)$. This assures us that $u_x(z)$, $u_y(z)$ exist at every point z of Ω. We define the *gradient of u at the point z* to be the vector

$$\nabla u(z) = \langle u_x(z), u_y(z) \rangle.$$

Some authors write grad $u(z)$ instead of $\nabla u(z)$. The upside-down delta ∇ is sometimes read "del."

Pictorially (see Figure 1.23), we think of the vector $\nabla u(z)$ as having its tail end fixed at the point z. This prompts a question: "What is so special about the direction in which the gradient vector points?" We will consider this below.

Examples of Gradient Vectors

1. Let $\Omega = \mathbb{R}^2$, $u(x, y) = x^2 y$, $z_0 = (1, 1)$. What is $\nabla u(z_0)$?
   We have $u_x = 2xy$, $u_y = x^2$, so that $u_x(1, 1) = 2$, $u_y(1, 1) = 1$. It follows that $\nabla u(1, 1) = \langle 2, 1 \rangle$.

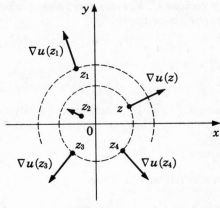

Figure   1.23

Gradients of $u(x, y) = \frac{1}{2}(x^2 + y^2)$

2. Let u be as above, but now let $z_1 = (-2, 3)$. Then we see that $\nabla u(-2, 3) = \langle -12, 4 \rangle$.

3. Let u be as above. For which z is $\nabla u(z) = \langle 0, 0 \rangle$? You should convince yourself that $\nabla u(z) = \langle 0, 0 \rangle$ if and only if z has the form $(0, y)$. Compare the y-axis.

Two things remain to be done: we must relate the gradient to the directional derivative, and also we must interpret the direction of the gradient.

The Gradient and the Directional Derivative

This is a simple observation. We know already that if $V = \langle k_1, k_2 \rangle$ is a unit vector, then the directional derivative has the nice form (recall Theorem 5, Paragraph 1.3.2)

$$u'(z_0; V) = u_x(z_0)k_1 + u_y(z_0)k_2.$$

Now we observe that the right-hand side here looks like the dot product of two vectors. In fact, we see

$$u'(z_0; V) = \nabla u(z_0) \cdot V.$$

In words, the rate of change of u at z_0 in the V-direction may be obtained by dotting the gradient vector with the unit vector V.

On the Gradient Direction

The dot product formula just obtained stimulates further thought. We recall from Section 1.2 that

$$U \cdot V = |U| \, |V| \cos \theta$$

for any vectors U, V. Here, θ is the angle between U and V. Hence, the paragraphs above imply

$$u'(z_0; V) = \nabla u(z_0) \cdot V = |\nabla u(z_0)|\, |V| \cos \theta.$$

But $|V| = 1$ (unit vector!), so that

$$u'(z_0; V) = |\nabla u(z_0)| \cos \theta.$$

We leave it to you to conclude this analysis by proving the next statement (interesting, but not crucial to what follows).

THEOREM 6

Let $u \in \mathscr{C}^1(\Omega)$, $z_0 \in \Omega$. Then

   (i) *the gradient vector $\nabla u(z_0)$ points from z_0 in the direction of maximum rate of increase of the value of u;*
   (ii) *the rate of increase (directional derivative) in the gradient direction is equal to the length $|\nabla u(z_0)|$;*
   (iii) *for any unit vector V, the directional derivative satisfies*

$$- |\nabla u(z_0)| \leq u'(z_0; V) \leq |\nabla u(z_0)|.$$

Hint: Look at the cosine. For (i) use the fact that $\cos \theta = 1$ if and only if $\theta = 0$. Otherwise, $\cos \theta < 1$. What does this say about $\nabla u(z_0)$ and V?

Moral

If the temperature at each point z (Figure 1.24) in the plane were equal to $u(z)$, and if you were a heat-seeking bug standing at the point

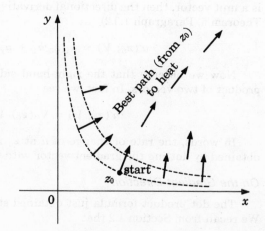

Figure   1.24

Gradients of temperature $u(x, y) = xy$

z_0, then in order to move toward higher temperatures most efficiently (in terms of degrees of increase in temperature per distance traveled) you should begin traveling away from z_0 in the direction pointed out by $\nabla u(z_0)$ (and at every point z follow $\nabla u(z)$).

Thought question: What if $\nabla u(z_0) = \langle 0, 0 \rangle$ (and hence has no direction)?

Exercises to Paragraph 1.3.3

1. Compute the gradient $\nabla u(z)$, where $u(z) = x - x^2 + 3y^2$ and z is given by
   (a) $z = (x, y)$ (typical point),
   (b) $z = (0, 0)$,
   (c) $z = (1, 2)$,
   (d) $z = (1/2, 0)$.

2. Do Exercise 1 to Paragraph 1.3.2, using the gradient.

3. (a) Let $r = r(z) = |z|$ as usual. Argue, using Theorem 6, that at each point $z_0 \neq 0$, the gradient $\nabla r(z_0)$ points from z_0 directly away from the origin (the direction of greatest rate of increase of r).
   (b) Verify your argument in (a) by actually computing ∇r.
   (c) In which direction is $\nabla \theta(z_0) = \nabla \arg z_0$ pointing from z_0? Assume here that z_0 does not lie on the nonpositive x-axis.

4. The function $u \in \mathscr{C}^1(\Omega)$ has a *local maximum* at z_0 in Ω if $u(z_0) \geq u(z)$ for all z in some open disc centered at z_0. Prove that if u has a local maximum at z_0, then $\nabla u(z_0) = \langle 0, 0 \rangle$, the zero vector. Compare the condition $f'(x_0) = 0$ from ordinary calculus.

5. Prove Theorem 6.

1.3.4 The Outward Normal Derivative $\partial u/\partial n$

Most of the directional derivatives we discuss will in fact be outward normal derivatives. This concept will enable us to develop one of our basic themes: the relation of the behavior of a function u inside a domain Ω to its behavior on the boundary $\partial \Omega$.

Here is the standard situation. Ω is a Jordan domain (see Section 1.2) and z_0 is a point on $\partial \Omega$. The outward normal vector $N(z_0)$ points out of Ω; see Figure 1.25. Suppose $u(z)$ is a function that is continuously differentiable in some larger domain containing Ω and $\partial \Omega$. We ask: "What is the rate of change of the value $u = u(z)$ as the variable z crosses the boundary $\partial \Omega$ at the point $z = z_0$ in the normal direction $N(z_0)$?"

We know the answer already. It is the directional derivative $u'(z_0; N(z_0))$. We call this value the *outward normal derivative of u at z_0 on $\partial \Omega$*. It is frequently denoted $(\partial u/\partial n)(z_0)$, with Ω and $\partial \Omega$ understood but not explicitly mentioned. Thus, we know that

$$\frac{\partial u}{\partial n}(z_0) = u'(z_0; N(z_0)) = \nabla u(z_0) \cdot N(z_0).$$

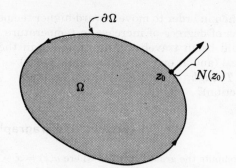

Figure   1.25

$$\frac{\partial u}{\partial n}(z_0) = \text{derivative of } u \text{ in } N(z_0) \text{ direction}$$

Sample Computations of $\partial u/\partial n$

1. Let $u(x, y) = x^2 y \in \mathscr{C}^1(\mathbb{R}^2)$ and let $\Omega = D(0; 1)$, the unit disc. Given $z_0 = (1/\sqrt{2}, 1/\sqrt{2})$ on the circle $\partial\Omega$. What is $(\partial u/\partial n)(z_0)$?

We compute $\nabla u(z_0) = \langle 1, \frac{1}{2}\rangle$. We note also that

$$N(z_0) = \langle 1/\sqrt{2}, 1/\sqrt{2}\rangle.$$

(Draw a picture!) Thus, we compute

$$\frac{\partial u}{\partial n}(z_0) = \left\langle 1, \frac{1}{2}\right\rangle \cdot \left\langle \frac{1}{\sqrt{2}}, \frac{1}{\sqrt{2}}\right\rangle = \frac{3}{2\sqrt{2}}.$$

It follows that $u(x, y)$ is increasing as (x, y) moves across the point $(1/\sqrt{2}, 1/\sqrt{2})$ in the radial direction.

2. Now we treat $(\partial u/\partial n)(z)$ as a function of the variable z on $\partial\Omega$. For definiteness, let u and Ω be as in computation 1. At a point $z = (x, y)$ on the unit circle, we observe (picture!) that the outward normal derivative satisfies

$$N(z) = N(x, y) = \langle x, y\rangle.$$

Note that this *is* a unit vector, since $x^2 + y^2 = 1$.

Also, we have for $u(x, y) = x^2 y$:

$$\nabla u(z) = \nabla u(x, y) = \langle 2xy, x^2\rangle.$$

It follows by taking a dot product that

$$\frac{\partial u}{\partial n}(z) = \frac{\partial u}{\partial n}(x, y) = 3x^2 y = 3y(1 - y^2),$$

since $x^2 = 1 - y^2$ on $\partial\Omega$.

If we choose to parametrize the circle by the usual angle θ, $\alpha(\theta) = (x, y) = (\cos\theta, \sin\theta)$, we obtain $\partial u/\partial n$ as a function of θ:

$$\frac{\partial u}{\partial n}(z) = \frac{\partial u}{\partial n}(\alpha(\theta)) = 3\cos^2\theta\sin\theta.$$

Preview

We will frequently treat $(\partial u/\partial n)(z)$ as a *function* defined on the curves that comprise $\partial\Omega$. This function reflects the tendency of $u(z)$ to change across the full boundary $\partial\Omega$.

Exercises to Paragraph 1.3.4

1. Let $\Gamma = \Gamma_0 + \Gamma_1 + \Gamma_2 + \Gamma_3$ be the (positively oriented) boundary of the unit square, as described in Exercise 1, Paragraph 1.2.6. Prove that the outward normal derivative $(\partial u/\partial n)(z)$ is given by
   (a) $-(\partial u/\partial y)(z)$, if $z \in \Gamma_0$;
   (b) $(\partial u/\partial x)(z)$, if $z \in \Gamma_1$;
   (c) $(\partial u/\partial y)(z)$, if $z \in \Gamma_2$;
   (d) $-(\partial u/\partial x)(z)$, if $z \in \Gamma_3$.

2. Let $\Omega = D(0; 1)$, so that $\partial\Omega = C(0; 1)$. Verify that the outward normal derivative $(\partial u/\partial n)(z)$, $z \in \partial\Omega$, is actually the radial derivative $(\partial u/\partial r)(z)$ in this case. Thus, if u is given in polar coordinates, $u = u(r, \theta)$, the outward normal derivative is obtained by differentiating with respect to the variable r.

3. Let Ω be the unit disc as in Exercise 2. Compute the outward normal derivative $(\partial u/\partial n)(z)$ by expressing the given u in polar coordinates and applying Exercise 2. Since $r = 1$ on $\partial\Omega$, $(\partial u/\partial n)(1, \theta)$ is a function of θ alone.
   (a) $u(z) = x$.
   (b) $u(z) = |z|^2$.
   (c) $u(z) = \arg z$, $(-\pi < \theta < \pi)$.
   (d) $u(z) = xy$.

4. Let Ω be the unit disc as in Exercises 2 and 3, and consider $u(z) = u(x, y) = x$. Without computing, argue that $(\partial u/\partial n)(z) = 0$ if z is $(0, 1)$ or $(0, -1)$, $(\partial u/\partial n)(z) > 0$ if z lies to the right of the y-axis $(x > 0)$, and $(\partial u/\partial n)(z) < 0$ if z lies to the left of the y-axis. The point here is that $u(z) = x$ increases as z moves from left to right. *Note:* $|z| = 1$.

5. (a) Suppose Ω is the unbounded domain determined by $|z| > 1$, so that $\partial\Omega$ is the unit circle that is given a clockwise orientation. Verify that $(\partial u/\partial n)(z) = -(\partial u/\partial r)(z)$ on $\partial\Omega$.
   (b) Given $u(z) = \ln |z|$, $z \neq 0$, compute $(\partial u/\partial n)(z)$ on $\partial\Omega$ as above.

6. Let Ω be the unit disc with positively oriented boundary as usual.
   (a) Construct a function $u(z)$ defined for all $z \neq 0$ such that $(\partial u/\partial n)(z) < 0$ for all z on $\partial\Omega$ and $\lim_{z\to 0} u(z) = \infty$.
   (b) Construct a function $v(z)$ defined for all z without exception such that $(\partial u/\partial n)(z) < 0$ for all z on $\partial\Omega$.
   (c) Construct a function $u(z)$ such that $(\partial u/\partial n)(z) = 0$ for z on the x- or y-axis, $(\partial u/\partial n)(z) > 0$ for z in the first or third quadrant, $(\partial u/\partial n)(z) < 0$ for z in the second or fourth quadrant. Of course $|z| = 1$ in all this.

7. *Some review questions.* Standard situation: Ω is a Jordan domain with boundary $\partial\Omega$ parametrized by $z(s) = (x(s), y(s))$, where the parameter s is arc length. The function $u(z)$ is in $\mathscr{C}^1(\Omega^+)$. Answer true or false.
   (a) For $z \in \Omega$, $\nabla u(z) = \langle u_y(z), u_x(z) \rangle$.
   (b) At a point $z(s)$ on $\partial\Omega$, $(\partial u/\partial n)(z(s)) = \nabla u(z(s)) \cdot N(z(s))$, where $N(z(s))$ is the outward normal vector at $z(s)$.

(c) The outward normal vector $N(z(s))$ equals $\langle y'(s), -x'(s) \rangle$.

(d) If Ω is the unit disc, then $N(z(s))$ points toward the origin of the z-plane.

(e) $(\partial u/\partial n)(z(s)) = u_x(z(s))y'(s) - u_y(z(s))x'(s)$.

(f) If we consider $u(z(s))$ as a function of s, then $(d/ds)u(z(s)) = \nabla u(z(s)) \cdot T(s)$, where, as usual, $T(s) = T(z(s))$ is the unit tangent vector at the point determined by s, that is, at $z(s)$.

(g) $(\partial u/\partial n)(z(s))$ is a vector in the plane.

1.3.5 Derivatives of Higher Order; the Laplacian

So far we have discussed the directional derivative, the gradient, and the outward normal derivative. Note that all of these are, in a sense, *first* derivatives. Starting at the middle of the next section, we will be using only functions that have many more derivatives. (By the way do you know any functions—say, $y = f(x)$—that do not have an infinite number of derivatives $f'(x), f''(x), \ldots$ at each x of the interval of definition?)

We recall that $(\partial^2 u/\partial x^2)$ or u_{xx} denotes the second partial derivative $(\partial/\partial x)(\partial u/\partial x)$ $(\partial^2 u/\partial x\,\partial y)$ or u_{xy} denotes $(\partial/\partial y)(\partial u/\partial x)$, and likewise for the partial derivatives of u_y. Thus, the higher derivatives are defined by iteration. The following folk theorem is used often.

EQUALITY OF MIXED PARTIALS

Let $u_{xy}(z)$ and $u_{yx}(z)$ exist for all z in Ω. If they are continuous functions of z, then they are equal; $u_{xy}(z) = u_{yx}(z)$ for all z.

Note: This theorem is false without some continuity for u_{xy} or u_{yx}. However, most of the functions we encounter will have continuous higher partial derivatives, so there is no cause for concern.

The Sets $\mathscr{C}^k(\Omega)$

In the spirit of $\mathscr{C}(\Omega)$ and $\mathscr{C}^1(\Omega)$, let us define $\mathscr{C}^k(\Omega)$, $k = 1, 2, \ldots$, to be the set of functions defined on Ω for which all mixed partial derivatives through the kth order exist and are continuous.

Note that

$$\mathscr{C}(\Omega) \supset \mathscr{C}^1(\Omega) \supset \mathscr{C}^2(\Omega) \supset \cdots \supset \mathscr{C}^\infty(\Omega).$$

Here, $\mathscr{C}^\infty(\Omega)$ consists of those functions $u(x, y)$ *all* of whose mixed partials exist and are continuous throughout Ω.

Announcement: The Laplacian

We will be very interested in the generalized second derivative of $u(x, y)$ given by the function $u_{xx} + u_{yy}$. This is called the *Laplacian* of u

(after the eighteenth century French mathematician Laplace) and denoted Δu; thus,

$$\Delta u(z) = \Delta u(x, y) = u_{xx}(x, y) + u_{yy}(x, y).$$

A function u in $\mathscr{C}^2(\Omega)$ is *harmonic in* Ω if and only if $\Delta u(z) = 0$ for all z in Ω.

Exercises to Paragraph 1.3.5

1. Which of the following functions $u(z)$ are harmonic?
   (a) $2x - 3y$.
   (b) $e^x \cos y$.
   (c) $e^x \cos x$.
   (d) $\ln |z|$.
   (e) $x^2 + y^2$.
   (f) $x^2 - y^2$.
2. (a) Verify that $f(x) = |x|$ is not differentiable at $x = 0$. Note that the graph $y = |x|$ has a corner at $x = 0$.
   (b) Verify that $f_1(x) = \int_{-1}^{x} |t| \, dt$ is continuously differentiable, but has no second derivative at $x = 0$.
   (c) Verify that $f_n(x) = \int_{-1}^{x} f_{n-1}(t) \, dt$ is n times continuously differentiable, but lacks an $(n + 1)$-st derivative at $x = 0$. Here, f_n is defined by iteration.
   (d) Use (c) to prove that $\mathscr{C}^{k+1}(\Omega)$ is a proper subset of $\mathscr{C}^k(\Omega)$ for every domain Ω.

Section 1.4 INTEGRAL CALCULUS IN THE PLANE

1.4.0 Introduction

We turn now from differentiating to integrating. We will discuss two integrals, the "one-dimensional" line integral

$$\int_{\Gamma} [p(x, y) \, dx + q(x, y) \, dy]$$

and the "two-dimensional" double integral

$$\iint_{\Omega} u(x, y) \, dx \, dy.$$

The first of these will be defined in terms of the usual integral of calculus, using the fact that Γ is a parametrized curve.

We have already completed most of the topology (open sets, domains) and curve theory that we will need. Much of it will be used in our presentation of Green's Theorem. This will be the central result of this section and the fact that makes the integration theory of harmonic and complex analytic functions work.

1.4.1 Line Integrals

Let Γ be any piecewise-smooth curve in $\mathbb{R}^2$, and let $p = p(x, y)$, $q = q(x, y)$ be continuous real-valued functions defined in some set containing Γ. We wish to give a meaning to

$$I = \int_{\Gamma} [p\ dx + q\ dy].$$

To do this, note that a typical point (x, y) on Γ is of the form $(x, y) = \alpha(t) = (\alpha_1(t), \alpha_2(t))$, where $\alpha: [a, b] \to \Gamma$ is a parametrization of the curve Γ. Since $x = \alpha_1(t)$, $y = \alpha_2(t)$ on Γ, we have from calculus that $dx = \alpha'_1(t)\ dt$, $dy = \alpha'_2(t)\ dt$. Let us rewrite the integral above with these replacements:

$$I_\alpha = \int_{t=a}^{t=b} [p(\alpha(t))\alpha'_1(t)\ dt + q(\alpha(t))\alpha'_2(t)\ dt].$$

Check and see that this is now a one-variable definite integral (on the t-axis) of the type we learned at our mother's knee. Were we given the functions p, q, and the parametrization $\alpha(t)$ explicitly, we could hope to compute I_α as a definite integral and get a number.

We are tempted to define $I = I_\alpha$ except for one thing. Suppose we had chosen a *different* parametrization for the curve Γ, perhaps the arc length parametrization $\sigma(s)$. Would we get the same number? Does $I_\sigma = I_\alpha$? If so, it would be reasonable to *define I* as I_α $(= I_\sigma = \cdots)$.

The following fact resolves our quandary.

THEOREM 7

Let α and β be equivalent parametrizations of the curve Γ. Then $I_\alpha = I_\beta$.

Note: In Section 1.2, $\alpha: [a_1, b_1] \to \Gamma$ and $\beta: [a_2, b_2] \to \Gamma$ were defined to be equivalent if and only if $\alpha(t) = \beta(E(t))$, where $E: [a_1, b_1] \to [a_2, b_2]$ is a one-one-onto continuous piecewise differentiable function with $E'(t) > 0$ for all t in $[a_1, b_1]$. As usual, it suffices to consider α smooth, E differentiable.

Proof: Say $\alpha(t) = \beta(E(t))$, $s = E(t)$. Thus, $ds = E'(t)\ dt$. Then we have

$$I_\alpha = \int_{t=a_1}^{t=b_1} [p(\alpha(t))\alpha'_1(t)\ dt + q(\alpha(t))\alpha'_2(t)\ dt]$$

$$= \int_{t=a_1}^{t=b_1} [p(\beta(E(t)))(\beta_1(E(t)))'\ dt + q(\beta(E(t)))(\beta_2(E(t)))'\ dt]$$

$$= \int_{t=a_1}^{t=b_1} [p(\beta(E(t)))\beta'_1(E(t))E'(t)\ dt + q(\beta(E(t)))\beta'_2(E(t))E'(t)\ dt]$$

$$= \int_{s=a_2}^{s=b_2} [p(\beta(s))\beta'_1(s)\ ds + q(\beta(s))\beta'_2(s)\ ds]$$

$$= I_\beta.$$

Done.

Note: This is a glorious exercise in the Chain Rule.

Since equivalent parametrizations yield the same definite integral, we define $I = \int_\Gamma [p\ dx + q\ dy]$ to be I_α, as above.

Observation

Since $x = \alpha_1(t)$, $y = \alpha_2(t)$, we have

$$\frac{dx}{dt}\ dt = \alpha'_1(t)\ dt,$$

$$\frac{dy}{dt}\ dt = \alpha'_2(t)\ dt.$$

A good way to remember the definition of I follows from this. It is (if t varies from $t = a$ to $t = b$)

$$I = \int_\Gamma [p\ dx + q\ dy] = \int_a^b \left(p\frac{dx}{dt} + q\frac{dy}{dt} \right) dt.$$

Caution: Even though the notation $\int_\Gamma [p\ dx + q\ dy]$ does not indicate that an explicit parametrization α is given, we will suppose that a parametrization is given, and we will work with this parametrization and those equivalent to it. A less ambiguous notation would be $\int_\alpha [p\ dx + q\ dy]$, with $\int_\alpha = \int_\beta$ if α and β are equivalent. Inequivalent parametrizations of Γ may, of course, yield different values of the line integral.

Example

Let $p(x, y) = xy$, $q(x, y) = 3$ (constant), and let $\alpha: [0, \pi] \to \Gamma$ parametrize the unit semicircle

$$\Gamma = \{z = (x, y) \mid |z| = 1, y \geq 0\}$$

via $\alpha(\theta) = (\cos \theta, \sin \theta)$; see Figure 1.26. What is $\int_\Gamma [p\ dx + q\ dy]$?

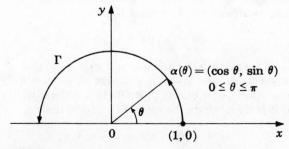

Figure   1.26

We have

$$
\int_{\Gamma} [p \, dx + q \, dy] = \int_{\theta=a}^{\theta=b} [p(\alpha(\theta))\alpha'_1(\theta) \, d\theta + q(\alpha(\theta))\alpha'_2(\theta) \, d\theta]
$$

$$
= \int_0^{\pi} [-\cos\theta \sin^2\theta \, d\theta + 3\cos\theta \, d\theta]
$$

$$
= \left[-\frac{\sin^3\theta}{3} + 3\sin\theta \right]_{\theta=0}^{\theta=\pi}
$$

$$
= 0.
$$

Let's compute this integral again. But now we choose another parametrization of Γ which is equivalent to α. We'll verify that the integral is again equal to zero, as promised by Theorem 7.

Let $\beta: [-1, 1] \to \Gamma$ be given by $\beta(t) = (-t, \sqrt{1 - t^2})$. You should check that $\beta(t)$ traverses the semicircle Γ in a counterclockwise manner as t increases from $t = -1$ to $t = 1$. Can you find $E: [0, \pi] \to [-1, 1]$ such that $t = E(\theta)$? Note $-t = \cos\theta = x$.

Now we compute I_β directly. We have

$$
I_\beta = \int_{t=-1}^{t=1} [p(\beta(t))\beta'_1(t) \, dt + q(\beta(t))\beta'_2(t) \, dt]
$$

$$
= \int_{t=-1}^{t=1} [-t\sqrt{1 - t^2}\,(-1) \, dt - 3t(1 - t^2)^{-\frac{1}{2}} \, dt]
$$

$$
= \frac{1}{2} \left[-\frac{1}{3}(1 - t^2)^{\frac{3}{2}} + 3(1 - t^2)^{\frac{1}{2}} \right]_{t=-1}^{t=1}
$$

$$
= 0.
$$

Thus, $I_\alpha = I_\beta$, verifying Theorem 7.

The Line Integral of an Exact Differential

We are about to compute $\int_{\Gamma} p \, dx + q \, dy$ for an important class of integrands.

Our expressions $p \, dx + q \, dy$ are known as "differentials," "first-order differentials," or "one-forms." If we are given a function $u = u(x, y)$ in $\mathscr{C}^1(\Omega)$, then its differential is defined to be

$$
du = u_x \, dx + u_y \, dy.
$$

For example, if $u(x, y) = x^2 y$, then $du = 2xy \, dx + x^2 \, dy$.

Now suppose we are given a differential $p \, dx + q \, dy$ with the functions $p = p(x, y)$, $q = q(x, y)$ in $\mathscr{C}(\Omega)$. This differential is said to be *exact in Ω* if it is equal to du for some u in $\mathscr{C}^1(\Omega)$; that is,

$$
p \, dx + q \, dy = du = u_x \, dx + u_y \, dy.
$$

In other words, p and q are not random but are related by the conditions $p = u_x, q = u_y$.

An interesting question (see exercises) is, "Given $p\,dx + q\,dy$, is it exact (in some Ω)?" We'll encounter this again.

Exact differentials, being derivatives (in a sense), have nice integral properties, as the following result shows. It should remind you of the Fundamental Theorem of Calculus.

THEOREM 8

Let Γ be a piecewise-smooth curve from z_0 to z_1 lying inside a domain Ω. Let $u \in \mathscr{C}^1(\Omega)$. Then

$$\int_\Gamma du = u(z_1) - u(z_0).$$

Note: Since the right-hand side here depends only on z_0, z_1, but not on Γ, the integral is independent of the curve joining z_0 to z_1.

Proof: Let $\alpha: [a, b] \to \Gamma$ be the parametrization. Then

$$\int_\Gamma du = \int_a^b [u_x(\alpha(t))\alpha'_1(t)\,dt + u_y(\alpha(t))\alpha'_2(t)\,dt]$$

$$= \int_a^b \frac{d}{dt}\,u(\alpha(t))\,dt$$

$$= u(\alpha(b)) - u(\alpha(a))$$

$$= u(z_1) - u(z_0).$$

Note that this boils down to the Chain Rule for $u(\alpha(t))$ and the Fundamental Theorem of Calculus for $(d/dt)u(\alpha(t))$. Done.

Question

Let Γ be a loop, $z_1 = z_0$, inside Ω. If $u \in \mathscr{C}^1(\Omega)$, what is $\int_\Gamma du$?

Exercises to Paragraph 1.4.1

1. Compute the following line integrals directly from the definition. In each case, Γ is the unit circle parametrized by $\alpha(t) = (\cos t, \sin t)$, $0 \le t \le 2\pi$, as usual.
   (a) $\int_\Gamma [x\,dx - y\,dy]$.
   (b) $\int_\Gamma x(dx + dy)$.
   (c) $\int_\Gamma y\,dx$.
   (d) $\int_\Gamma dy$.
2. (a) Prove that if the differential $p\,dx + q\,dy$ is exact in Ω (that is, $p\,dx + q\,dy = du$) and $u \in \mathscr{C}^2(\Omega)$, then $p_y = q_x$.
   (b) Do you think the converse is true? (A delicate question.)

3. Which of the following differentials are exact in $\Omega = \mathbb{R}^2$? Find u if $p \, dx + q \, dy = du$.
   (a) $y \, dx + x \, dy$.
   (b) dx.
   (c) $y \, dx - x \, dy$.
   (d) $e^x \cos y \, dx - e^x \sin y \, dy$.

4. (a) Let $u \in \mathscr{C}^1(\Omega)$. Prove that if Γ is a closed loop lying inside Ω, then $\int_\Gamma du = 0$.
   (b) To which of the integrals of Exercise 1 does this apply?

5. Let Ω be the plane $\mathbb{R}^2$ with the nonpositive x-axis deleted. Recall $\theta = \arg z \in \mathscr{C}^2(\Omega)$, once we agree that $-\pi < \theta < \pi$. See the exercises to Paragraph 1.3.1.
   (a) Verify $d\theta = (x^2 + y^2)^{-1} (-y \, dx + x \, dy)$.
   (b) Observe that the functions $-y(x^2 + y^2)^{-1}, x(x^2 + y^2)^{-1}$ are defined in $\mathbb{R}^2 - \{0\}$ and hence on the unit circle Γ.
   (c) Verify $\int_\Gamma (x^2 + y^2)^{-1} (-y \, dx + x \, dy) = 2\pi \neq 0$. Note that $|z| = 1$.
   (d) Why doesn't (c) contradict the result of Exercise 4(a)?
   (e) How does (c) reflect the multivaluedness of $\arg z$? This example is worthy of meditation.

6. (a) Find $u(x, y)$ such that $ye^x \, dx + e^x \, dy = du$.
   (b) Compute $\int_\Gamma [ye^x \, dx + e^x \, dy]$, where Γ is the graph of $y = \sin x$ parametrized by x, $0 \le x \le \pi/2$. *Hint:* Make the most of (a).

7. (a) Prove that if Γ is horizontal [vertical], then $dy = 0$ [$dx = 0$] on Γ. This may simplify certain integrations.
   (b) Let $\Gamma = \Gamma_1 + \Gamma_2$, where Γ_1 is the horizontal segment from $z_0 = (x_0, y_0)$ to $z_1 = (x_1, y_0)$ and Γ_2 is the vertical segment from z_1 to $z_2 = (x_1, y_1)$. Prove that

$$\int_\Gamma \left[p(x, y) \, dx + q(x, y) \, dy \right] = \int_{x_0}^{x_1} p(x, y_0) \, dx + \int_{y_0}^{y_1} q(x_1, y) \, dy,$$

provided the parametrization of Γ_1 is equivalent to one of the natural parameters $\pm x$ and likewise for Γ_2 and $\pm y$ (as is usually the case).

8. Prove the following algebraic properties of the line integral by reducing to the case of ordinary integrals.
   (a) $(\int_\Gamma [p_1 \, dx + q_1 \, dy]) + (\int_\Gamma [p_2 \, dx + q_2 \, dy]) = \int_\Gamma [(p_1 + p_2) dx + (q_1 + q_2) \, dy]$.
   (b) $\int_\Gamma ap \, dx + bq \, dy = a \int_\Gamma p \, dx + b \int_\Gamma q \, dy$, $(a, b, \in \mathbb{R})$.

9. *Proving* $\int_{-\Gamma} [p \, dx + q \, dy] = - \int_\Gamma [p \, dx + q \, dy]$.
   (a) Recall that if Γ is parametrized by $\alpha(t)$, $a \le t \le b$, then $-\Gamma$ denotes the reverse journey that is given by $\gamma(\tau) = \alpha(t(\tau))$, where $t = t(\tau) = a - (\tau - b)$ and $a \le \tau \le b$. Refer to the exercises to Paragraph 1.2.3.
   (b) It suffices to prove $\int_{-\Gamma} p \, dx = - \int_\Gamma p \, dx$. Verify that

$$\int_{-\Gamma} p(x, y) \, dx = \int_{\tau=a}^{\tau=b} p(\gamma(\tau)) \frac{d\gamma_1(\tau)}{d\tau} \, d\tau$$

where, as usual, $(x, y) = \gamma(\tau) = (\gamma_1(\tau), \gamma_2(\tau))$.
   (c) Verify that the right-hand integral in (b) equals

$$\int_{t=b}^{t=a} p(\alpha(t)) \frac{d\alpha_1(t)}{dt} \, dt.$$

Hint: Use $\gamma(\tau) = \gamma(\tau(t)) = \alpha(t)$ and the chain rule.

(d) Observe that the integral in (c) equals

$$-\int_\Gamma p(x, y)\, dx$$

as desired. Note that this is nothing more than a long-winded variation on the familiar relation

$$\int_b^a f(x)\, dx = -\int_a^b f(x)\, dx$$

of ordinary calculus.

1.4.2 Double Integrals, Iterated Integrals

Now we turn from one-dimensional (line) integrals to two-dimensional (area) integrals. We trust you have seen some of these already.

The Idea of a Double Integral

Suppose that Ω is a bounded domain (a disc, say) in $\mathbb{R}^2$, and let $u = u(x, y)$ be a continuous function defined on $\overline{\Omega}$, $u \in \mathscr{C}(\overline{\Omega})$. If $u(x, y) \geq 0$ for all $(x, y) \in \overline{\Omega}$, then the double integral

$$\iint\limits_\Omega u\, dA$$

will be the volume (a nonnegative number) of the right cylinder whose base (in the plane) is $\overline{\Omega}$ and whose height above (x, y) is $u(x, y)$; see Figure 1.27. If $u(x, y)$ is negative for some (x, y), then minor adjustments must be made to this description.

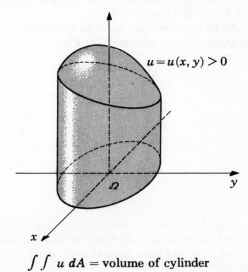

Figure   1.27

$$\iint\limits_\Omega u\, dA = \text{volume of cylinder}$$

On the Definition

This interpretation in terms of volume is not a definition of

$$\iint_{\Omega} u \, dA$$

because we have not defined "volume." The usual definition is in terms of a limit. Thus, let us subdivide $\overline{\Omega}$ into pieces $\Delta A_1, \ldots, \Delta A_n$ (where ΔA_j also stands for the area of the jth region). Let z_j be a point in the jth piece ΔA_j and form the sum

$$S_n = \sum_{j=1}^{n} u(z_j) \, \Delta A_j.$$

It can be shown that for simple types of domain Ω, pieces ΔA_j, and continuous functions $u(x, y)$, a sequence $S_1, S_2, S_3, \ldots, S_n, \ldots$ of such sums (numbers!) will approach a definite limit, provided the "diameters" of all pieces ΔA_j approach zero as n approaches ∞. Moreover, this limit will depend only on u and Ω, not on the choice of subdivisions. We call this limit the *double integral* of u *over* Ω, denoted

$$\iint_{\Omega} u \, dA \qquad \text{or} \qquad \iint_{\Omega} u \, dx \, dy.$$

Computation: Iterated Integrals

Once again we find ourselves with a reasonable definition, but with no means of computing. Happily, however, many double integrals may be computed as "iterated" integrals: two one-dimensional integrations performed successively.

An example should serve to recall this method. Given that Ω is the triangular region pictured in Figure 1.28 and that $u(x, y) = xy$, we have that

$$\begin{aligned}
\iint_{\Omega} u \, dx \, dy &= \int_{x=0}^{x=1} \left[\int_{y=0}^{y=x} u \, dy \right] dx \\
&= \int_{x=0}^{x=1} \left[\frac{xy^2}{2} \right]_{y=0}^{y=x} dx \\
&= \int_{x=0}^{x=1} \frac{x^3}{2} \, dx \\
&= \frac{1}{8}.
\end{aligned}$$

Note that we "integrated out" the variable y first, obtaining an integral in x alone, and then integrated out the variable x to get a number. It would have been slightly more complicated to integrate x out first.

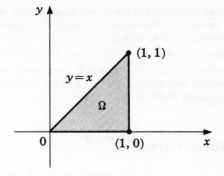

Figure   1.28

Exercises to Paragraph 1.4.2

1. Let Ω be the unit square with corners $(0, 0)$, $(1, 0)$, $(1,1)$, $(0,1)$. Compute the following double integrals:
   (a) $\iint_\Omega dx\, dy$,
   (b) $\iint_\Omega xy\, dx\, dy$.

2. Let $u, v \in \mathcal{C}(\overline{\Omega})$. True or false?
   (a) $\iint_\Omega dx\, dy = $ area of Ω.
   (b) $u \geq v$ on Ω implies $\iint_\Omega u\, dx\, dy \geq \iint_\Omega v\, dx\, dy$.
   (c) $u > 0$ throughout Ω implies $\iint_\Omega u\, dx\, dy > 0$.
   (d) The sign of $\iint_\Omega u\, dx\, dy$ depends on the orientation of $\partial\Omega$.
   (e) If $|u(z)| \leq M$ for all $z \in \overline{\Omega}$, then $|\iint_\Omega u\, dx\, dy| \leq M$ (area of Ω).

3. Let $\Omega = \{(x, y) \mid x_0 < x < x_1, y_0 < y < y_1\}$, a rectangle with sides parallel to the coordinate axes. Suppose $p(x, y) \in \mathcal{C}^1(\Omega^+)$, where Ω^+ is a domain containing $\overline{\Omega}$.
   (a) Prove $\iint_\Omega p_y\, dx\, dy = \int_{x_0}^{x_1} (p(x, y_1) - p(x, y_0))\, dx$.
   (b) Prove that the right-hand side of (a) equals $-\int_{\partial\Omega} p\, dx$, provided the path $\partial\Omega$ is given a positive (counterclockwise) orientation in the usual manner.

   This is the essence of Green's Theorem for a rectangle. See **Paragraph 1.4.3** for interpretation.

4. Let D be the disc $D(0; R)$, and let r, θ be the usual polar coordinates. Suppose $u \in \mathcal{C}(\overline{D})$.
   (a) Verify

$$\iint_D u(x, y)\, dx\, dy = \iint_D u(r\cos\theta, r\sin\theta)r\, dr\, d\theta$$

$$= \int_{\theta=0}^{\theta=2\pi} \left\{ \int_{r=0}^{r=R} u(r\cos\theta, r\sin\theta)r\, dr \right\} d\theta.$$

   (b) Verify that we may interchange the order of integration in the last integral of (a).
   (c) Suppose $u(x, y) = \varphi(x^2 + y^2)$ for some function $\varphi(t)$. Verify that

$$\iint_D u\, dx\, dy = 2\pi \int_0^R \varphi(r^2)r\, dr.$$

Note here that the integrand depends only on distance r but not on angle θ. An example of this would be $u(x, y) =$ temperature at (x, y) due to a point-heat source situated at the origin 0.

5. Compute $\iint_D (x^2 + y^2)\, dx\, dy$ where $D = D(0; 1)$. *Hint:* Use Exercise 4(c).

6. Compute $\iint_\Omega xy\, dx\, dy$, where Ω is the triangle of the example in the text, by integrating first with respect to x. Your result should, of course, agree with that obtained in the text.

1.4.3 Green's Theorem

This is the most important result of Chapter 1. The statements of Green's Theorem and the three Green's Identities (coming) will involve most of the notions we have discussed so far: domains, arc length, the gradient, the outward normal derivative, line integrals, double integrals.

What is Green's Theorem? It is the two-dimensional version of the familiar Fundamental Theorem of Calculus. We recall that this latter theorem says

$$F(b) - F(a) = \int_a^b F'(x)\, dx.$$

Think of it this way: The left-hand side is concerned with the behavior of the function F on the boundary (two points!) of the interval, while the right-hand side is concerned with the behavior of the "differential" $dF = F'(x)\, dx$ on the entire interval (domain of $F(x)$).

Just so, Green's Theorem will tell us that

$$\int_{\partial\Omega} \left[p\, dx + q\, dy \right] = \iint_\Omega (q_x - p_y)\, dx\, dy.$$

Again, the left-hand side is concerned with the behavior of $p\, dx + q\, dy$ on a boundary (curve!), while the right-hand side is concerned with the behavior of the "differential" $(q_x - p_y)\, dx\, dy$ over the entire domain. We will make a more complete statement of Green's Theorem in a moment.

History: The English scientist George Green included this theorem in a paper published in 1828. However, others had known of it earlier, notably Lagrange and Gauss.

Comment

We will state and apply a general version of Green's Theorem. However, we will give a proof only in a special case, albeit an instructive one, namely, the case where the domain is a rectangle. Most of the difficulty of the general proof is caused by the wildly curving boundaries that a Jordan domain may have. A proof of an even more general version is given in Apostol's book, *Mathematical Analysis*.

Before stating the theorem, we introduce another notation. If Ω is a domain with boundary $\partial\Omega$ as usual, we let Ω^+ denote a larger domain containing both Ω and $\partial\Omega$. Our purpose is this: We wish to be sure that our functions are nicely behaved on $\partial\Omega$ as well as inside Ω. Hence, we will play it safe and require that our functions be well behaved on the larger domain Ω^+.

GREEN'S THEOREM

Let Ω be a k-connected Jordan domain, and suppose $p(x, y)$, $q(x, y)$ are functions in $\mathscr{C}^1(\Omega^+)$, where Ω^+ is some domain containing Ω and $\partial\Omega$. Then

$$\int_{\partial\Omega} [p\ dx + q\ dy] = \iint_{\Omega} (q_x - p_y)\ dx\ dy.$$

Proof: It suffices to prove the two independent equalities

$$\int_{\partial\Omega} p\ dx = -\iint_{\Omega} p_y\ dx\ dy,$$

$$\int_{\partial\Omega} q\ dy = \iint_{\Omega} q_x\ dx\ dy.$$

The minus sign in the first equality derives from the orientation. We shall prove this first equality and leave the second as an exercise. The two are entirely similar.

As mentioned above, we consider only the special case of a rectangle:

$$\Omega = \{(x, y) \mid x_0 < x < x_1, y_0 < y < y_1\}$$

Thus $\partial\Omega$ consists of the parametrized line segments Γ_1, Γ_2, Γ_3, Γ_4 in Figure 1.29. Note the counterclockwise orientation.

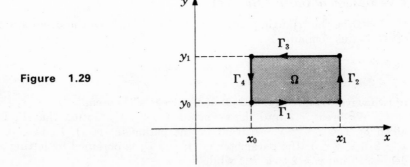

Figure 1.29

We have

$$-\iint_\Omega p_y \, dx \, dy = -\int_{x_0}^{x_1} \left(\int_{y_0}^{y_1} p_y \, dy \right) dx$$

$$= -\int_{x_0}^{x_1} (p(x, y_1) - p(x, y_0)) \, dx$$

$$= \int_{x_0}^{x_1} p(x, y_0) \, dx + \int_{x_1}^{x_0} p(x, y_1) \, dx$$

$$= \int_{\Gamma_1} p(x, y) \, dx + \int_{\Gamma_3} p(x, y) \, dx$$

$$= \int_{\partial\Omega} p(x, y) \, dx,$$

where the reasons for the five equal signs are first, iteration of the integral; second, Fundamental Theorem of Calculus for the integrand p_y considered as a function of y alone; third, basic algebra of the usual integral; fourth, notation; fifth, $dx = 0$ on the vertical sides Γ_2 and Γ_4. This finishes our proof.

Comments

1. Green's Theorem will enable us to obtain further integral relations between behavior inside Ω and behavior on $\partial\Omega$.

2. Our style of proof works for the more general convex domain whose boundary consists of curves that can be expressed as the graphs of functions, as pictured in Figure 1.30.

3. The one nontrivial idea occurring in our proof is the Fundamental Theorem of Calculus for functions of one variable.

4. We will use Green's Theorem freely on Jordan domains Ω. Remember that for us a Jordan domain may have holes (the annulus) and that Green's Theorem concerns the integral $\int_{\partial\Omega}$ over the full boundary of Ω, a total of k Jordan curves if Ω is k-connected.

A Verification of Green's Theorem

Let Ω be the triangle pictured in Figure 1.31. Note the orientation of $\partial\Omega$. Let us compute

$$\int_{\partial\Omega} [y \, dx - x \, dy]$$

in two ways: (1) directly, and (2) using Green's Theorem.

1. We break $\partial\Omega$ into the segments Γ_1, Γ_2, Γ_3, noting that Γ_1 is given by $1 \le x \le 2, y = 0$ (that is, the parameter is x); Γ_2 is given $x = 2, 0 \le y \le 1$ (the parameter is y); and Γ_3 is obtained by letting x decrease from $x = 2$ to $x = 1$ while $y = x - 1$.

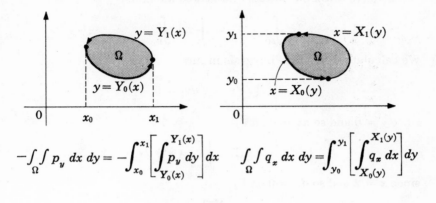

$$-\iint_{\Omega} p_y \, dx \, dy = -\int_{x_0}^{x_1}\left[\int_{Y_0(x)}^{Y_1(x)} p_y \, dy\right]dx \qquad \iint_{\Omega} q_x \, dx \, dy = \int_{y_0}^{y_1}\left[\int_{X_0(y)}^{X_1(y)} q_x \, dx\right]dy$$

$$\overline{\Omega} = \overline{\Omega}_1 \cup \overline{\Omega}_2$$

Typical decomposition into convex domains:

$$\iint_{\Omega} = \iint_{\Omega_1} + \iint_{\Omega_2}$$

Figure   1.30

Figure   1.31

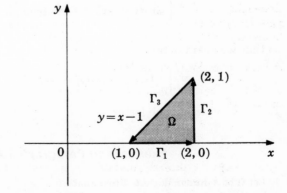

Now we break our integral into three integrals:

$$\int_{\partial\Omega} = \int_{\Gamma_1} + \int_{\Gamma_2} + \int_{\Gamma_3}$$

We calculate these three integrals in turn.

$$\int_{\Gamma_1} [y \, dx - x \, dy] = 0,$$

since $y = 0$ and so $dy = 0$ on Γ_1.

$$\int_{\Gamma_2} [y \, dx - x \, dy] = \int_{y=0}^{y=1} (-2) \, dy = -2,$$

since $x = 2$ and so $dx = 0$ on Γ_2.

$$\int_{\Gamma_3} [y \, dx - x \, dy] = \int_{x=2}^{x=1} [(x - 1) \, dx - x \, dx],$$

since $dy = d(x - 1) = dx$ on Γ_3. This becomes

$$-\int_{x=1}^{x=2} (-1) \, dx = 2 - 1 = 1.$$

Putting all this together, we obtain with our bare hands

$$\int_{\partial\Omega} [y \, dx - x \, dy] = 0 - 2 + 1 = -1.$$

2. Green's Theorem tells us, since $q_x = -1, p_y = 1$, that

$$\int_{\partial\Omega} [y \, dx - x \, dy] = \iint_\Omega (-1 - 1) \, dx \, dy = \iint_\Omega (-2) \, dx \, dy.$$

But we note that

$$\iint_\Omega dx \, dy = \text{area of the triangle } \Omega = \tfrac{1}{2}$$

so that we again obtain

$$\int_{\partial\Omega} [y \, dx - x \, dy] = -2 \left(\tfrac{1}{2}\right) = -1.$$

This method was much faster.

Exercises to Paragraph 1.4.3

1. Let Ω be a Jordan domain. Prove that
   (a) $(\tfrac{1}{2}) \int_{\partial\Omega} [-y \, dx + x \, dy] = $ area of Ω,
   (b) $\int_{\partial\Omega} x \, dy = $ area of Ω.
      Thus the area may be determined by integrating around the boundary!

2. Verify that Green's Theorem yields a short proof of $\int_\Gamma du = 0$ in the special case that $u \in \mathscr{C}^2(\Omega)$ and both the loop Γ and the points enclosed by Γ lie within Ω. Compare equality of mixed partial derivatives.

   Of course we've proved this already in somewhat more general circumstances.

3. Prove $\int_{\partial\Omega} q\, dy = \int\int_\Omega q_x\, dx\, dy$ (half of Green's Theorem) in the case Ω is a rectangle with sides parallel to the coordinates axes. Use the method of the text.

4. Prove that $\int_\Gamma [p(x)\, dx + q(y)\, dy] = 0$, provided $p(x)$, $q(y)$ (functions of x and y, respectively), are continuously differentiable on and inside the closed loop Γ.

5. *A physical interpretation of Green's Theorem.* Let $V(x, y) = \langle p(x, y), q(x, y)\rangle$ be a *vector field*, that is, a rule that affixes to the point (x, y) the vector $V(x, y)$. Suppose $p, q \in \mathscr{C}^1(\Omega^+)$ (usual convention with $\Omega^+ \supset \overline{\Omega} \supset \Omega$).

   (a) Verify $\int_{\partial\Omega} [p\,dx + q\,dy] = \int_{\partial\Omega} V \cdot T\, ds$, where $T = T(s)$ is the unit tangent to the curves comprising $\partial\Omega$ as usual, and the integral on the right is taken over all such boundary curves.

   If we think of $V(x, y)$ pointing out a fluid flow over the plane, then $\int_{\partial\Omega} V \cdot T\, ds$, the *circulation*, indicates the net amount of flow around the curves $\partial\Omega$.

   (b) We define $q_x - p_y$ to be the *curl* of $V = \langle p, q\rangle$, written curl V. The number curl $V(x, y)$ measures the rotational tendencies of the flow V at the point (x, y). Prove (mathematically *and* physically) that

   $$\text{Circulation around } \partial\Omega = \iint_\Omega \text{curl } V\, dx\, dy.$$

6. Prove Green's Theorem for the convex domain Ω whose boundary (a loop) consists of four curves, each of which can be expressed as the graph of a function of x or y, as mentioned in the text (see Comment 2).

   Note that the disc and triangle are included here.

7. *What guarantees* $\int_\Gamma [p\, dx + q\, dy] = 0$? Let Γ be a simple closed curve in the domain Ω, and suppose the functions p, q are in $\mathscr{C}(\Omega)$. Consider the following hypotheses:

   (i) The interior of Γ is contained in Ω;
   (ii) Ω is simply connected (no holes);
   (iii) $p\, dx + q\, dy$ is exact in Ω;
   (iv) $p, q \in \mathscr{C}^1(\Omega)$ and $p_y = q_x$ throughout Ω.

   (a) Only hypothesis (iii) is sufficient to guarantee that the integral in question vanishes. Give a full justification of this statement, including proofs and counterexamples.

   (b) Verify that either geometric hypothesis (i) or (ii), taken with either condition (iii) or (iv), together suffice to guarantee that the integral vanishes.

8. *What guarantees* $p\, dx + q\, dy$ *is exact?* Let $p, q \in \mathscr{C}(\Omega)$. Consider the following conditions:

   (i) $p, q \in \mathscr{C}^1(\Omega)$ and $p_y = q_x$ throughout Ω;
   (ii) If Γ is any closed curve in Ω, then $\int_\Gamma [p\, dx + q\, dy] = 0$;
   (iii) If Γ is any simple closed curve in Ω whose interior also lies in Ω, then $\int_\Gamma [p\, dx + q\, dy] = 0$;
   (iv) For every pair z_0, z_1 of points in Ω, the line integral of $p\, dx + q\, dy$ is the same over every curve from z_0 to z_1 which lies in Ω ("independence of path").

(a) Which of these conditions is alone sufficient to guarantee that $p \, dx + q \, dy = du$ for some $u \in \mathscr{C}^1(\Omega)$? Give proof or counterexample for each condition. *Helpful:* Exercise 5 to Paragraph 1.4.1.

(b) Suppose now we give Ω some structure, namely, we grant that Ω is simply connected. Which of (i), (ii), (iii), (iv) now guarantee the existence of u?

(c) *For meditation:* Some "likely" differentials fail to be exact on Ω if Ω is not simply connected.

9. *The Inside-Outside Theorem.* Let Ω be a k-connected Jordan domain and $u(x, y) \in \mathscr{C}^2(\Omega^+)$. Then

$$\int_{\partial\Omega} \frac{\partial u}{\partial n} \, ds = \iint_{\Omega} \Delta u \, dx \, dy.$$

Hint: Write $(\partial u/\partial n) \, ds$ in the form $p \, dx + q \, dy$ and apply Green's Theorem. See Paragraph 1.4.4 also, and Exercise 9 there.

1.4.4. Green's Identities

Green's Theorem may be used to simplify the computation of some line integrals. There are many other applications, however. The first of these will be the three Green's Identities. These formulas may appear somewhat bizarre, but they are very useful. The first two are all but immediate consequences of Green's Theorem. Green's III is a bit deeper (it works only in conjunction with a particular function, the logarithm) and takes a bit more work to establish.

Throughout our discussion, Ω is a Jordan domain and the functions $p(x, y)$, $q(x, y)$ are in $\mathscr{C}^2(\Omega^+)$; that is, all second partials exist and are continuous throughout some domain larger than $\overline{\Omega} = \Omega \cup \partial\Omega$. Hence, we need not worry about existence of derivatives at the boundary, etc.

We have, as usual, $\nabla p(x, y) = \langle p_x(x, y), p_y(x, y) \rangle$, the gradient of p (a vector depending on the point (x, y)), and $\Delta p(x, y) = p_{xx}(x, y) + p_{yy}(x, y)$, the Laplacian of p (a continuous function).

It is sometimes helpful to write

$$\nabla = \left\langle \frac{\partial}{\partial x}, \frac{\partial}{\partial y} \right\rangle,$$

the gradient differential operator, so that

$$\nabla \cdot \nabla = \left\langle \frac{\partial}{\partial x}, \frac{\partial}{\partial y} \right\rangle \cdot \left\langle \frac{\partial}{\partial x}, \frac{\partial}{\partial y} \right\rangle$$

$$= \frac{\partial^2}{\partial x^2} + \frac{\partial^2}{\partial y^2}.$$

But this is the Laplacian operator, $\Delta = \nabla \cdot \nabla$; this last is also written ∇^2, so that we sometimes see $\Delta = \nabla^2$.

An Important Integral

We will be discussing the integral

$$\int_{\partial\Omega} \frac{\partial q}{\partial n}\, ds.$$

Our notation here appears confused. The symbol $\int_{\partial\Omega}$ indicates a line integral over the Jordan curves that comprise $\partial\Omega$, whereas the integrand looks like an ordinary function of s (in contrast with $A\, dx + B\, dy$). Let us recall, therefore, that if $\sigma(s) = (\sigma_1(s), \sigma_2(s))$ parametrizes one of the curves of $\partial\Omega$ by arc length, then

$$N = N(s) = \langle \sigma'_2(s),\ -\sigma'_1(s) \rangle$$

(refer to Section 1.2). Hence,

$$\frac{\partial q}{\partial n}\, ds = \nabla q \cdot N\, ds$$

$$= \langle q_x, q_y \rangle \cdot \langle \sigma'_2(s),\ -\sigma'_1(s) \rangle\, ds$$
$$= q_x \sigma'_2(s)\, ds - q_y \sigma'_1(s)\, ds$$
$$= q_x\, dy - q_y\, dx.$$

Thus we agree that

$$\int_{\partial\Omega} \frac{\partial q}{\partial n}\, ds \qquad \text{means} \qquad \int_{\partial\Omega} [q_x\, dy - q_y\, dx],$$

an honest line integral! Our hybrid notation will prove worthwhile, however.

THEOREM (GREEN'S I)

Let Ω be a k-connected Jordan domain and $p, q \in \mathscr{C}^2(\Omega^+)$. Then

$$\iint_{\Omega} \nabla p \cdot \nabla q\, dx\, dy = \int_{\partial\Omega} p\, \frac{\partial q}{\partial n}\, ds - \iint_{\Omega} p\, \Delta q\, dx\, dy.$$

Meaning: This theorem is a two-dimensional version of integration by parts from ordinary calculus. Recall that

$$\int_a^b u\, dv = uv \Big]_a^b - \int_a^b v\, du.$$

Now correspond $u \leftrightarrow \nabla q$, $dv \leftrightarrow \nabla p$, whence $v \leftrightarrow p$, $du \leftrightarrow \nabla \cdot \nabla q = \Delta q$.

Proof: The idea is to use Green's Theorem on the line integral in the statement. Unwinding this integral yields

$$\int_{\partial\Omega} p\,\frac{\partial q}{\partial n}\,ds = \int_{\partial\Omega} [pq_x\,dy - pq_y\,dx]$$

$$= \int_{\partial\Omega} [(-pq_y)\,dx + (pq_x)\,dy]$$

$$= \iint_{\Omega} [p_x q_x + pq_{xx} + p_y q_y + pq_{yy}]\,dx\,dy$$

$$= \iint_{\Omega} [\nabla p \cdot \nabla q + p\,\Delta q]\,dx\,dy,$$

which is equivalent to Green's I. Done.

THEOREM (GREEN'S II)

Let Ω be a k-connected Jordan domain and $p, q \in \mathscr{C}^2(\Omega^+)$. Then

$$\int_{\partial\Omega} \left(p\,\frac{\partial q}{\partial n} - q\,\frac{\partial p}{\partial n}\right) ds = \iint_{\Omega} (p\,\Delta q - q\,\Delta p)\,dx\,dy.$$

Meaning: Symmetrized Green's I.

Proof: Since the left-hand side of Green's I is symmetric in p and q and the right-hand side is not, we may interchange p and q to get another formula. Subtracting the second formula from the first yields

$$0 = \int_{\partial\Omega} \left(p\,\frac{\partial q}{\partial n} - q\,\frac{\partial p}{\partial n}\right) ds - \iint_{\Omega} (p\,\Delta q - q\,\Delta p)\,dx\,dy,$$

which is Green's II. Done.

Before going on to Green's III, we pause to note a very easy consequence of Green's I. We will use it once or twice crucially. It appeared earlier in the exercises to Paragraph 1.4.3.

THE INSIDE-OUTSIDE THEOREM

Let Ω be a k-connected Jordan domain and $q \in \mathscr{C}^2(\Omega^+)$. Then

$$\int_{\partial\Omega} \frac{\partial q}{\partial n}\,ds = \iint_{\Omega} \Delta q\,dx\,dy.$$

Proof: Let $p = 1$ (constant) in Green's I. Done.

Comment on the Inside-Outside Theorem

The outward normal derivative $\partial q/\partial n$ measures the rate of change of q across the boundary $\partial\Omega$ at a fixed point on the boundary. Integrating this rate of change over the full boundary $\partial\Omega$ gives some sort of "net change" of q across $\partial\Omega$. This is the "outside" part of the theorem. We are told that this net change of q across $\partial\Omega$ can be obtained by integrating Δq over the "inside," that is, over Ω.

Towards Green's III

Now we come to Green's III and thereby raise a theme central to these chapters: determining interior behavior by behavior on the boundary. To be more explicit, we want to know $q(\zeta)$ for all $\zeta \in \Omega$ (a Jordan domain, of course) when we are given only (i) $q(\zeta)$ exists and is twice continuously differentiable on Ω and $\partial\Omega$, (ii) $\Delta q(z)$ is known for all $z \in \Omega$, (iii) $q(z)$ is known for all $z \in \partial\Omega$ (boundary values of q), (iv) $\nabla q(z)$ is known for all $z \in \partial\Omega$. Green's III assures us that $q(\zeta)$ is determined by the data (i)–(iv), and in fact we have the remarkable formula

$$q(\zeta) = \frac{1}{2\pi} \iint\limits_{\Omega} \ln r \, \Delta q(z) \, dx \, dy - \frac{1}{2\pi} \int_{\partial\Omega} \left(\ln r \, \frac{\partial q}{\partial n}(z) - q(z) \frac{\partial \ln r}{\partial n} \right) ds,$$

where $r = |z - \zeta|$. In this formula, $z = (x, y)$ and all the z's, x's, and y's are "integrated out," leaving a number that depends on ζ. This number is $q(\zeta)$.

The function $\ln r$ is, of course, the natural logarithm of r. We need some information about $\ln r$ in order to prove Green's III. Consider

LEMMA 9

Fix $\zeta \in \mathbb{R}^2$. Let $r = |z - \zeta|$ for variable $z \in \mathbb{R}^2$. Then

  (i) *$\ln r$ is constant on circles $C(\zeta; r)$ centered at ζ;*
  (ii) *$\ln r$ is harmonic as a function of $z = (x, y)$ in the punctured plane $\mathbb{R}^2 - \{\zeta\}$; that is, $\Delta \ln r = 0$.*

Proof: (i) This is true because r itself (the radius) is constant on circles.

(ii) This is a straightforward exercise in taking partial derivatives. To begin, if $z = (x, y)$ and $\zeta = (\xi, \eta)$, then $r = [(x - \xi)^2 + (y - \eta)^2]^{1/2}$. From basic calculus, therefore, $\ln r = \frac{1}{2} \ln [(x - \xi)^2 + (y - \eta)^2]$. We leave it to you to show that the Laplacian of this function vanishes. Done.

Now we may establish our main result.

THEOREM (GREEN'S III)

*Let Ω be a k-connected Jordan domain with $\zeta \in \Omega$ and let $q \in \mathscr{C}^2(\Omega^+)$.
Let $z = (x, y)$ and $r = |z - \zeta|$. Then the value $q(\zeta)$ is given by*

$$q(\zeta) = \frac{1}{2\pi} \int\int_{\Omega} \ln r \, \Delta q(z) \, dx \, dy$$

$$- \frac{1}{2\pi} \int_{\partial\Omega} \left(\ln r \, \frac{\partial q}{\partial n}(z) - q(z) \frac{\partial \ln r}{\partial n} \right) ds.$$

Proof: Our plan is to apply Green's II with $p = \ln r$ in the domain $\Omega - \bar{D}(\zeta; \varepsilon)$, where $\bar{D}(\zeta; \varepsilon)$ is the *closed* disc of radius $\varepsilon > 0$. Then we will let ε tend to zero and show that everything works. Note that we can't apply Green's II to the full domain Ω because $\ln r$ is not defined if $z = \zeta$ (whence $r = 0$).

(a) We orient $C(\zeta; \varepsilon)$, the circular boundary of $\bar{D}(\zeta; \varepsilon)$, so that the outward (that is, outward from $\Omega - \bar{D}(\zeta; \varepsilon)$)) normal points into the disc $\bar{D}(\zeta; \varepsilon)$ directly toward ζ. See Figure 1.32.

Figure 1.32

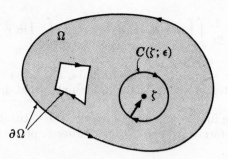

(b) Hence, taking the outward normal derivative in the direction of N gives -1 times the "radial" derivative, that is, the derivative with respect to $r(=$ the distance from $\zeta)$.

(c) From this follows

$$\frac{\partial p}{\partial n} = -\frac{\partial p}{\partial r} = -\frac{1}{r},$$

which equals $-(1/\varepsilon)$ on $C(\zeta; \varepsilon)$. Note here that $p = \ln r$.

(d) Writing Green's II and using $\Delta p = \Delta \ln r = 0$, $\partial p/\partial n = -(1/\varepsilon)$ on $C(\zeta; \varepsilon)$, $ds = \varepsilon \, d\theta$ on $C(\zeta; \varepsilon)$ (polar coordinates centered at ζ), we have

$$\int_{\partial\Omega} \left(\ln r \, \frac{\partial q}{\partial n} - q \frac{\partial \ln r}{\partial n} \right) ds + \int_0^{2\pi} \left(\ln \varepsilon \, \frac{\partial q}{\partial n} + q \frac{1}{\varepsilon} \right) \varepsilon \, d\theta$$

$$= \int\int_{\Omega - \bar{D}(\zeta;\varepsilon)} \ln r \, \Delta q \, dx \, dy.$$

Here we have used the fact that the boundary of $\Omega - \overline{D}(\zeta; \varepsilon)$, for small enough ε, is the union of $\partial\Omega$ and $C(\zeta; \varepsilon)$. Thus we get integrals around $\partial\Omega$ and the circle $C(\zeta; \varepsilon)$.

(e) Now we examine

$$\int_0^{2\pi} \left(\ln \varepsilon \, \frac{\partial q}{\partial n} + q \, \frac{1}{\varepsilon} \right) \varepsilon \, d\theta.$$

We integrate each term separately. First, we have the crucial cancellation of $1/\varepsilon$ with ε (thanks to $\ln r$), giving $\int_0^{2\pi} q \, d\theta$. Here, q is regarded as a function on $C(\zeta; \varepsilon)$, so $q = q(\varepsilon, \theta)$ in *polar coordinates* centered at ζ. Since the integral involves θ only (ε is fixed), we have from the Average Value Theorem of integral calculus that there exists an angle θ_ε (depending on ε) such that

$$\int_0^{2\pi} q \, d\theta = 2\pi \cdot q(\varepsilon, \theta_\varepsilon), \qquad 0 \le \theta_\varepsilon \le 2\pi.$$

(Recall that the Average Value Theorem says that if $g(x)$ is continuous, then $\int_a^b g(x) \, dx = g(x^*)(b - a)$, where x^* is some point satisfying $a < x^* < b$. Geometrically, this says that the net area under the graph of g is equal to the area of a rectangle whose height is $g(x^*)$, the "average value" of $g(x)$ on the interval.) See Figure 1.33.

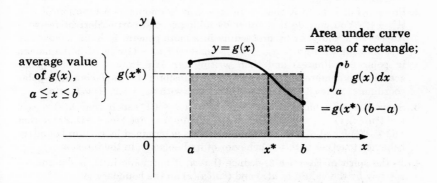

Average-Value Theorem of Integral Calculus

Figure   1.33

It follows that as ε tends to zero, $2\pi q(\varepsilon, \theta_\varepsilon)$ approaches $2\pi q(\zeta)$, since q is continuous and ζ is determined by the condition $\varepsilon = 0$.

(f) What of the other half, namely, $\int_0^{2\pi} \varepsilon \ln \varepsilon \, (\partial q / \partial n) \, d\theta$? This equals

$$\varepsilon \ln \varepsilon \int_0^{2\pi} \frac{\partial q}{\partial n} \, d\theta$$

for fixed ε. But as ε tends to zero, it can be shown that $\varepsilon \ln \varepsilon$ also tends to zero (use L'Hospital's Rule).

Hence, the equation in (d) reduces to

$$2\pi q(\zeta) = \lim_{\varepsilon \to 0} \iint_{\Omega - \overline{D}(\zeta;\varepsilon)} \ln r \, \Delta q \, dx \, dy - \int_{\partial\Omega} \left(\ln r \frac{\partial q}{\partial n} - q \frac{\partial \ln r}{\partial n} \right) ds.$$

We conclude that the limit on the double integral must exist (since everything else in the equation does) and it must equal

$$\iint_{\Omega} \ln r \, \Delta q \, dx \, dy.$$

This proves Green's III. Done.

Green's III is primarily of theoretical importance. We'll use it as a steppingstone to the Poisson Integral Formula in Chapter 2.

Exercises to Paragraph 1.4.4

1. Suppose p, q are harmonic, $\Delta p = \Delta q = 0$ (see Section 1.3) throughout Ω^+. Prove that
   (a) $\int_{\partial\Omega} p(\partial q/\partial n) \, ds = \int_{\partial\Omega} q(\partial p/\partial n) \, ds$,
   (b) $\int_{\partial\Omega} (\partial q/\partial n) \, ds = 0$.

2. Prove (ii) of Lemma 9, that is, $\ln |z - \zeta|$ is harmonic as a function of z in $\mathbb{R}^2 - \{\zeta\}$. You may do this either by taking second partial derivatives with respect to x and y, or by proceeding in a more general fashion, namely, by first showing that $\Delta u = u_{xx} + u_{yy} = u_{rr} + (1/r)u_r + (1/r^2)u_{\theta\theta}$ (the Laplacian in polar coordinates) and then writing $u(r, \theta) = \ln |z - \zeta| = \ln r$ (polar coordinates centered at ζ). You should transform the Laplacian into polar coordinates at least once in your life. Start with $u_x = u_r r_x + u_\theta \theta_x$, etc.

3. Suppose Ω is a Jordan domain, and $u, v \in \mathscr{C}^2(\Omega^+)$ such that $u(z) = v(z)$ and $(\partial u/\partial n)(z) = (\partial v/\partial n)(z)$ for $z \in \partial\Omega$, and $\Delta u(\zeta) = \Delta v(\zeta)$ for $\zeta \in \Omega$. Prove that $u(\zeta) = v(\zeta)$ for all $\zeta \in \Omega$. Thus, a function is determined by certain boundary behavior together with the behavior of its Laplacian in the interior.

4. In the spirit of Exercise 3, deduce that u, if harmonic in Ω, is determined entirely by the values of $u(z)$ and $(\partial u/\partial n)(z)$ on the boundary $\partial\Omega$.

   We comment that Chapter 2 will show that a harmonic function is fully determined by the values $u(z)$ on $\partial\Omega$ alone; that is, $(\partial u/\partial n)(z)$ also depends on the values of u on $\partial\Omega$ and the fact that $\Delta u \equiv 0$. It follows that we should not expect to specify the "boundary values" $u(z)$ and $(\partial u/\partial n)(z)$, $z \in \partial\Omega$, at will, and then concoct a harmonic function u in Ω which assumes those values. In general, we must be very careful in using Green's III to *construct* functions on Ω.

5. Prove $\varepsilon \ln \varepsilon \to 0$ as $\varepsilon \to 0$. *Hint:* Consider $\ln \varepsilon / (1/\varepsilon)$.

6. Name four famous theorems of ordinary one-variable calculus that were applied in this chapter.

7. Write down a line integral over the boundary $\partial\Omega$ which shows how a function $u(\zeta)$ harmonic in Ω^+ is determined inside Ω by information relating to $\partial\Omega$. Use one of the Green's Identities.

8. Let Ω be the disc $D(0; R)$ and suppose u is harmonic in Ω^+. Use your integral representation of Exercise 1.7 to prove the Circumferential Mean-Value Theorem (here $z = z(s)$ on $C = \partial\Omega$):

$$u(0) = \frac{1}{2\pi R} \int_C u(z) \, ds = \frac{1}{2\pi} \int_0^{2\pi} u(R \cos\theta, R \sin\theta) \, d\theta.$$

Thus, the value $u(0)$ at the center is an average or "mean" of its values on the rim.

Deduce that the integrals must be independent of the radius R because the value $u(0)$ surely is.

9. *A direct proof of the Inside-Outside Theorem.* Obtain this formula using Green's Theorem only, without mention of the Identities. We have Ω a Jordan domain with boundary curve (or curves) given by $z = z(s) = (x(s), y(s))$ as usual (arc length parametrization). Also, $N(z(s))$ is the outward normal vector to Ω at the point $z(s)$, and $u = u(z) \in \mathscr{C}^2(\Omega^+)$.
   (a) Verify that $(\partial u/\partial n)(z(s)) = \nabla u(z(s)) \cdot N(z(s))$.
   (b) Verify $N(z(s)) = \langle y'(s), -x'(s) \rangle$.
   (c) Use (a) to write $(\partial u/\partial n)(z(s)) \, ds$ in the form $p \, dx + q \, dy$.
   (d) Apply Green's Theorem to conclude the Inside-Outside Theorem:

$$\int_{\partial\Omega} \frac{\partial u}{\partial n} \, ds = \iint_\Omega \Delta u \, dx \, dy.$$

10. *The Divergence Theorem for vector fields.* This theorem generalizes the Inside-Outside Theorem to an arbitrary vector field $V(z) = \langle A(z), B(z) \rangle$ (see Exercise 5 to Paragraph 1.4.3), not merely the gradient of u. Here, $A(z), B(z)$ are functions in $\mathscr{C}^1(\Omega^+)$ and we define the *divergence* of V (written div V) to be the function

$$\text{div } V(z) = \nabla \cdot V(z) = A_x(z) + B_y(z).$$

   (a) Write $V(z(s)) \cdot N(z(s)) \, ds$ in the form $p \, dx + q \, dy$.
   (b) Apply Green's Theorem to conclude that

$$\int_{\partial\Omega} V \cdot N \, ds = \iint_\Omega \text{div } V \, dx \, dy.$$

   (c) Convince yourself that the left-hand integral measures the net flow of the vector field $V(z)$ across the full boundary $\partial\Omega$.
   (d) Likewise, div $V(z)$ is a number that measures the tendency of V to flow away (diverge) from the vicinity of the point z. Look at some special cases to convince yourself of this.
   (e) Convince yourself, making reference to (c) and (d), that the Divergence Theorem in (b) is reasonable on physical grounds.

2

Harmonic Functions in the Plane

Section 2.1 SOME BASIC PROPERTIES

Now we single out a certain class of functions for special consideration. A function $u = u(z) = u(x, y)$ is *harmonic* in Ω if and only if $u \in \mathscr{C}^2(\Omega)$ and also u satisfies Laplace's equation in Ω; that is,

$$\Delta u(x, y) = u_{xx}(x, y) + u_{yy}(x, y) = 0$$

for all (x, y) in Ω. As usual, Ω is a domain (not necessarily bounded) in $\mathbb{R}^2$. Harmonic functions have readily understandable physical interpretations (see Section 2.2) as well as nice mathematical properties and are also prominent in the study of another important class of functions, the analytic functions of a complex variable (Chapter 3).

Examples of Harmonic Functions

Example 1. Any constant function, in particular the function that is identically zero, is harmonic in $\Omega = \mathbb{R}^2$.

Example 2. The functions $u(x, y) = x$, $v(x, y) = y$ are harmonic in $\mathbb{R}^2$, since all second derivatives are zero.

Example 3. The functions $x^2 - y^2$ and xy are harmonic in $\mathbb{R}^2$, as you should verify. However, $x^2 + y^2$ is *not* harmonic.

Example 4. The functions $e^x \sin y$ and $e^x \cos y$ are both harmonic in $\mathbb{R}^2$.

Example 5. The sum $u + v$ of two harmonic functions in Ω is harmonic in Ω (proof?) as is the product cu of a harmonic function u by a scalar c. Thus, $2x + 3y$ and $2 - x + y$ are harmonic. However, the product of two harmonic functions may fail to be harmonic (consider x^2).

Example 6. In contrast to the above, the function $u(z) = \ln r = \ln |z - \zeta|$ is not defined at $z = \zeta$ (some given point of $\mathbb{R}^2$) and hence it cannot be harmonic in *all* of $\mathbb{R}^2$. It is a straightforward exercise in taking second derivatives to verify that $u(z) = u(x, y)$ is harmonic in the punctured domain $\mathbb{R}^2 - \{\zeta\}$. We used this function in our discussion of Green's III in Chapter 1. We'll use it again.

Now let us apply some of the general integral theorems of Chapter 1 to harmonic functions. Hence, Ω will be a Jordan domain with boundary $\partial\Omega$, and Ω^+ is a larger domain containing Ω and $\partial\Omega$. Our functions will be nicely behaved in all of Ω^+, hence certainly in Ω and its boundary.

THEOREM 1

Let Ω be a k-connected Jordan domain and u a harmonic function in $\mathscr{C}^2(\Omega^+)$. Then

(i) $\int_{\partial\Omega} (\partial u/\partial n)(z)\, ds = 0$;

(ii) *for $\zeta \in \Omega$, $z \in \partial\Omega$, $r = |z - \zeta|$ as usual, we have*

$$u(\zeta) = -\frac{1}{2\pi} \int_{\partial\Omega} \left(\ln r\, \frac{\partial u}{\partial n}(z) - u(z)\, \frac{\partial \ln r}{\partial n} \right) ds.$$

Proof: Just apply $\Delta u = 0$ to the Inside-Outside Theorem and Green's III, respectively. Done.

Comments

1. The first formula says that for a harmonic function, the "net flux" across the entire boundary $\partial\Omega$ is zero. We will discuss this at greater length in Section 2.2.

2. The second formula says that a harmonic function is determined by its "boundary behavior." That is, if we know that u exists and is harmonic in Ω^+ and if we know also $u(z)$ and $(\partial u/\partial n)(z)$ for z on $\partial\Omega$, then we can by integration obtain $u(\zeta)$ for ζ *inside* Ω. This uses the fact that once the curve (or curves) $\partial\Omega$ is given and ζ is selected inside Ω, then the function $\ln r = \ln|z - \zeta|$ is well known. "Determination by boundary values" is a remarkable property of harmonic functions. We will have much more to say on this also. In fact, we'll see that $u(\zeta)$ is obtainable without $(\partial u/\partial n)(z)$.

3. In both formulas (i) and (ii) we understand that $z = z(s)$; that is, each curve comprising a part of $\partial\Omega$ is parametrized by arc length s. Note, too, that r in formula (ii) is a function of s as well as of z.

Exercises to Section 2.1

1. Which of the following functions are harmonic in Ω?
   (a) $4x - 3y$, $\Omega = \mathbb{R}^2$;
   (b) e^{xy}, $\Omega = \mathbb{R}^2$;
   (c) $2x^2 - 3y^2$, $\Omega = \mathbb{R}^2$;
   (d) $\ln |z|$, $\Omega = \mathbb{R}^2 - \{0\}$;
   (e) $e^x \cos x$, $\Omega = \mathbb{R}^2$;
   (f) $x^3 - 3xy^2$, $\Omega = \mathbb{R}^2$.

2. Suppose $u(x, y) = e^x g(y)$ is harmonic in $\mathbb{R}^2$. Show that $g(y)$ must be of the form $A \cos y + B \sin y$ for real A, B.

3. Suppose u, v are functions defined in Ω such that u, $v \in \mathscr{C}^2(\Omega)$ and, moreover, $u_x = v_y$, $u_y = -v_x$ (the Cauchy–Riemann equations). Prove that u and v are each harmonic; that is, show $u_{xx} + u_{yy} = 0$, and likewise for v.

4. Given that u is harmonic in a domain Ω containing the closed disc $\bar{D}(0; R)$. Deduce from Theorem 1 that $\int_0^{2\pi} (\partial u/\partial r)(R, \theta) \, d\theta = 0$, where r, θ denote polar coordinates.

5. A naive student deduces from Green's III that every function $q(x, y)$ in $\mathscr{C}^2(\Omega^+)$, harmonic or not, is determined inside Ω by its behavior on the boundary $\partial\Omega$. Which term of the identity has he overlooked?

6. The functions $u(\zeta) \equiv 0$ and $v(\zeta) = \ln |\zeta|$ agree on the unit circle $C(0; 1)$ and are harmonic in $\mathbb{R}^2 - \{0\}$, yet are clearly different functions. Why does this not contradict the fact that harmonic functions are determined by "behavior on the boundary?" *Hint:* Can we represent $\ln |\zeta|$ for $|\zeta| < 1$ by Green's III?

7. Suppose u is harmonic in Ω and, moreover, $u \in \mathscr{C}^3(\Omega)$ so that u_x, $u_y \in \mathscr{C}^2(\Omega)$. Are u_x, u_y harmonic?

8. Verify formula (i) of Theorem 1 for $u(x, y) = x^2 - y^2$, $\Omega = D(0; R)$, as follows:
   (a) In $\mathbb{R}^2$, $u(x, y) = u(r \cos \theta, r \sin \theta) = r^2 \cos 2\theta$.
   (b) $\partial u/\partial n = \partial u/\partial r = 2r \cos 2\theta$ for positively oriented circles ($r = $ constant) centered at the origin.
   (c) Fixing $r = R$, show $\partial u/\partial r = 2R \cos (2s/R)$. *Hint: $s = R\theta$.*
   (d) Deduce $\int_{\partial\Omega} (\partial u/\partial n) \, ds = \int_0^{2\pi R} 2R \cos (2s/R) \, ds$, which vanishes.

Section 2.2 HARMONIC FUNCTIONS AS STEADY-STATE TEMPERATURES

2.2.0 Introduction

The purpose of this section is to show that a harmonic function defined in a domain is the same thing as a steady-state temperature in the domain. We know what a harmonic function is ($\Delta u = 0$). So now we must explain the meaning of "steady-state temperature." This will be very informal kitchen-stove physics.

2.2.1 Steady-state Temperatures

Suppose Ω is a Jordan domain with boundary $\partial\Omega$ as usual. We assume that $\bar{\Omega} = \Omega \cup \partial\Omega$ is made of some heat-conducting substance (a metal plate, say). We further suppose that at each point z of $\partial\Omega$, heat is either being fed into the plate $\overline{\Omega}$ or removed from it. The temperature at the point z is denoted $u(z)$. This is a real number—positive, negative, or zero—but will not be referred to any scale such as Celsius. We suppose that as z varies around the curve (or curves) $\partial\Omega$, the function $u(z)$ varies continuously. See Figure 2.1.

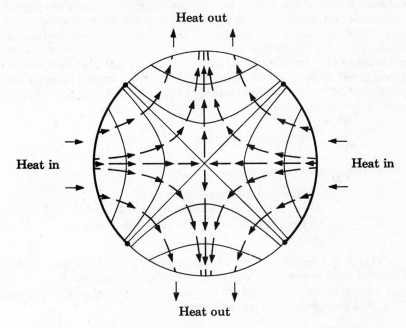

Unbroken curves: curves of constant temperature (isotherms)
Arrows: paths of heat flow

A steady-state heat flow

Figure   2.1

Of course, heat is flowing in the entire plate $\overline{\Omega}$. Roughly speaking, it is flowing across Ω from hotter spots on the boundary to relatively cooler spots on the boundary. The temperature may vary from point to point in $\overline{\Omega}$, but we make the additional very important assumption that at any fixed point ζ the temperature does not vary with time. In this case (temperature $u(\zeta)$ is independent of time) we say that *steady-state* conditions prevail, that $u(\zeta)$ is a *steady-state* (or *stationary*) temperature in $\overline{\Omega}$, etc. Our aim is to show that $\Delta u = 0$ throughout Ω.

Hence, we want to transmute this situation into a mathematical statement about the function u. We assert that, in the steady-state case, the amount of heat (a notion we do not define!) flowing into any small portion of Ω in a given span of time is the same as the amount of heat flowing out of that portion. Otherwise, that portion would heat up (or cool off) and the temperature at points inside the portion of plate would change in time, contradicting the steady-state hypothesis.

Suppose Γ is a small simple closed curve (a loop) inside Ω whose interior lies entirely inside Ω. In the steady-state case, the net amount of heat flowing across Γ is zero (heat in $-$ heat out $= 0$). But rate of heat flow across a given point z on Γ depends essentially on $(\partial u/\partial n)(z)$, the outward normal derivative of temperature at z. An example to convince you of this: If $(\partial u/\partial n)(z)$ is a large positive number, then the temperature just outside Γ near z is a good deal higher than the temperature just inside Γ near z, and hence heat is flowing into the interior of Γ at the point z (from hot to cold). Since we agree that the net heat flow across Γ is zero, we must conclude that the net temperature change ("temperature flux") accumulated around the entire curve Γ is zero. But this gives us an integral equation:

$$\text{Net temperature flux} = \int_\Gamma \frac{\partial u}{\partial n}\, ds = 0.$$

In summary, we have argued as follows:

u = steady-state temperature in $\overline{\Omega}$
$\Rightarrow$ net heat flow across every loop Γ is zero
$\Rightarrow$ net temperature flux across Γ is zero
$\Rightarrow \int_\Gamma (\partial u/\partial n)\, ds = 0$.

This last formula is reminiscent of Theorem 1. However, we still have not shown that a steady-state temperature satisfies $\Delta u = 0$. We will attend to this last link in the chain of reasoning in Paragraph 2.2.2. It is concerned with integration, not physics.

Exercises to Paragraph 2.2.1

1. *One-dimensional heat flow.* Regard the real interval $a \le x \le b$ as a heated wire, with the temperature at x given by some function $u(x)$. Suppose $u(x)$ does not vary in time.
   (a) Convince yourself that $u(a) > u(b)$ implies that heat is flowing at a steady rate from a to b.
   (b) Moreover, the graph of $y = u(x)$ is a straight line, whence u satisfies the "one-dimensional Laplace equation" $u'' = u_{xx} = 0$.
2. *The heat-flow vector.* This is another mathematical tool (not used in this section) for describing the temperature $u(x, y)$ in a plane domain Ω.
   (a) Recall that the gradient $\nabla u(z)$ points from z in the direction of maximal rate of increase of temperature u.
   (b) Argue that, at the point z in Ω, heat is flowing in the direction pointed out by the "heat-flow" vector $V(z) = -\nabla u(z)$ (note minus sign).
   (c) Verify that the temperature u is harmonic if and only if the "divergence" div V of the heat-flow vector is identically zero. (If $V(z) = \langle p(z), q(z)\rangle$ is

any vector field, we define div $V =$ div $V(z) = p_x(z) + q_y(z)$, a function of z.) Language: A steady-state heat flow is "divergence-free." Heat is not being dissipated at any point.

3. *Logarithmic temperature distribution.* Let $u(z) = -\ln |z|$ give the temperature at point z. We know it's harmonic. We deal here with temperature in the entire plane rather than in a bounded domain.

(a) Observe that the temperature at the origin is infinite (an infinite source of heat).

(b) Observe that the temperature is the same at all points of any circle centered at the origin.

(c) Find all points with zero temperature.

(d) Verify that the net flux of temperature across the unit circle is -2π. *Note:* $\int_\Gamma (\partial u/\partial n) \, ds$.

(e) Relate the negativity of the net temperature flux across the unit circle to the fact that heat is flowing away from the origin and out of the unit disc at a steady rate.

(f) Why doesn't all the heat eventually leave the domain $0 < |z| < 1$?

(g) Verify that the net temperature flux across *any* circle $|z| = R$ equals -2π, independent of R (compare (e)) and justify this independence of R on elementary physical grounds.

In this exercise we have translated mathematics into physics. Contrast the following exercise.

4. *An infinite heat source.* Suppose the origin is an infinite source of heat in the z-plane and that the heat flows radially outward equally in all directions. Suppose further that steady-state conditions prevail in the sense that the total amount of heat flowing each second across any circle $|z| = R$ is the same, independent of the radius R (and so temperature at a point does not vary in time). Show that the temperature function *must* be of the form $u(z) = A \ln |z| + B$ with $A < 0$, as follows: Note that we are here translating physics into mathematics. Contrast the preceding exercise.

(a) Argue that $u(z)$ satisfies $\lim_{z\to 0} u(z) = +\infty$ (compare $u(0) = +\infty$).

(b) Argue that $u_\theta = 0$ so that u depends on $r = |z|$ alone.

(c) Argue that $\int_0^{2\pi} u_r(R)R \, d\theta$ is independent of R.

(d) Deduce that $u_r(r)$ is a constant multiple of r^{-1}.

(e) Conclude that u is of the form $A \ln r + B$, as claimed.

Moral: The logarithm and the source of a steady-state heat flow are even more closely related than you may have guessed from Exercise 3. But after seeing Green's III and Theorem 1 in Section 2.1 you should agree that the logarithm is special.

5. *Source and sink.* Let the temperature $u(\zeta) = \ln |\zeta - \zeta_0| - \ln |\zeta - \zeta_1|$, where ζ_0, ζ_1 are distinct.

(a) Verify that u is harmonic except at ζ_0, ζ_1.

(b) Locate the source and sink of heat in this temperature distribution.

(c) Describe the set of points of zero temperature.

6. *One-dimensional vs. two-dimensional.*

(a) Verify that the steady-state temperature $f = f(r)$ in a wire $a \le r \le b$ such that $f(a) = 0, f(b) = 1$ is given by $f(r) = (r - a)/(b - a)$. Compare Exercise 1.

(b) Verify that the steady-state temperature $u = u(r, \theta)$ in the annular plate $a \le r \le b, 0 \le \theta \le 2\pi$ (with $a > 0$), which satisfies $u(a, \theta) = 0, u(b, \theta) = 1$

(constant on each boundary circle) is given by $u(r, \theta) = (\ln r - \ln a)/(\ln b - \ln a)$. Note independence of θ.

Moral: The temperature distribution for the annulus, although independent of θ, is *not* obtained merely by rotating the distribution for the wire. Note that $(r - a)/(b - a)$ is not harmonic in the plane.

(c) Can you prove the uniqueness assertion implicit in our wording of (b)?

7. Argue on physical grounds that if there are no sources or sinks of heat in the closed disc $|z| \le 1$ and if the temperature at every point does not vary with time and, moreover, equals zero everywhere on the rim $|z| = 1$, then in fact the temperature is zero everywhere in the disc.

8. *The heat equation.* Granted that the temperature $u = u(x, y, t)$ at the point (x, y) and time t satisfies the *heat equation* $u_t = u_{xx} + u_{yy}$. Argue that if the temperature is actually steady-state, then in fact u satisfies Laplace's equation. *Hint:* Steady-state = time-independent.

9. *Potential functions.* A vector field $V(z) = \langle p(z), q(z) \rangle$, with $p(z), q(z)$ functions in $\mathscr{C}(\Omega)$, is *conservative* if it is actually a gradient, that is, if there exists a function ψ in $\mathscr{C}^1(\Omega)$ such that $V(z) = \nabla\psi(z)$. In this case, ψ is said to be the *potential* for the field V.

(a) Prove that the heat-flow vector of Exercise 2 is conservative by observing that $\psi(z) = -u(z)$, where u is the temperature function, serves as a potential function.

(b) Prove that $V(z)$ is conservative in Ω if and only if the integral $\int_\Gamma V(z) \cdot T(z)\, ds = 0$ for every closed loop Γ inside Ω. Here, $z = z(s)$ is the arc length parametrization of Γ and T is the unit tangent vector. *Hint:* If $V = \nabla\psi$, then $p\, dx + q\, dy = d\psi$, an exact differential.

(c) Conclude that if V is conservative, $V = \nabla\psi$, then

$$\int_{z_0}^{z_1} V(z) \cdot T(z)\, ds = \psi(z_1) - \psi(z_0),$$

where the integral is taken along any curve in Ω from z_0 to z_1. Thus, *the flow from z_0 to z_1 depends only on the difference in potential at the two points.* Conservative fields are desirable in part because functions are often easier to handle than vector fields.

10. *Which vector fields are conservative?* Let $V(z) = \langle p(z), q(z) \rangle$ with $p, q \in \mathscr{C}^1(\Omega)$. Define curl $V(z) = q_x(z) - p_y(z)$, as in Exercise 5 to Paragraph 1.4.3.

(a) Prove that V conservative implies curl $V = 0$ throughout Ω.

(b) Let Ω be an open disc. Prove that V is conservative if and only if curl $V = 0$ throughout Ω. *Hint:* Green's Theorem and Exercise 9(b).

(c) Let Ω' be the punctured disc $0 < |z| < 1$, and

$$V(z) = \left\langle \frac{-y}{x^2 + y^2}, \frac{x}{x^2 + y^2} \right\rangle$$

in Ω'. Prove curl $V = 0$ in Ω' but V is not conservative in Ω'.

Moral: On domains with holes (not simply connected), necessary conditions for conservatism need not be sufficient. The same is true for exactness of differentials.

2.2.2 A Characterization of Harmonic Functions

This is an extension of the first part of Theorem 1.

THEOREM 2

Let Ω be a plane domain and $u \in \mathscr{C}^2(\Omega)$. Then u is harmonic in $\Omega \Leftrightarrow$ for every Jordan curve Γ inside Ω whose interior lies inside Ω we have

$$\int_\Gamma \frac{\partial u}{\partial n}\, ds = 0.$$

Proof: ($\Rightarrow$) This is a minor variation of (i) of Theorem 1.

($\Leftarrow$) This will follow from the so-called "Bump Principle." We want to show $\Delta u(\zeta) = 0$ for all $\zeta \in \Omega$.

Assume, on the contrary, that $\Delta u(\zeta) > 0$ for some $\zeta \in \Omega$ (same idea if $\Delta u(\zeta) < 0$). But Δu is continuous, so we conclude that $\Delta u(z) > 0$ for all points z in some small open area around the point ζ. That is, its graph has a "bump" above ζ. Let Γ be a small circle around ζ inside this open area, and let D be the disc whose boundary is Γ. Then

$$\iint_D \Delta u\, dx\, dy > 0$$

because $\Delta u > 0$ in D. By the Inside-Outside Theorem, therefore, the number $\int_\Gamma (\partial u/\partial n)\, ds > 0$. This contradiction shows $\Delta u(\zeta) = 0$ and completes the proof of Theorem 2.

The Steady-State Interpretation Completed

Using Theorem 2 and Paragraph 2.2.1, we conclude that *a steady-state temperature $u = u(z)$ inside a Jordan domain is harmonic. Conversely, we may interpret a given harmonic function as a steady-state temperature.* What we have done is deduce from physical considerations the partial differential equation of steady-state heat flow, namely, Laplace's equation,

$$\Delta u = 0.$$

Now we may use our intuition about temperature to guess at or interpret theorems about harmonic functions. See Paragraph 2.2.3 for some guesses.

For example, "determination of a harmonic function by boundary behavior" (see (ii) of Theorem 1) is now quite believable: The temperature inside a plate is determined by the heat input that is maintained along the edge (boundary) of the plate (assuming no heat sources or sinks inside the plate).

More on the Bump Principle

This is worth singling out for special mention. It enables us to deduce information about a continuous function from information about its integrals. Note that it depends on continuity, not on differentiability. One statement of this basic idea is the following:

BUMP PRINCIPLE

Let Ω be a domain and q a continuous function on Ω (that is, $q \in \mathscr{C}(\Omega)$) and also $q(z) \geq 0$ for all $z \in \Omega$. Then q is identically zero, $q \equiv 0$, in $\Omega \Leftrightarrow$

$$\iint_{\Omega} q \, dx \, dy = 0.$$

For if the graph of a nonnegative continuous function has no "bump," the function must be zero everywhere. You should make your own private peace with this.

What is the analogous principle for continuous functions $f(x)$ of ordinary calculus?

Exercises to Paragraph 2.2.2

1. Let Ω be a Jordan domain and $q \in \mathscr{C}(\Omega^+)$. True or false?
   (a) If q vanishes everywhere on $\partial\Omega$, then q vanishes throughout Ω.
   (b) If $\iint_{\Omega} q \, dx \, dy = 0$, then q vanishes identically on Ω.
   (c) If $\iint_{D} q \, dx \, dy = 0$ for every open disc D contained in Ω, then q vanishes identically in Ω.
   (d) If q vanishes identically in Ω, then q vanishes on $\partial\Omega$ as well.
   (e) If $z_0 \in \Omega$, $q(z_0) > 0$, then $\iint_{D} q \, dx \, dy > 0$ for some $D = D(z_0; r)$.
2. State and prove a "Bump Principle" for real functions $y = g(x) \geq 0$ defined and continuous on an interval $a \leq x \leq b$. What can you prove about $g(x)$ if $\int_a^b g(x) \, dx = 0$?

2.2.3 Some Conjectures About Harmonic Functions

The analogy with steady-state temperature prompts three educated guesses about existence and behavior of harmonic functions.

Conjecture 1. On Existence: Let $D = D(0; r)$ be the disc of radius r and $C = C(0; r)$ its circular boundary, $C = \partial D$. Given a real-valued continuous function g defined on the circle C only. Does there exist a function $u = u(x, y)$ which is (i) continuous in $\overline{D} = D \cup C$, (ii) harmonic in D, and (iii) agrees with g, $u(z) = g(z)$, at boundary points $z \in C$? Thus, we ask if we can "pull" the domain of definition of g into the entire disc $\overline{D}$ so that the extended function (now called u) is harmonic in D. This is a certain kind of *boundary value problem*, the *Dirichlet problem* for the disc D.

Here is a physical "solution" to the Dirichlet problem. We construct u by heating up the rim C of the disc so that the temperature at each point z on C is maintained at the given value $g(z)$. Heat will flow from the rim inward and after a long time steady-state conditions will prevail. Now let $u(\zeta)$ be the temperature (use a thermometer!) at the point ζ inside D. This defines a continuous (why?) function u in all of $\overline{D}$, harmonic (why?) in D and equal to g on C. Of course we have given a physical proof here, not a mathematical proof. We will have more to say on the Dirichlet problem later.

Conjecture 2. On Maxima and Minima: Suppose we have a harmonic function $u = u(z)$ defined on $\overline{D}$ as above. It is reasonable that if the temperature u is not constant throughout $\overline{D}$, then the hottest and coldest points of $\overline{D}$ all occur on the boundary C. In other words, we cannot have a hot interior (in D) point surrounded by cooler points when the sources (and sinks) of heat are all on the boundary C. See Figures 2.2

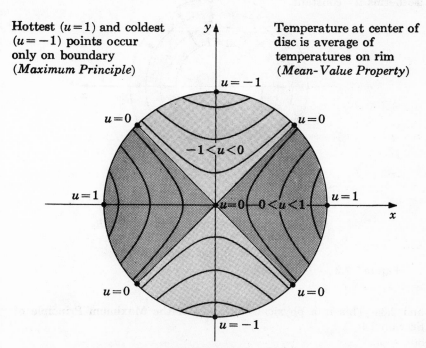

Steady-state temperature $u(x, y) = x^2 - y^2$
Isotherms in the disc.
(Compare figure 2.1)

Figure   2.2

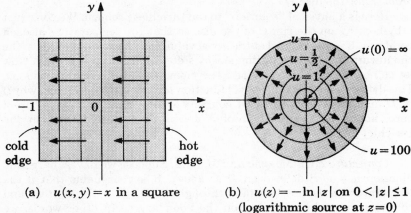

(a) $u(x, y) = x$ in a square

(b) $u(z) = -\ln |z|$ on $0 < |z| \leq 1$
(logarithmic source at $z = 0$)

Note: Lines of heat flow $(\rightarrow)$ are perpendicular to isotherms $u = $ constant.

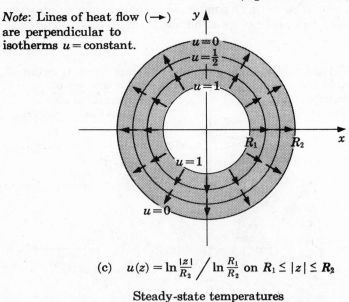

(c) $u(z) = \ln \dfrac{|z|}{R_2} \bigg/ \ln \dfrac{R_1}{R_2}$ on $R_1 \leq |z| \leq R_2$

Steady-state temperatures

Figure   2.3

and 2.3a. This is a physical statement of the Maximum Principle of Section 2.4.

   Conjecture 3. Mean Values: Though it may be less evident on purely physical grounds, it is not unreasonable to expect that the **temperature at the exact center of the disc** D **is some sort of average or mean of the temperatures throughout the disc** (or again, of the **temperatures around the rim** C). We will discuss such mean-value properties in Section 2.3.

A Comment on Our Approach

We are in the process of learning about harmonic functions by examining Laplace's equation $\Delta u = 0$ and steady-state heat flow. Note that

  (i) we have very few examples of harmonic functions yet (though we believe they are plentiful by Conjecture 1 above);
 (ii) we are not studying these functions from the vantage point of explicit formulas (such as $e^x \sin y$);
(iii) we have not endeavored to seek explicit harmonic functions by solving $\Delta u = 0$ (as one does when studying elementary differential equations).

What are we doing then? A little bit of the *qualitative* theory of partial differential equations: studying the *behavior* of solutions of $\Delta u = 0$ without knowing the solutions explicitly.

The preceding paragraph prompts a question: Which technical theorem or equation has enabled us to study the behavior of solutions of $\Delta u = 0$ without knowing explicit formulas for these solutions?

Exercises to Paragraph 2.2.3

1. Suppose u and v are harmonic in a domain Ω containing the closed disc $|z| \leq 1$. True or false?
   (a) If $u(z) > v(z)$ whenever $|z| = 1$, then $u(z) > v(z)$ for all z inside the disc as well.
   (b) If $u(z) = v(z)$ whenever $|z| = 1$, then $u(z) = v(z)$ for all z inside the disc as well.
   (c) If $u(0) \geq u(z)$ whenever $|z| = 1$, then $u(z)$ is constant for $|z| \leq 1$.
   (d) If $u(z) = 0$ whenever $|z| = 1$, then $u(z) = A \ln|z|$ for some constant $A \neq 0$.
   (e) If $\Omega = \mathbb{R}^2$ and $u(z) > 0$ for all z, then $u(z)$ is constant.
   (f) If $\Omega = \mathbb{R}^2$ and $\lim_{z \to \infty} u(z)$ exists (and is finite), then u is a constant function.

   Hint: Use your physical intuition (temperatures!). We have not yet developed sufficient machinery to give complete proofs here of those statements that are true.
2. Answer the question raised at the end of Paragraph 2.2.3. How have we been able to use $\Delta u = 0$?

Section 2.3 MEAN-VALUE PROPERTIES OF HARMONIC FUNCTIONS

Now we make mathematical sense out of Conjecture 3 of Section 2.2.

CIRCUMFERENTIAL MEAN-VALUE THEOREM

Let u be harmonic in a domain Ω, let ζ be a point of Ω, and suppose the closed disc $D = D(\zeta; R)$ is contained inside Ω. Then the value $u(\zeta)$

is the arithmetic mean of the values $u(z)$ taken on the circle $C = \partial D$, $|z - \zeta| = R$. That is,

$$u(\zeta) = \frac{1}{2\pi R} \int_C u(z) \, ds = \frac{1}{2\pi} \int_0^{2\pi} u(R, \theta) \, d\theta,$$

where (r, θ) denote polar coordinates centered at the point ζ (Figure 2.4).

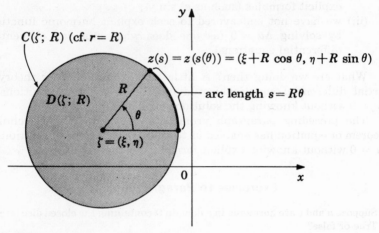

Polar coordinates centered at ζ

Figure 2.4

Note: The equality of the two integrals follows immediately from $s = R\theta$ (when θ is measured in radians). The first integral is a mean or average because it is a limit of the following averaging process: Cut the circle C into pieces $C_1, C_2, \dots, C_n$ and then form the sum $\sum u(z_k) \, \Delta s_k$, where z_k is on C_k and Δs_k is the length of C_k. Finally, divide this sum by the total circumference $2\pi R$.

First Proof (Using the Inside-Outside Theorem): First we show that the value of the integral in the statement is a constant independent of R.

Let r satisfy $0 < r \le R$ and consider any circle $C = C(\zeta; r)$. Since u is harmonic, we have

$$\begin{aligned}
0 &= \int_C \frac{\partial u}{\partial n} \, ds \\
&= \int_0^{2\pi} \frac{\partial u}{\partial r} (r, \theta) r \, d\theta \\
&= r \frac{d}{dr} \int_0^{2\pi} u(r, \theta) \, d\theta.
\end{aligned}$$

To obtain the second equality here, we used the fact that $\partial/\partial n$ is the same as $\partial/\partial r$ on discs. The third equality follows from differentiation under the integral sign (useful; see Appendix A2.1). It follows that the integral in the statement has zero derivative with respect to r and is therefore a constant independent of r.

Now we evaluate this constant in two ways to get the theorem. On the one hand, when $r = R$, the constant equals $\int_0^{2\pi} u(R, \theta)\, d\theta$ (part of the statement of theorem!). Now we let r tend to zero. By the Average Value Theorem of integral calculus, for each r it is true that

$$\int_0^{2\pi} u(r, \theta)\, d\theta = 2\pi u(r, \theta_r),$$

where θ_r is some angle depending on r. Now as r tends to zero, the right-hand side (and therefore the left-hand side) clearly tends to $2\pi u(\zeta)$. Thus,

$$\int_0^{2\pi} u(R, \theta)\, d\theta = 2\pi u(\zeta)$$

as claimed. Done.

Second Proof (Using Green's III): We have that

$$u(\zeta) = -\frac{1}{2\pi} \int_C \left(\ln r \frac{\partial u}{\partial n} - u \frac{\partial \ln r}{\partial n} \right) ds$$

$$= -\frac{1}{2\pi} \left(\ln R \int_C \frac{\partial u}{\partial n}\, ds - \int_C u \frac{1}{R}\, ds \right),$$

since $r = R$ on C. But the first integral here vanishes (why?) and the second is

$$\int_0^{2\pi} u(R, \theta)\, d\theta,$$

since $ds = R\, d\theta$ on C. Done.

Questions

1. The second proof is much shorter. Why?
2. Which standard tools of calculus were used in the first proof? See Appendix A2.1.

The following fact is also plausible on physical grounds.

SOLID MEAN-VALUE THEOREM

Let u be harmonic in a domain Ω, let ζ be a point of Ω, and suppose the closed disc $\overline{D} = \overline{D}(\zeta; R)$ is contained in Ω. Then the value $u(\zeta)$ is given by

$$u(\zeta) = \frac{1}{\pi R^2} \iint_D u\, dx\, dy.$$

Proof: For each r with $0 < r \le R$ we know that

$$u(\zeta) = \frac{1}{2\pi} \int_0^{2\pi} u(r, \theta)\, d\theta.$$

Now multiply both sides by $r\, dr$ and integrate from $r = 0$ to $r = R$. The left side gives $u(\zeta)\,(R^2/2)$. The right side is the iterated integral

$$\frac{1}{2\pi} \int_0^R \int_0^{2\pi} u(r, \theta) r\, dr\, d\theta.$$

But $r\, dr\, d\theta$ is the element of area in polar coordinates. Thus, the iterated integral is the same as

$$\frac{1}{2\pi} \iint_D u\, dx\, dy.$$

This proves the theorem.

Comment

You might attempt to verify these theorems for some specific harmonic functions (see exercises).

More on Mean-Value Properties

Now we show that the relation between harmonic functions and the circumferential mean-value property is even more intimate than we first suspected. We will now see that *only* harmonic functions have the mean-value property throughout a domain Ω. Thus, we could actually use this property as a *definition* of harmonic function. Compare the "net flux equals zero" property of Theorem 2; this also characterizes harmonic functions.

First we need some precise language. Let Ω be any domain. We say that $u \in \mathscr{C}^2(\Omega)$ has the *circumferential mean-value property* in Ω if and only if, for each disc $D = D(\zeta; R)$, such that $\overline{D} \subset \Omega$, we have

$$u(\zeta) = \frac{1}{2\pi R} \int_C u\, ds,$$

where C is the circle ∂D.

THEOREM 3

Let $u \in \mathscr{C}^2(\Omega)$ where Ω is any domain. Then u is harmonic in $\Omega \Leftrightarrow u$ has the circumferential mean-value property in Ω.

Proof: ($\Rightarrow$) Done already.

($\Leftarrow$) We are given (using polar coordinates centered at ζ)

$$u(\zeta) = \frac{1}{2\pi} \int_0^{2\pi} u(r, \theta)\, d\theta$$

for all r (provided $D(\zeta; r)$ lies inside Ω, of course). Differentiating in r and multiplying by r yields

$$0 = r \frac{d}{dr} \int_0^{2\pi} u(r, \theta) \, d\theta$$

$$= \int_0^{2\pi} r \frac{\partial u}{\partial r} (r, \theta) \, d\theta$$

$$= \int_{C(\zeta;r)} \frac{\partial u}{\partial n} \, ds$$

$$= \iint_{D(\zeta;r)} \Delta u \, dx \, dy.$$

Since this is true for all discs $D(\zeta; r)$ inside Ω, the Bump Principle allows us to conclude that the integrand Δu is zero. Done.

Comment

The final step in the preceding proof is a repetition of what we did in proving Theorem 2 and could be omitted by appealing to that theorem.

Exercises to Section 2.3

1. *Some standard notation.* Details as in the statement of the Circumferential Mean-Value Theorem. This exercise shows how the statement is affected by choice of parametrization.
   (a) Convince yourself that the integral $\int_C u(z) \, ds$ means $\int_0^{2\pi R} u(z(s)) \, ds$, where $z(s)$ parametrizes the circle C by arc length.
   (b) Use $C = C(\zeta; R)$ and $\zeta = (\xi, \eta)$ to verify that the arc length parametrization of C is given by $z(s) = (x(s), y(s)) = (\xi + R \cos(s/R), \eta + R \sin(s/R))$, with $0 \le s \le 2\pi R$. *Hint:* $s = R\theta$ on $C(\zeta; R)$.
   (c) If we parametrize C by the usual angle θ rather than by s, then $z(s) = z(s(\theta)) = (\xi + R \cos \theta, \eta + R \sin \theta)$ with $0 \le \theta \le 2\pi$.
   (d) If we agree that r, θ denote polar coordinates centered at ζ, then C is determined by $r = R$ and we save space by writing $u(R, \theta)$ instead of $u(\xi + R \cos \theta, \eta + R \sin \theta)$.

   We remark that these explicit parametrizations are not necessary in the proof of the theorem.

2. Evaluate $(1/2\pi R) \int_C u \, ds$ for $u = u(z)$, z on $C = C(\zeta; R)$, as follows:
   (a) $u(z) = xy$, $\zeta = (1, 2)$, $R = 40$;
   (b) $u(z) = e^x \cos y$, $\zeta = 0$, $R = 1$;
   (c) $u(z) = \ln|z|$, $\zeta = (1, 0)$, $R = 1/2$;
   (d) $u(z) = x + y$, $\zeta = (1, -1)$, $R = 5$.

   Hint: Don't integrate.

3. Evaluate $(1/2\pi) \int_0^{2\pi} u(\xi + R \cos \theta, \eta + R \sin \theta) \, d\theta$ for $u(x, y)$, R and (ξ, η) as follows:

     (a) $u(x, y) = xy$, $(\xi, \eta) = (1, 2)$, $R = 40$;

     (b) $u(x, y) = e^x \cos y$, $(\xi, \eta) = (0, 0)$, $R = 1$.

     Hint: Compare Exercise 2.

4. Let r, θ denote polar coordinates centered at ζ. Evaluate $(1/2\pi) \int_0^{2\pi} u(R, \theta) \, d\theta$ for u, ζ, $r = R$ as follows:

     (a) $u(z) = \ln|z|$, $\zeta = (1, 0)$, $R = 1/2$;

     (b) $u(z) = x + y$, $\zeta = (1, -1)$, $R = 5$.

     Hint: No need to change $u(z)$ to polar coordinates nor to integrate.

5. Evaluate $\iint\limits_{D} u(x, y) \, dx \, dy$ for u and $D = D(\zeta; R)$ as follows:

     (a) $u(x, y) = xy$, $\zeta = (1, 2)$, $R = 4$;

     (b) $u(z) = \ln|z|$, $\zeta = (1, 0)$, $R = 1/2$.

6. Let $u \in \mathscr{C}^2(\Omega)$, where Ω is a domain containing the closed disc $|z| \leq 1$. Suppose $u(0) > u(z)$ for $0 < |z| \leq 1$ so that $z = 0$ is a local maximum for u.

     (a) Prove $u(0) > (1/2\pi) \int_0^{2\pi} u(\cos \theta, \sin \theta) \, d\theta$.

     (b) Deduce that u fails to be harmonic at some (perhaps all) points of Ω.

     (c) Suppose the function $u(z)$ here denotes the temperature in Ω (a heated plate) at a certain moment of time. Assuming that heat is neither added nor subtracted, what should happen at $z = 0$ (where presently $u(0)$ is relatively high) as time goes by?

7. Are harmonic functions characterized (among those in $\mathscr{C}^2(\Omega)$) by the Solid Mean-Value Property? Define this property carefully before you answer.

8. What is the Average-Value Theorem of integral calculus referred to in the first proof of this section? Note that we used it in the proof of Green's III in Chapter 1 also.

Section 2.4 THE MAXIMUM PRINCIPLE

2.4.0. Introduction

In Section 2.2 we conjectured that a nonconstant harmonic function could not assume its highest value in the interior of an open domain Ω; the highest (and lowest) value of $u(x, y)$ are attained only if (x, y) is a point of $\partial\Omega$ but not of Ω itself. We will establish this result now, utilizing the mean-value results of Section 2.3.

2.4.1 The Strong Maximum Principle

Let Ω be a plane domain (open, connected, not necessarily bounded) and let $u : \Omega \to \mathbb{R}$. We say that u *assumes its maximum* in Ω if and only if there exists at least one point $\zeta \in \Omega$ and real number c such that $u(\zeta) = c$ and, moreover, $u(z) \leq c$ for all $z \in \Omega$. There is an entirely similar definition of u assuming its minimum.

For example, the function $u(x, y) = 1 - x^2 - y^2$ assumes its maximum in $\mathbb{R}^2$ at $(x, y) = (0, 0)$, whereas the function $v(x, y) = x$ does not assume its maximum in $\mathbb{R}^2$; there is no point with highest x-coordinate.

Now we assert

STRONG MAXIMUM PRINCIPLE

A nonconstant harmonic function in a domain Ω does not assume its maximum or its minimum on Ω.

Proof: Assume the harmonic function u assumes its maximum at $\zeta \in \Omega$, $u(\zeta) = c$. If u is not constant in Ω, then the set of $z \in \Omega$ such that $u(z) < c$ is nonempty and open (continuity!). Consequently, the set of points z for which $u(z) = c$ is not open also, or else Ω would be disconnected. Thus, there is a point $\zeta_1 \in \Omega$ such that $u(\zeta_1) = c$ and a disc $\overline{D}(\zeta_1; r)$ contained in Ω with boundary $C(\zeta_1; r)$ containing at least one point z with $u(z) < c$. Since u is continuous, however, there exists an *arc* on the circle $C(\zeta_1; r)$ such that $u(z) < c$ for all points z on this arc. It follows (using a Bump Principle on the arc and polar coordinates r, θ centered at ζ_1) that

$$\frac{1}{2\pi} \int_0^{2\pi} u(r, \theta)\, d\theta < \frac{1}{2\pi} \int_0^{2\pi} u(\zeta_1)\, d\theta = u(\zeta_1).$$

But this is a contradiction, since u is harmonic.

The same sort of proof works in the case of a minimum, or one can note that $-u$ is also harmonic and that its maximum would occur at the minimum of u. This completes the proof.

This theorem is called "strong" because it makes no restriction on the boundedness of Ω nor on the behavior of u on the boundary $\partial\Omega$.

Exercises to Paragraph 2.4.1

1. What property of harmonic functions was contradicted to conclude the proof of the Strong Maximum Principle?
2. Emphasize the importance of the connectedness of Ω by finding a nonconstant function u defined on the open set $\Omega_1 \cup \Omega_2$ (where Ω_1, Ω_2 are disjoint domains) such that $\Delta u = 0$ everywhere, yet u assumes both its maximum and its minimum on $\Omega_1 \cup \Omega_2$. *Hint:* Construct a continuous function that assumes only two values.

    Note that connectedness was utilized in the proof of the Strong Maximum Principle.
3. Give an informal physical argument why a steady-state (time-independent) temperature could not assume its maximum at any point inside a heated disc unless heat was being added there continually. *Hint:* What would happen to the heat concentrated at the hottest point as time passed?

2.4.2 The Weak Maximum Principle

We can say even more if Ω is a bounded domain (contained in same disc). Then its closure $\overline{\Omega} = \Omega \cup \partial\Omega$ is closed and bounded ("compact"). It can be shown that if u is continuous on the compact set $\overline{\Omega}$, then u assumes its maximum and minimum somewhere on $\overline{\Omega}$. This follows immediately from the fact that this set of all values $u(z)$, $z \in \overline{\Omega}$, is a closed interval (end points included) of finite length in the continuum of real numbers. You might try to prove this, using the fact that $\overline{\Omega}$ is connected.

Now, since a harmonic function u is continuous, it must assume extreme values on $\overline{\Omega}$. The next result assures us that these extreme values are assumed at certain points on $\partial\Omega$ only, unless u is constant on $\overline{\Omega}$.

WEAK MAXIMUM PRINCIPLE

Let Ω be a bounded domain with u continuous on $\overline{\Omega}$ and harmonic in Ω. Then either u is constant on $\overline{\Omega}$ or u assumes its maximum and minimum values on $\partial\Omega$ only.

Note: Two claims here. Extreme values are assumed on $\partial\Omega$, but not anywhere else in $\overline{\Omega}$.

Proof: Left to you. Use the Strong Maximum Principle plus the comments above.

As an example, note that the harmonic function $u(x, y) = y$ assumes its maximum and minimum values on the boundary $C(0; 1)$ of the closed unit disc $\overline{D}(0; 1)$, in fact, at the points $(0, 1)$ and $(0, -1)$, respectively. For (x, y) in the open disc $D(0; 1)$, we have $-1 < u(x, y) < 1$.

Exercises to Paragraph 2.4.2

1. *Some real analysis.* We make occasional, but crucial, use of this fact: If S is a closed and bounded (that is, compact) subset of the plane and u is a continuous (not necessarily harmonic) function defined on S, then u assumes its maximum and minimum values on S. (For us, of course, $S = \overline{\Omega}$, where Ω is a bounded domain.) This may be proved as follows (you may wish to consult a reference here):

   (a) Prove that S is compact if and only if every infinite subset of S has a point of accumulation which lies in S.

   (b) The image set $u(S)$ on the real line cannot be unbounded. *Hint:* Unboundedness implies a sequence of values $u(z_1)$, $u(z_2)$,... tending to $\pm\infty$ in $\mathbb{R}$. But $z_1, z_2,...$ has a point of accumulation (say, z^*), in S. What of $u(z^*)$?

   (c) The image set $u(S)$ must contain its least upper bound and greatest lower bound (the maximum and minimum values of u). *Hint:* Choose a sequence $u(z_1)$, $u(z_2)$,... tending to the least upper bound of $u(S)$ and argue as in (b).

2. Locate the maximum and minimum values of the following functions on $\overline{\Omega}$. Do they occur only on the boundary?
   (a) $u(x, y) = xy$, $\overline{\Omega} = \overline{D}(0; 1)$;
   (b) $v(x, y) = x^2 - y^2$, $\overline{\Omega} = \overline{D}(0; 1)$;
   (c) $u(z) = \ln|z|$, $1 \le |z| \le 2$;
   (d) $u(z) = e^x \cos y$, $|z| \le 1$.
3. Let $\overline{\Omega}$ be the closed upper half-plane $y \ge 0$.
   (a) Verify that the harmonic function $u(x, y) = x + y$ does not assume its maximum or minimum in $\overline{\Omega}$.
   (b) Does this contradict the Weak Maximum Principle? Explain.
4. Convince yourself (that is, prove!) that the following statement is equivalent to the Weak Maximum Principle: If Ω is a bounded domain and u is a nonconstant function continuous on $\overline{\Omega}$ and harmonic in Ω, then for every point ζ in Ω, we have (in obvious notation)

$$\min_{z \in \delta\Omega} u(z) < u(\zeta) < \max_{z \in \partial\Omega} u(z).$$

2.4.3 Application

Certain uniqueness theorems follow readily from the preceding work. Here is a sample.

THEOREM 4

Let Ω be a bounded domain and let u, v be functions continuous on $\overline{\Omega}$, and equal on $\partial\Omega$; that is, $u(z) = v(z)$ for all $z \in \partial\Omega$. Then $u(\zeta) = v(\zeta)$ for all ζ inside Ω as well.

Proof: The function $u - v$ is harmonic in Ω and vanishes on Ω. Since its extreme values are equal to zero, we must have $u - v = 0$ everywhere in $\overline{\Omega}$. Done.

This result shows that there is at most one harmonic function on a bounded domain Ω (not necessarily a Jordan domain, by the way) which is continuous on $\overline{\Omega}$ and assumes given values on $\partial\Omega$. Thus, *boundary behavior is decisive*. This takes care of the "uniqueness" part of the so-called harmonic boundary-value problem or Dirichlet problem on a bounded domain Ω. The question of existence is generally more difficult. See Appendix A2 for the special case where Ω is an open disc.

A good exercise: Translate the proof of Theorem 4 into the language of temperatures.

Exercises to Paragraph 2.4.3

1. Let $D = D(0; 1)$, the unit disc, with boundary C parametrized by the angle θ as usual; that is, $z(\theta) = (\cos \theta, \sin \theta)$.
   (a) Verify that $u(x, y) = xy$ assumes the boundary values $f(\theta) = \frac{1}{2} \sin 2\theta$ on C; that is, $u(z(\theta)) = f(\theta)$.

(b) Find a harmonic function $v(x, y)$ in D which agrees with $g(\theta) = \cos 2\theta$ on C.

(c) Are the harmonic functions $u(x, y)$ and $v(x, y)$ uniquely determined in D by the boundary values $f(\theta)$ and $g(\theta)$, respectively?

2. Why may the following pairs u, v of harmonic functions differ in Ω without contradicting Theorem 4? Note that $u = v$ everywhere on $\partial\Omega$.

   (a) $u \equiv 0$, $v(x, y) = y$, Ω = the upper half-plane.

   (b) $u \equiv 0$, $v(z) = \ln|z|$, Ω = the unit disc.

3. The harmonic version of Green's III (see Theorem 1) assured us that the harmonic function u in the Jordan domain Ω was determined by the set of values $u(z)$ and $(\partial u/\partial n)(z)$, $z \in \partial\Omega$.

   (a) How does Theorem 4 improve matters in this regard?

   (b) Can you think of a formula for $u(\zeta)$, $\zeta \in \Omega$, in terms of $u(z)$, $z \in \partial\Omega$, but not involving $(\partial u/\partial n)(z)$?

Section 2.5 HARNACK'S INEQUALITY AND LIOUVILLE'S THEOREM

2.5.0. Introduction

Roughly speaking, the Weak Maximum Principle asserted that if Ω was "small," and u harmonic on Ω and continuous on $\overline{\Omega}$, then the set of values $u(z)$, $z \in \Omega$, was also "small"—a single point or an open interval of finite length.

Now we investigate the opposite extreme: the case $\Omega = \mathbb{R}^2$, so that Ω is as unbounded as possible. A harmonic function u is said to be *entire* if its domain of definition is $\mathbb{R}^2$. Liouville's Theorem will tell us that the set of values $u(z)$, $z \in \mathbb{R}^2$, is either a single number (the degenerate case u = constant) or consists of all real numbers, provided u is harmonic. We will see a similar result when we study complex analytic functions.

Note that we state no theorems about the set of values $u(z)$, $z \in \Omega$, in the general case that Ω is unbounded but not all of $\mathbb{R}^2$ (except for the Strong Maximum Principle, of course).

2.5.1. Harnack's Inequality

This is a prerequisite to our proof of Liouville's Theorem.

HARNACK'S INEQUALITY

Let $D = D(z_0; R)$ be an open disc and let u be harmonic on D such that $u(z) \geq 0$ for all $z \in D$. Then for all $z \in D$, we have

$$0 \leq u(z) \leq \left(\frac{R}{R - |z - z_0|}\right)^2 u(z_0).$$

Note: This inequality bounds the growth of the value $u(z)$ as z moves away from the center z_0. For instance, if $|z - z_0| = \frac{1}{2}R$ so that z is halfway to ∂D, then we are told that $u(z)$ is not more than four times larger than the value $u(z_0)$. Given the lower bound $0 \le u(z)$, we get an upper bound.

Figure   2.5

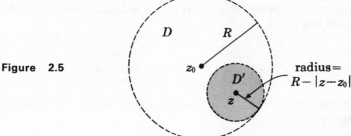

Proof: Apply the Solid Mean-Value Theorem to the small disc D' centered at z (Figure 2.5) with radius $R - |z - z_0|$ to obtain

$$0 \le u(z) = \frac{1}{\pi(R - |z - z_0|)^2} \iint\limits_{D'} u \, dx \, dy.$$

Now, since $u \ge 0$ in D, which contains D', we see that

$$\iint\limits_{D'} u \, dx \, dy \le \iint\limits_{D} u \, dx \, dy = \pi R^2 u(z_0),$$

where the last equality again follows from the Solid Mean-Value Theorem. Together, these inequalities give

$$0 \le u(z) \le \frac{R^2}{(R - |z - z_0|)^2} u(z_0),$$

as claimed. Done.

Comment

This is noteworthy because here the value $u(z)$ is not bounded in terms of the derivatives (compare rate of change) of the function u, but only in terms of $u(z_0)$ itself. Of course this is so only because u is harmonic; for nonharmonic functions, the same sort of theorem would almost certainly involve the derivatives of u.

Exercises to Paragraph 2.5.1

1. Use Harnack's Inequality to prove that if u is harmonic in $D(z_0; R)$, with $u(z) \ge 0$ there, and $u(z_0) = 0$, then in fact $u(z) = 0$ throughout $D(z_0; R)$.

2. Verify that Exercise 1 is the Minimum Principle for the harmonic function u.
3. Use Harnack's Inequality to prove the Minimum Principle for an arbitrary harmonic function u in an arbitrary domain Ω. *Hint:* A minimum in Ω is a minimum in some $D(z_0; R)$. Also, if $u(z_0) = c_0$ at the minimum z_0, then $u(z) - c_0$ is nonnegative near z_0.

    We remark that both Harnack's Inequality and the Minimum Principle followed from mean-value considerations. Thus, the present exercise gives no radically new method for obtaining the Minimum (and Maximum) Principle.
4. Does there exist a nonnegative harmonic function $u(z)$ in the open disc $D(0; 1)$ such that $\lim_{z \to 1} u(z) = \infty$?

2.5.2 Liouville's Theorem

Now we apply Harnack's Inequality to obtain the result mentioned in Paragraph 2.5.0.

STRONG LIOUVILLE THEOREM

If u is entire harmonic and is bounded either from above or from below, then u is a constant function.

Proof: Suppose u is bounded from below so that $u(z) \geq a$ for some number a. Then $v = u - a \geq 0$ is also harmonic and v is constant if and only if u is constant. Hence, we may as well suppose $a = 0$, $u \geq 0$ (as required for Harnack's Inequality).

Now choose any two points $z, \zeta \in \mathbb{R}^2$. For any radius $R > |z - \zeta|$, we have that

$$0 \leq u(z) \leq \left(\frac{R}{R - |z - \zeta|} \right)^2 u(\zeta).$$

Now let R approach ∞; this is allowed because u is entire! Then

$$\left(\frac{R}{R - |z - \zeta|} \right)^2 \to 1,$$

so that $u(z) \leq u(\zeta)$. Interchanging z and ζ in the argument, we get $u(\zeta) \leq u(z)$ also. Thus, $u(z) = u(\zeta)$ for any z, ζ, whence u is constant.

If, on the other hand, u is bounded from above, then $-u$ is bounded from below. We leave the rest to you. Done.

Comment

If $u = u(x, y)$ is a steady-state temperature on the entire plane $\mathbb{R}^2$ and is not constant, then there must be points in the plane at any given temperature, no matter how high or low. For example, let $u(x, y) = xy$.

Exercises to Paragraph 2.5.2

1. A naive student applies Harnack's Inequality to an arbitrary entire harmonic function u, obtains $u(z) = u(\zeta)$ for all z, ζ, and concludes that all entire harmonic functions are constant. We know this is false.
   (a) Try to reconstruct his doomed argument. Be naive!
   (b) Point out the mistake in (a).

2. (a) Give a quick proof that the real-valued function $\sin xy$ is not harmonic. *Hint:* It's defined for all (x, y).
   (b) Suppose the harmonic function $u(x, y)$ satisfies $|u(x, y)| \leq |\sin xy|$ for all (x, y). What is $u(x, y)$?

Appendix 2.1 ON DIFFERENTIATION UNDER THE INTEGRAL SIGN AND AN APPLICATION: HARMONIC FUNCTIONS ARE $\mathscr{C}^\infty$

We revisit calculus in $\mathbb{R}^2$. In our first proof of the Circumferential Mean-Value Property, we were confronted with the question

$$\frac{d}{dr} \int_0^{2\pi} u(r, \theta)\, d\theta \overset{?}{=} \int_0^{2\pi} \frac{\partial u}{\partial r}(r, \theta)\, d\theta.$$

That is, the integral $\int_0^{2\pi} u(r, \theta)\, d\theta$ is a function of r; can its first derivative be computed by computing the integral of $(\partial u/\partial r)(r, \theta)\, d\theta$? We claimed at the time that the answer is "yes."

Both differentiation and integration are limit processes. The present question partakes of the general mathematical problem: "When can two limit processes be interchanged without altering the result?" Some cases where this interchange is possible:

(1)
$$\frac{\partial^2 u}{\partial x\, \partial y} = \frac{\partial^2 u}{\partial y\, \partial x},$$

provided both partial derivatives are continuous;

(2)
$$\int \left[\int u\, dx \right] dy = \int \left[\int u\, dy \right] dx,$$

provided certain conditions on u and the domain of integration are met.

Now we state and prove a reasonably general result on differentiating under the integral sign. This will justify what we did in our proof of the Circumferential Mean-Value Property and will also be applied to the question of smoothness (differentiability) of harmonic functions.

THEOREM

Let R be the closed rectangle $a \leq s \leq b, c \leq t \leq d$ in the st-plane, and suppose the functions $F(s, t)$ and its partial derivative $F_t(s, t)$ are continuous on some open set containing R. Then $\int_a^b F(s, t)\, ds$ is a differentiable function of t for $c < t < d$ and, in fact,

$$\frac{d}{dt} \int_b^a F(s, t)\, ds = \int_a^b F_t(s, t)\, ds.$$

Proof: It is helpful to define

$$f(t) = \int_a^b F(s, t)\, ds, \qquad g(t) = \int_a^b F_t(s, t)\, ds.$$

These are continuous functions in $a \leq t \leq b$ because the integrands are continuous. The idea of the proof is to show

$$\int_c^\tau g(t)\, dt = f(\tau) + \text{constant}$$

whenever $c < \tau < d$. For, having this, the familiar Fundamental Theorem of Calculus would allow us to conclude that $f'(\tau) = g(\tau)$, as desired.

Now we have

$$\int_c^\tau g(t)\, dt = \int_c^\tau \int_a^b F_t(s, t)\, ds\, dt$$

$$= \int_a^b \int_c^\tau F_t(s, t)\, dt\, ds$$

$$= \int_a^b \{F(s, \tau) - F(s, c)\}\, ds$$

$$= f(\tau) - \int_a^b F(s, c)\, ds.$$

Here we have interchanged the order of integration and actually integrated F_t with respect to t.

Now note that the second term, $-\int_a^b F(s, c)\, ds$, is a constant independent of τ. By our remarks at the start of the proof, we are done.

A Verification

We differentiate under the integral sign generally as part of a proof. However, an explicit computation may be instructive. For $0 \leq s \leq 1, 0 \leq t \leq 1$, let

$$F(s, t) = t^2 + st$$

so that

$$F_t(s, t) = 2t + s.$$

On the one hand,

$$\int_0^1 F(s, t)\, ds = \int_0^1 (t^2 + st)\, ds = t^2 + \tfrac{1}{2}t,$$

whence the derivative of the integral is

$$\frac{d}{dt} \int_0^1 F(s, t)\, ds = 2t + \tfrac{1}{2}.$$

On the other hand, the integral of the derivative is

$$\int_0^1 F_t(s, t)\, ds = \int_0^1 (2t + s)\, ds = 2t + \tfrac{1}{2},$$

which equals $(d/dt) \int_0^1 F(s, t)\, ds$ found just above. This verifies the theorem.

Comment

There is an obvious generalization to cases where the integrand is of the form $F(s, t, \lambda)\, ds$ and the derivative in question involves $(\partial/\partial t)$, $(\partial/\partial \lambda)$, or combinations of these.

Application

This will be somewhat remarkable. We have defined the function u to be harmonic on Ω if and only if $u \in \mathscr{C}^2(\Omega)$ and $\Delta u = 0$. Thus, we *require* only continuous second partials and a relation between them. Now consider

THEOREM

If u is harmonic on Ω, then $u \in \mathscr{C}^\infty(\Omega)$; that is, all mixed partial derivatives of u exist and are continuous throughout Ω.

Proof: Green's III for harmonic u tells us that

$$u(x, y) = -\frac{1}{2\pi} \int_{\partial D} \left(\ln r \, \frac{\partial u}{\partial n}(\sigma(s)) - u(\sigma(s)) \frac{\partial \ln r}{\partial n} \right) ds,$$

where D is some Jordan domain inside Ω which contains (x, y), $\sigma(s)$ is a point on ∂D, and $r = |(x, y) - \sigma(s)|$. Note that the arc length variable s gets integrated out in this formula, leaving a function of x, y only. Now

$$r = r(x, y, s) = \sqrt{(x - \sigma_1(s))^2 + (y - \sigma_2(s))^2}$$

is infinitely differentiable in both x and y for $x \neq \sigma_1(s)$, $y \neq \sigma_2(s)$. Since the logarithm is also infinitely differentiable for $r > 0$, we may attack the formula for $u(x, y)$ with $(\partial/\partial x)$ or $(\partial/\partial y)$ as often as we wish. Done.

Example

$$u_{xxy}(x, y) = -\frac{1}{2\pi} \int_{\partial D} \left\{ (\ln r)_{xxy} \frac{\partial u}{\partial n}(\sigma(s)) - u(\sigma(s)) \left(\frac{\partial \ln r}{\partial n} \right)_{xxy} \right\} ds.$$

Since the integrand exists and is continuous, the integral u_{xxy} exists also. And so on.

Moral

Solutions of partial differential equations may be surprisingly smooth.

Comment

Even more is true. A harmonic function $u(x, y)$ may be expanded in convergent power series,

$$u(x, y) = \sum c_{mn} x^m y^n.$$

This will follow from our work in Chapter 5.

Appendix 2.2 THE DIRICHLET PROBLEM FOR THE DISC

A2.0 Introduction

This will be a leisurely discussion, stressing ideas (some ideas) rather than proofs or details.

As usual $D = D(0; R)$ is the open disc of radius $R > 0$ centered at the origin. Its boundary is the circle $C = C(0; R)$. Suppose we are given a continuous function g defined on the circle. The Dirichlet problem comprises three parts:

(1) Does there exist a function u continuous on the closed disc $\overline{D}$ and harmonic in D which agrees with the given g on C? This is the *Existence Problem*.

(2) Is the function u uniquely determined by g? (*The Uniqueness Problem*.)

(3) How do we represent u by a formula involving only g? (*The Representation Problem*.)

Happily, we know from Theorem 4 that if u exists, it is unique. This settles uniqueness (2). The other questions are not yet so immediate.

A2.1 The Representation Problem Begun

Suppose we know that the function u exists and is equal to g on C. Then Green's III for harmonic functions tells us that

$$u(\zeta) = \frac{-1}{2\pi} \int_C \left(\ln|z - \zeta| \frac{\partial u}{\partial n}(z) - u(z) \frac{\partial}{\partial n} \ln|z - \zeta| \right) ds_z,$$

where ds_z is written to remind us that $z \in C$ is the variable on the path of integration and is integrated out, leaving a function of $\zeta \in D$. This formula doesn't satisfy us, however, because it requires knowledge of $(\partial u/\partial n)(z)$ as well as of $u(z)$, whereas our physical intuition tells us that this is superfluous: Only the boundary temperature is needed to determine the steady-state temperature inside. Hence, what we want for the Representation Problem (3) is an "improved" Green's III, one that doesn't involve $(\partial u/\partial n)(z)$.

Here is another reason for seeking an improved Green's III. Suppose we had it:

$$u(\zeta) = [\text{formula involving } \zeta \in D, z \in C, u(z)].$$

Then we could toy with this as follows: Given any continuous $g(z)$, $z \in C$, we could replace the expression $u(z)$ in the formula by $g(z)$ and obtain a function $v(\zeta)$. We could then ask:

(i) Is v a harmonic function of $\zeta \in D$?
(ii) Are the boundary values of v actually given by the original function g? That is, as ζ approaches $z \in C$, does $v(\zeta)$ approach $g(z)$? If both answers are "Yes," then we have solved the Existence Problem (1). Hence, an improved Green's III may be the key to both major questions.

A2.2 Comments on Green's III

Note that if H is any function continuous in $\overline{D}$ and harmonic D, then Green's II tells us (assuming u harmonic as above) that

$$-\frac{1}{2\pi} \int_C \left(H(z) \frac{\partial u}{\partial n}(z) - u(z) \frac{\partial H}{\partial n}(z) \right) ds_z = 0.$$

Subtracting this from Green's III (see above) yields

$$u(\zeta) = -\frac{1}{2\pi} \int_C \left\{ [\ln|z - \zeta| - H(z)] \frac{\partial u}{\partial n}(z) \right.$$
$$\left. - u(z) \frac{\partial}{\partial n} (\ln|z - \zeta| - H(z)) \right\} ds_z.$$

Nothing profound here. But now suppose the harmonic function H has the further property that $H(z) = \ln|z - \zeta|$ for all $z \in C$. In that case the term involving $(\partial u/\partial n)(z)$ evaporates, and we are left with

$$u(\zeta) = \frac{-1}{2\pi} \int_C u(z) \frac{\partial}{\partial n} (H(z) - \ln|z - \zeta|) \, ds_z.$$

This is it! A formula involving $u(z)$ only—except for the mysterious harmonic function H and the known function $\ln|z - \zeta|$, of course. It is H we must now track down.

A2.3 The Green's Function

First some language. Let us write $H(z, \zeta)$ for the function H above. As usual, z is but one variable, yet H very likely depends on ζ in some way also, since (for example) H was supposed to agree with $\ln|z - \zeta|$ if $z \in C$ and ζ is any point inside C.

Now we write

$$G(z, \zeta) = H(z, \zeta) - \ln|z - \zeta|,$$

the so-called *Green's function* of D. Note that $G(z, \zeta)$ satisfies

    (i) $G(z, \zeta) = 0$, for $z \in C$, $\zeta \in D$, and
    (ii) $G(z, \zeta)$ differs from $-\ln|z - \zeta|$ by a function harmonic (in z) on all of D and continuous in $\overline{D}$.

These two properties define G because they were used to define H above. It follows that

    (iii) $G(z, \zeta)$ is harmonic for all $z \in D$ except at $z = \zeta$, since as z approaches ζ, the number $G(z, \zeta)$ approaches ∞.

Here is a physical interpretation of the Green's function. Since $G(z, \zeta)$ is harmonic on $D - \{\zeta\}$, we regard it as a steady-state temperature there, as follows: There is a point source of heat at ζ. The temperature is infinite there. Heat flows away from ζ and hence toward $C = \partial D$. The temperature drops as we approach C and is zero on C. From this description it is intuitively clear that $G(z, \zeta) > 0$ for $z \in D - \{\zeta\}$. Can you prove this?

Having the notion of the Green's function, we may state

THEOREM

Let u be continuous in $\overline{D}$ and harmonic in D. Let $G(z, \zeta)$ be the Green's function for D. Then for all $\zeta \in D$ we have

$$u(\zeta) = -\frac{1}{2\pi} \int_{\partial D} u(z) \frac{\partial}{\partial n} G(z, \zeta) \, ds_z \,.$$

Thus, the Representation Problem is solved, provided we know the Green's function for D. We remark also that the above discussion applies to any Jordan domain D, not only the disc. Of course, different domains would have quite different Green's functions (see the definition).

It is not yet clear to us whether the disc D (or any other Jordan domain) does have a Green's function. Actually, for quite general domains, a Green's function is known to exist. However, it is not known explicitly (no formula) for most domains and it is sometimes next to impossible to calculate.

Happily, in the special case $D = D(0; R)$ the Green's function can be discovered by elementary considerations in a few pages. We will content ourselves with a brief sketch.

We want $G(z, \zeta) = H(z, \zeta) - \ln|z - \zeta|$. The idea in getting $H(z, \zeta)$ is this: Try $H(z, \zeta) = \ln c|z - \zeta^*|$, where ζ^* is a point in $\mathbb{R}^2$ *outside* the disc (so that H is harmonic inside the disc), c and ζ^* depend only on ζ (not on z), and $\ln c|z - \zeta^*| - \ln|z - \zeta| = 0$ whenever $z \in C$. The physical interpretation is this: Just as we imagined ζ to be a source of heat, we choose ζ^* as a heat sink (temperature is $-\infty$) toward which the heat from ζ flows. More specifically, we choose ζ^* just far enough away from ζ so that the temperature at every point of C is zero. See Figure 2.6.

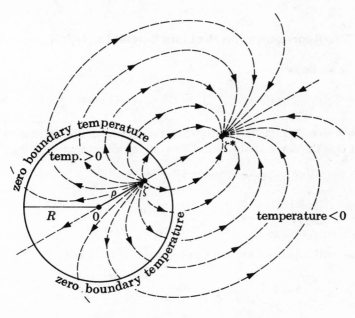

Locating the sink $\zeta^* = \left(\dfrac{R}{\rho}\right)^2 \zeta$

Figure   2.6

The constant c helps balance things. The problem of locating ζ^*, given ζ, may be solved using some classical geometry combined with common sense physics. The explicit solution is this. If $\zeta = (\rho, \varphi)$ in polar coordinates and R is the radius of the disc, then it can be shown that

$$c = \frac{\rho}{R} \quad \text{and} \quad \zeta^* = \left(\frac{R}{\rho}\right)^2 \zeta = \left(\left(\frac{R}{\rho}\right)^2 \rho, \theta\right).$$

Note that $|\zeta^*| = (R/\rho)^2 |\zeta| = (R^2/\rho) > R$ so that ζ^* is outside the disc $D(0; R)$. It follows that $H(z, \zeta)$ is harmonic in ζ for each $z \in D$.

Accepting the above, we may now write down the Green's function explicitly. We use polar coordinates, of course, with $z = (r, \theta)$, $\zeta = (\rho, \varphi)$ and $0 \le r, \rho \le R$. Then

$$G(z, \zeta) = \ln \left(\frac{\rho}{R} \left| z - \left(\frac{R}{\rho} \right)^2 \zeta \right| \right) - \ln|z - \zeta|$$

$$= \frac{1}{2} \ln \frac{R^2 - 2\rho r \cos(\theta - \varphi) + \rho^2 r^2 R^{-2}}{r^2 - 2\rho r \cos(\theta - \varphi) + \rho^2}$$

is the Green's function for $D(0; R)$. Note that $G(z, \zeta) = 0$ if $|z| = R$, as required. Once again kitchen-stove physics has helped us get an answer.

A.2.4 The Representation Problem Solved for D(0,R)

Now we know

$$u(\zeta) = -\frac{1}{2\pi} \int_C u(z) \frac{\partial}{\partial n} G(z, \zeta) \, ds_z$$

and we know $G(z, \zeta)$. Recalling that the outward normal derivative around C is the same as the radial derivative, we compute

$$\frac{\partial G}{\partial r}(z, \zeta) = \frac{-\rho \cos(\theta - \varphi) + \rho^2 r R^{-2}}{R^2 - 2\rho r \cos(\theta - \varphi) + \rho^2 r^2 R^{-2}}$$

$$- \frac{r - \rho \cos(\theta - \varphi)}{r^2 - 2\rho r \cos(\theta - \varphi) + \rho^2}.$$

Since we will integrate around C, we set $r = R$ to obtain

$$\frac{\partial G}{\partial r}(z, \zeta) \bigg|_{z \in C} = \frac{1}{R} \frac{\rho^2 - R^2}{R^2 - 2\rho R \cos(\theta - \varphi) + \rho^2}.$$

It is customary to introduce the notation

$$\mathscr{P}(R, \theta; \rho, \varphi) = \frac{R^2 - \rho^2}{R^2 - 2\rho R \cos(\theta - \varphi) + \rho^2}.$$

This is the *Poisson kernal*. We note

$$\mathscr{P}(R, \theta; \rho, \varphi) = -R \frac{\partial G}{\partial n}(z, \zeta) \qquad (z \in C).$$

It is standard practice to use the Poisson kernel, rather than the outward normal derivative of the Green's function, in the statements of theorems. At last we have solved the Representation Problem:

Poisson Integral Formula

Let u be continuous in $\overline{D}(0; R)$ and harmonic in $D(0; R)$. Then for $\zeta \in D$ we have

$$u(\zeta) = u(\rho, \varphi) = \frac{1}{2\pi} \int_0^{2\pi} u(R, \theta)\, \mathscr{P}(R, \theta; \rho, \varphi)\, d\theta$$

$$= \frac{R^2 - \rho^2}{2\pi} \int_0^{2\pi} \frac{u(R, \theta)}{R^2 - 2\rho R \cos(\theta - \varphi) + \rho^2}\, d\theta.$$

Note: $u(R, \theta)$ is a boundary value $(r = R)$ of $u = u(r, \theta)$.

Proof: The main work has been done. You may fill in the details.

Comment

We have not used the qualitative theorems about harmonic functions (e.g., Mean-Value, Maximum Principle) in this Appendix. All we have used is Green's III and some special properties of the logarithm. You may now apply the Poisson Integral Formula to prove results about harmonic functions. For instance, you should be able to derive a one-line proof of the Circumferential Mean-Value Property. Also, can you prove that $u \equiv 0$ is the only harmonic function vanishing on the entire circle C and defined everywhere in the interior D? Having this, can you prove that a harmonic function is uniquely determined by its boundary values on C? (No Maximum Principle required in this proof!)

A2.5 The Existence Problem Solved

Now we consider the situation where we are given $g = g(\theta)$ defined on the circular rim C of the disc D. We seek a continuous function u on $\overline{D}$ harmonic in D and agreeing with g on C. This can be done by altering the Poisson Integral Formula as follows:

Theorem

Let g be as above. Then the function u defined in $D = D(0; R)$ by

$$u(\rho, \varphi) = \frac{1}{2\pi} \int_0^{2\pi} g(\theta)\mathscr{P}(R, \theta; \rho, \varphi)\, d\theta$$

is continuous in $\overline{D}$, harmonic in D and has boundary values given by g; that is, as ρ approaches R, $u(\rho, \varphi)$ approaches $g(\varphi)$.

Comments on the Proof

1. Convince yourself that it is *not* immediately clear that the boundary values of u as defined by the integral are given by g. A proof requires close examination of the Poisson kernel.

2. Continuity of $u(\rho, \varphi)$ inside D again raises the problem of the interchange of limit operations. Again we omit details.

3. The fact $\Delta u(\rho, \varphi) = 0$ is easiest to prove. Since the Laplacian Δ involves differentiation with respect to the variables ρ and φ only, we may bring Δ under the integral sign as in Appendix A1:

$$\Delta u(\rho, \varphi) = \frac{1}{2\pi} \int_0^{2\pi} g(\theta)\, \Delta \mathscr{P}(R, \theta; \rho, \varphi)\, d\theta.$$

But the Poisson kernel is harmonic as a function of (ρ, φ). To see this, recall $\mathscr{P}(R, \theta; \rho, \varphi) = -R(\partial G/\partial r)(z, \zeta)$ with $z \in C$. Thus,

$$\Delta \mathscr{P}(R, \theta; \rho, \varphi) = -R\, \Delta\, \frac{\partial G}{\partial r}\, (z, \zeta)$$

$$= -R\, \frac{\partial}{\partial r}\, \{\Delta G(z, \zeta)\},$$

since Δ involves only ρ, φ, not r. But the Green's function is harmonic in $\zeta = (\rho, \varphi)$, provided $z = (R, \theta)$ lies on C (why?). Thus, $\Delta \mathscr{P}(R, \theta; \rho, \varphi) = 0$, whence $u(\rho, \varphi)$ is harmonic.

A2.6 An Alternate Approach: Fourier Series

This explanation will require only a superficial knowledge of Fourier series. We make no mention of the Green's function in this approach.

1. Given $g = g(\theta)$, we can write its Fourier series

$$A_0 + \sum_{n=1}^{\infty} (A_n \cos n\theta + B_n \sin n\theta).$$

The A's and B's are real numbers, the so-called *Fourier coefficients* of g, and are given by certain straightforward formulas. For example,

$$A_0 = \frac{1}{2\pi} \int_0^{2\pi} g(\theta)\, d\theta,$$

which (as we know) is the average value of $g(\theta)$ for $0 \leq \theta \leq 2\pi$.

2. To extend g into the interior of the disc D, we define, for $0 \leq \rho < R$ and $0 \leq \varphi \leq 2\pi$,

$$u(\rho, \varphi) = A_0 + \sum_{n=1}^{\infty} \left(\frac{\rho}{R}\right)^n (A_n \cos n\varphi + B_n \sin n\varphi),$$

where the A's and B's come from g. This pulls things in along radii and gives a function of two variables in D.

3. Informally, we see by comparing series that $\lim_{\rho \to R} u(\rho, \varphi) = g(\varphi)$, so that the boundary values of u *are* given by g, as desired.

4. Also, letting ρ approach zero shows that the value of u at the origin is A_0. But this is the circumferential mean of g on C, as we would expect if u were harmonic!

5. *Caution:* These informal remarks are not proofs. Nor is it at all clear that the function u is harmonic. To prove $\Delta u = 0$, one must differentiate the series for u, term by term (using the Laplacian in polar coordinates), and verify that each term $(\rho/R)^n (A_n \cos n\varphi + B_n \sin n\varphi)$ is harmonic. This raises obvious convergence questions, which we will not treat here.

6. We add that it is possible to obtain the Poisson Integral Representation for u by manipulating with the sines and cosines in the series for u given above. In fact, the integral sign in the Poisson formula is provided by the A_n and B_n, which are definite integrals.

7. We emphasize, however, that the key idea in the Fourier series approach to the Dirichlet problem is the trick of pulling the function g into the interior of D by introducing the factor $(\rho/R)^n$ into the nth term of the Fourier series for g. The official name for this is *Abel–Poisson summation*.

3

Complex Numbers and Complex Functions

Section 3.1 THE COMPLEX NUMBERS

3.1.0 Introduction

In this section we construct the set of complex numbers, identifying each such number with a point in the plane $\mathbb{R}^2$. This amounts to endowing the plane with a multiplication as well as an addition. This multiplication is of profound importance for later developments, as we shall see.

3.1.1 Basic Definitions

Let $\mathbb{C}$ denote the set of all elements z of the form

$$z = x + yi,$$

where $x, y \in \mathbb{R}$ and i is a symbol, to be interpreted below (it is not a **real** number). If $y = 0$, then we write $z = x$.

It follows from our definition that

$$z = x + yi \qquad \zeta = \xi + \eta i$$

are equal as elements of $\mathbb{C}$ if and only if

$$x = \xi \qquad y = \eta.$$

Note that here we continue to let x, y, ξ, η denote real numbers. One complex equation gives two real equations.

Some examples: $2 + 3i$, $\frac{1}{2} + 5i$, $0 + \pi i$ (usually written πi), $\frac{1}{2} + (-5)i$ (usually written $\frac{1}{2} - 5i$), $3 + 0i = 3$, $0 + 0i = 0$.

Addition and Subtraction

If $z = x + yi$ and $\zeta = \xi + \eta i$ are in $\mathbb{C}$, then we define their *sum* and *difference*, respectively, to be

$$z + \zeta = (x + \xi) + (y + \eta)i,$$
$$z - \zeta = (x - \xi) + (y - \eta)i,$$

and note that these are also in $\mathbb{C}$.

Multiplication

We know how to multiply two *real* numbers. Let us agree to multiply the symbol i as follows:

$$i \cdot i = i^2 = -1$$

(very important), and also for a and b real,

$$1i = i,$$
$$a(bi) = (ab)i,$$
$$ai = ia.$$

To multiply $z = x + yi$ by $\zeta = \xi + \eta i$, we repeatedly apply these rules by means of a "distributive law." Thus,

$$\begin{aligned} z\zeta &= (x + yi)(\xi + \eta i) = x(\xi + \eta i) + yi(\xi + \eta i) \\ &= x\xi + (x\eta + y\xi)i + y\eta i^2 \\ &= (x\xi - y\eta) + (x\eta + y\xi)i, \end{aligned}$$

since $i^2 = -1$. Thus, the product of two elements of $\mathbb{C}$ is again an element of $\mathbb{C}$. Note that $z\zeta = \zeta z$ always, just as in $\mathbb{R}$.

For example, $(2 + 3i)(-1 + 5i) = -17 + 7i$ (check it!).

Complex Numbers

The set $\mathbb{C}$ with addition and multiplication thus defined is called the set of *complex numbers*, and each $z = x + yi$ is a complex number.

Since we write $x + 0i$ simply as x, the set $\mathbb{R}$ of real numbers may be thought of as a subset of the set $\mathbb{C}$ of complex numbers, $\mathbb{R} \subset \mathbb{C}$. Note that

$$x\xi = (x + 0i)(\xi + 0i) = x\xi + 0i,$$

so that real multiplication and the newly defined complex multiplication are compatible.

Zero: We have $0 = 0 + 0i$. Note that, for all $z \in \mathbb{C}$,

$$z + 0 = z, \quad z - 0 = z, \quad z0 = 0z = 0.$$

More on i: The complex number i is not real, but its square -1 is. It is standard to write $i = \sqrt{-1}$, the "square root of minus one." We recall that only *nonnegative* real numbers have *real* square roots. The letter i may stand for "imaginary." If $z = x + iy$, then x and y are the *real* (Re) and *imaginary* (Im) parts of z, respectively. We denote this

$$x = \text{Re } z, \qquad y = \text{Im } z.$$

Exercises to Paragraph 3.1.1

1. Let $z = 2 - i$, $\zeta = 1 + 3i$. Compute
   (a) $z + \zeta$,
   (b) $\zeta - z$,
   (c) $z\zeta$.
2. Find a complex τ such that $z\tau = \zeta$ (with z, ζ as above). Is τ unique?
3. Use the complex number i to obtain both roots of the quadratic equations
   (a) $X^2 + 1 = 0$,
   (b) $X^2 + 9 = 0$.
4. For which integers n is i^n real? positive? negative? nonreal?
5. If z satisfies $z^2 + z + 1 = 0$, prove without computing z that $z^3 = 1$. *Hint:* Introduce z^3 by multiplying the given relation by z.
6. What are the real and imaginary parts of $z\zeta$ in Exercise 1(c)?

3.1.2 The Complex Plane

From now on we will identify the complex number $x + iy$ and the point (x, y) of $\mathbb{R}^2$. You should check that this identification is compatible with the additions in $\mathbb{C}$ and in $\mathbb{R}^2$. Hence we may now multiply two points in the plane and obtain a third point (this is *not* a dot product of vectors!). See Figures 3.1 and 3.5.

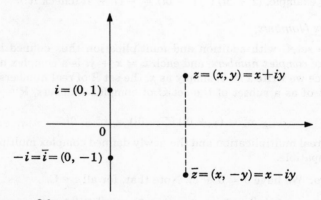

Figure 3.1

If $z = x + iy$, its *conjugate* is the complex number $\bar{z}$ given by

$$\bar{z} = x - iy.$$

It is obtained by reflecting z across the x-axis as in the Figure 3.1. Thus,

$$\overline{3 + 2i} = 3 - 2i, \bar{i} = -i, \overline{-4} = -4.$$

It is important that if $z = x + iy$, then

$$z\bar{z} = x^2 + y^2 = |z|^2,$$

where $|z|$ is the usual norm (distance from the origin) of a point in the plane. Thus, the multiplication ties in nicely with the geometry, thanks to the introduction of $\bar{z}$.

Let us carry this a bit farther. For $z, \zeta \in \mathbb{C}$, we have $\bar{z}, \bar{\zeta} \in \mathbb{C}$, so we may form the product $\bar{z}\,\bar{\zeta}$. On the other hand the product $z\zeta$, and hence its conjugate $\overline{z\zeta}$, are also in $\mathbb{C}$. We leave it to you to check (using $z = x + iy$, $\zeta = \xi + i\eta$, and the definition of conjugate) that

$$\bar{z}\,\bar{\zeta} = \overline{z\zeta}.$$

Thus, in multiplying and taking conjugates, it doesn't matter in which order we do things.

Having this, we observe that

$$|z\zeta|^2 = z\zeta\overline{z\zeta} = z\zeta\bar{z}\,\bar{\zeta} = z\bar{z}\zeta\bar{\zeta} = |z|^2|\zeta|^2.$$

Thus, by taking square roots, we conclude that "the norm of a product is the product of the norms,"

$$|z\zeta| = |z|\,|\zeta|.$$

Of course we knew this already for real numbers.

Exercises to Paragraph 3.1.2

1. Let $z = 2 - i$, $\zeta = 1 + 3i$. Locate the points z, ζ, $z + \zeta$, $z\zeta$, $\bar{z}$, $\bar{\zeta}$, $\bar{z}\bar{\zeta}$ in a sketch of the complex plane.

2. For z as above, compute $z\bar{z}$.

3. Sketch the set of all points z in the plane $\mathbb{C}$ which satisfy $z\bar{z} = 4$.

4. Let z be as in Exercise 1. Compute $|z|$, arg z (= the *argument* of z; that is, the counterclockwise angle in radians from the positive x-axis to the line segment from the origin to z). Now form the product iz and compute $|iz|$, arg iz. What does multiplication by i do to the point z geometrically? Illustrate your answer with a sketch.

5. Find a complex τ such that, for all z, the point τz is obtained by rotating the point z about the origin through one right angle (= $\pi/2$ radians) in the *clockwise* sense. *Hint:* See the preceding exercise.

3.1.3 Division by Complex Numbers

Now we show that the quotient ζ/z has a reasonable meaning as a complex number, provided $z \neq 0$. We will define the quotient as a product:

$$\frac{\zeta}{z} = \left(\frac{1}{z}\right)\zeta.$$

Hence, it suffices to show that $1/z$ determines a complex number (that is, something of the form $a + bi$ with a, b real).

Since $z \neq 0$, we have $\bar{z} \neq 0$, and hence it is not unreasonable to write $\bar{z}/\bar{z} = 1$. We see that $1/z$ should satisfy

$$\frac{1}{z} = \frac{1}{z} \cdot \frac{\bar{z}}{\bar{z}} = \frac{\bar{z}}{z\bar{z}} = \frac{\bar{z}}{|z|^2}$$

But the right-hand side here may now be dealt with! The denominator is a real number $|z|^2$, and we may use it to divide the real and imaginary parts of the numerator $\bar{z}$. Thus, let $z = x + iy$ so that $\bar{z} = x - iy$ and $|z|^2 = x^2 + y^2$. Then, as we just saw,

$$\frac{1}{z} = \frac{\bar{z}}{|z|^2} = \frac{x - iy}{x^2 + y^2} = \left(\frac{x}{x^2 + y^2}\right) - \left(\frac{y}{x^2 + y^2}\right)i.$$

This is of the form $a + bi$, a complex number.

For example, if $z = 3 - 2i$, then

$$\frac{1}{z} = \frac{3}{13} + \frac{2}{13}\,i.$$

Furthermore, if $\zeta = 1 + i$ and $z = 3 - 2i$, then

$$\frac{\zeta}{z} = \zeta\left(\frac{1}{z}\right) = (1 + i)\left(\frac{3}{13} + \frac{2}{13}\,i\right) = \frac{1}{13} + \frac{5}{13}\,i.$$

Summary

We may add, subtract, multiply, and divide (except by zero) in the complex numbers. Also,

$$|z\zeta| = |z|\,|\zeta|, \qquad |z| = \sqrt{z\bar{z}}, \qquad \frac{1}{z} = \frac{\bar{z}}{|z|^2}.$$

Exercises to Paragraph 3.1.3

1. Let $z = 2 - i$, $\zeta = 1 + 3i$. Compute:
   (a) $1/z$,
   (b) ζ/z.

2. Find τ such that $z\tau = \zeta$ (with z, ζ as above). See the exercises to **Paragraph 3.1.1.**
3. Give proof or counterexample: $|1/z| = 1/|z|$ for $z \neq 0$.
4. Write $1/i$ in the form $x + iy$ with x, y real.

Section 3.2 COMPLEX ANALYTIC FUNCTIONS

3.2.0 Introduction

Now we will introduce the notion of a complex-valued function and then single out the more special analytic functions for further consideration. We'll then construct some important analytic functions and see that they are built from harmonic functions, just as complex numbers are built from real numbers.

3.2.1 Preliminaries

We write $\mathbb{C}$ instead of $\mathbb{R}^2$ for the (complex) plane. Let Ω be a subset (usually a domain) of the plane $\mathbb{C}$. A *complex function* $f\colon \Omega \to \mathbb{C}$ is a rule that assigns to each $z \in \Omega$ a complex number $f(z)$. Full name: *complex-valued function of a complex variable*.

An example: polynomials such as $f(z) = z^2 + (1 + i)z - 7$ or $g(z) = z^3 - i$ (here, $\Omega = \mathbb{C}$). Note that we make use here of the multiplication in $\mathbb{C}$.

Another example: the "conjugation function" $F(z) = \bar{z}$.

Pictures

Since $\mathbb{C}$ ($= \mathbb{R}^2$) is two-dimensional, the graph of $w = f(z)$ would require a four-dimensional (z, w)-space. Hence, we do not graph complex-valued functions. It is standard, however, to draw the z-plane (xy-plane) containing Ω and the w-plane (uv-plane, target plane, image plane) and consider the function f as a geometric transformation or mapping that carries the domain Ω onto a subset (call it $f(\Omega)$) in the image plane. This is the point-of-view of Chapter 8. See Figure 3.2.

Real and Imaginary Parts

Let f be a complex function defined on Ω, and let us write $w = f(z)$ for $z \in \Omega$. Since w is complex, we may write it $w = u + iv$ with u, v real.

Now, as z varies in Ω, the number $w = f(z)$ varies, and hence the real and imaginary parts of w vary. It follows that $u = u(z)$ and $v = v(z)$ are real-valued functions of z and so we have a familiar-looking decomposition:

$$w = f(z) = u(z) + iv(z).$$

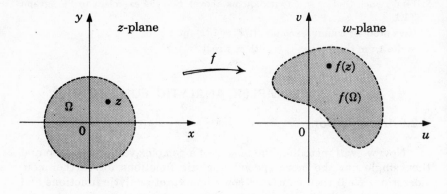

The complex function $w = f(z)$ as a mapping

Figure 3.2

The real-valued functions $u(z)$, $v(z)$ are called the *real* and *imaginary* parts respectively of the function f, in complete analogy with complex numbers. Thus, we may write

$$u = \operatorname{Re}\,(f), \qquad v = \operatorname{Im}\,(f).$$

You might check that if $f(z) = z^2$, then

$$\operatorname{Re}\,(f) = u(z) = x^2 - y^2, \qquad \operatorname{Im}\,(f) = v(z) = 2xy.$$

Limits in $\mathbb{C}$

Let $f \colon \Omega \to \mathbb{C}$ be a complex function, z_0 a point in the z-plane, w_0 in the w-plane $\mathbb{C}$. We write

$$\lim_{z \to z_0} f(z) = w_0$$

if and only if, given any "target" disc $D(w_0; \varepsilon)$ about the point w_0, *no matter how small the radius* $\varepsilon > 0$, there is a "confidence" disc $D(z_0; \delta)$, $\delta > 0$, which is such that if z is any point of $D(z_0; \delta) \cap \Omega$, then we may be confident that the value $f(z)$ lies in the target disc $D(w_0; \varepsilon)$. Of course the confidence radius δ will generally depend on the given target radius ε. See Figure 3.3.

Intuitively, $\lim_{z \to z_0} f(z) = w_0$ means that if z is a point moving on some path in the plane toward z_0, then the point $f(z)$ will move in the image plane along a path leading to the point w_0.

The following observations, whose proofs will be left to you, show that the taking of limits is compatible with the usual operations of algebra. Suppose w_0, w_1 are in $\mathbb{C}$ and

$$\lim_{z \to z_0} f(z) = w_0, \qquad \lim_{z \to z_0} g(z) = w_1.$$

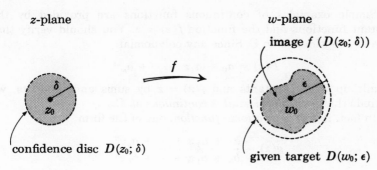

z-plane

w-plane

image $f\ (D(z_0; \delta))$

f

confidence disc $D(z_0; \delta)$

given target $D(w_0; \epsilon)$

Figure 3.3

Then we assert

$$\lim_{z \to z_0} (f(z) \pm g(z)) = \lim_{z \to z_0} f(z) \pm \lim_{z \to z_0} g(z) = w_0 \pm w_1,$$

$$\lim_{z \to z_0} f(z)g(z) = \lim_{z \to z_0} f(z) \lim_{z \to z_0} g(z) = w_0 w_1,$$

$$\lim_{z \to z_0} \frac{f(z)}{g(z)} = \frac{\lim_{z \to z_0} f(z)}{\lim_{z \to z_0} g(z)} = \frac{w_0}{w_1},$$

where in the last line we require $g(z)$ and w_1 to be different from zero.

We remark that if $f(z) = u(z) + iv(z)$ and $w_0 = u_0 + iv_0 \in \mathbb{C}$, then it can be proved that

$$\lim_{z \to z_0} f(z) = w_0$$

is equivalent to

$$\lim_{z \to z_0} u(z) = u_0, \qquad \lim_{z \to z_0} v(z) = v_0.$$

Continuity

The function $f: \Omega \to \mathbb{C}$ is *continuous* at z_0 if and only if

(i) $z_0 \in \Omega$, so that $f(z_0)$ is defined;
(ii) $\lim_{z \to z_0} f(z) = f(z_0)$.

Thus, the limit of $f(z)$ as z approaches z_0 is just what it should be.

A function is *continuous* on the domain Ω, provided it is continuous at each point of Ω.

Since limits are well behaved with respect to the standard algebraic operations, it can be shown that *the sum, difference, product, and quotient of two functions continuous at z_0 are also continuous at z_0* (provided, as usual, we do not divide by zero).

From our remarks above on the real and imaginary parts of f and limits, you should convince yourself that *f is continuous at z_0 if and only if the real-valued functions u and v are also, where $f(z) = u(z) + iv(z)$.*

Simple examples of continuous functions are provided by the constant functions and the function $f(z) = z$. You should verify that these are continuous on $\mathbb{C}$. Since any polynomial

$$P(z) = a_0 + a_1 z + \cdots + a_n z^n$$

is built up from constants and $f(z) = z$ by sums and products, we conclude that *every polynomial is continuous on* $\mathbb{C}$.

In fact, if $g(z)$ is a *rational function,* one of the form

$$g(z) = \frac{a_0 + a_1 z + \cdots + a_n z^n}{b_0 + b_1 z + \cdots + b_k z^k},$$

then $g(z)$ is certainly continuous at all points z_0 where it is defined. By the way, at what points is $g(z)$ *not* defined?

We leave it to you to decide whether the conjugation function $f(z) = \bar{z}$ is continuous.

Finally, we remark that most of the complex functions we are to meet will be continuous and, in fact, have the stronger property of analyticity. We turn to this now.

Exercises to Paragraph 3.2.1

1. Let $f(z) = z^3$. Compute the real and imaginary parts $u(x, y)$, $v(x, y)$ of f. Of course $z = x + iy$.

2. Likewise for $g(z) = 1/z$.

3. Let Ω be the unit disc $|z| < 1$. Sketch the sets Ω and $f(\Omega)$ (in the w-plane), where $w = f(z) = z - i$. Likewise where $f(z) = 2z + 3i$.

4. Formulate the definition of limit, using the terminology of absolute values rather than discs, by completing the following statement: $\lim_{z \to z_0} f(z) = w_0$ if and only if given any $\varepsilon > 0$ there exists a number $\delta > 0$ such that if $|z - z_0| < \delta$, then....
   This "arithmetic" formulation is more useful in certain numerical situations than the more picturesque statement involving discs.

5. Keeping Exercise 4 in mind, prove
$$\lim_{z \to z_0} (f(z) + g(z)) = \lim_{z \to z_0} f(z) + \lim_{z \to z_0} g(z),$$
as claimed in the text.

6. Prove that $f(z) =$ constant and $g(z) = z$ are continuous at all points of $\mathbb{C}$.

7. Is $f(z) = \bar{z}$ continuous? Give reasons.

8. Prove that a function $f \colon \Omega \to \mathbb{C}$ is continuous throughout Ω if and only if whenever S is an open subset of $\mathbb{C}$, then the set $f^{-1}(S) = \{z \in \Omega \mid f(z) \in S\}$ is also open. *Hint:* Use discs.

9. *An important property of continuous functions.* Prove that if $f(z_0) \neq 0$, then there is an open subset (neighborhood) containing z_0 such that $f(z) \neq 0$ for any z in the open subset. In words, if $f(z_0) \neq 0$, then $f(z) \neq 0$ for all z sufficiently close to z_0. *Hint:* Use Exercise 8.

10. (a) Recall from calculus the meaning of the *composite function* $w = f(g(\zeta))$, where $w = f(z)$, $z = g(\zeta)$.
(b) Prove that if g is continuous at ζ_0 and f is continuous at $z_0 = g(\zeta_0)$, then $f(g(\zeta))$ is continuous at ζ_0.

3.2.2 The Complex Derivative

Let $f: \Omega \to \mathbb{C}$, $z_0 \in \Omega$. Then the first derivative $f'(z_0)$ of f at z_0 is, by definition,

$$f'(z_0) = \lim_{z \to z_0} \frac{f(z) - f(z_0)}{z - z_0} \qquad (z \in \Omega),$$

provided this limit exists as a finite complex number. It is sometimes written as $(df/dz)(z_0)$.

We require $z \in \Omega$ in the definition so that $f(z)$ makes sense.

Note that the limit above must equal the same complex number $f'(z_0)$ *regardless of the manner in which the variable z approaches z_0.* Since z_0 is an interior point of Ω, the point z may approach z_0 from any direction, spiral inward, etc. The limit is required to be the same number $f'(z_0)$ in each case.

Comment

Since we don't graph complex functions, we don't speak of the tangent to a graph. In particular, $f'(z_0)$ is *not* to be thought of as the "slope" of anything. For now, it is simply a complex number. We will interpret it and compute it later. We'll soon discover, for example, that $f(z) = z^2$ implies $f'(z) = 2z$.

Some Properties of the Derivative

Since the derivative is defined as a limit and limits are, as we have seen, well behaved with respect to the elementary algebraic operations, we have the following identities reminiscent of ordinary real calculus:

$$(f(z) \pm g(z))' = f'(z) \pm g'(z) \qquad \text{(Sum Rule)}$$

$$(f(z)g(z))' = f'(z)g(z) + f(z)g'(z) \qquad \text{(Product Rule)}$$

$$\left(\frac{f(z)}{g(z)}\right)' = \frac{g(z)f'(z) - g'(z)f(z)}{[g(z)]^2} \qquad \text{(Quotient Rule)}$$

Also, if we are given $f: \Omega \to \mathbb{C}$ as usual and another function: $g: \Omega_1 \to \Omega$, $g(\zeta) = z$, then we may form the *composite function* $f \circ g: \Omega_1 \to \mathbb{C}$ defined by

$$(f \circ g)(\zeta) = f(g(\zeta)).$$

For example, if $f(z) = z^2 + 1$ and $g(\zeta) = \zeta - 4$, then $(f \circ g)(\zeta) = f(\zeta - 4) = (\zeta - 4)^2 + 1$. This composite is a function of ζ, as expected.

How does the derivative of $(f \circ g)(\zeta)$ relate to the derivatives of f and g? The answer is given by

THE CHAIN RULE

Let the composite function $(f \circ g)(\zeta)$ be defined as above, and suppose the derivatives $f'(z)$ and $g'(\zeta)$ exist throughout their respective domains. Then

$$(f \circ g)'(\zeta) = f'(g(\zeta))g'(\zeta).$$

That is, writing $z = g(\zeta)$,

$$\frac{d}{d\zeta}(f \circ g)(\zeta) = \frac{d}{dz} f(z) \frac{d}{d\zeta} g(\zeta) = f'(z)g'(\zeta).$$

Proof: We do this only in the special case $z = g(\zeta) = a\zeta + b, a \neq 0$. Then, at the point ζ_0, we have

$$\frac{f(g(\zeta)) - f(g(\zeta_0))}{\zeta - \zeta_0} = \frac{f(z) - f(z_0)}{z - z_0} \cdot \frac{z - z_0}{\zeta - \zeta_0},$$

writing $z_0 = g(\zeta_0)$. Now, using the facts that $z - z_0 = a(\zeta - \zeta_0)$ and that z is different from z_0 if ζ is different from ζ_0, we may let ζ approach ζ_0 on the right-hand side to obtain in the limit $f'(z_0)a$, that is, $f'(z_0)g'(\zeta_0)$ as claimed. Done.

We remark that the proof is somewhat more delicate if we contend with the possibility that $z = z_0$ (that is, $g(\zeta) = g(\zeta_0)$) for a sequence of points ζ tending toward ζ_0. See the Exercises.

Let's conclude this subsection by noting that if $f(z) = c$, a complex constant independent of z, then

$$\frac{f(z) - f(z_0)}{z - z_0} = \frac{c - c}{z - z_0} = 0$$

so that surely $f'(z_0) = 0$. *The derivative of a constant function is identically zero.*

Likewise, if $f(z) = az + b$, then $f'(z) = a$, as we noted in the proof of the Chain Rule.

Finally, if $f(z) = z^n$, where n is a nonnegative integer ($z^0 = 1$), repeated applications of the Product Rule (p. 111) and the fact that $(z)' = 1$ yield the not unexpected result

$$f'(z) = (z^n)' = nz^{n-1}.$$

Afterthought

We used the difference quotient

$$\frac{f(z) - f(z_0)}{z - z_0}$$

to define $f'(z_0)$. This is an extremely important instance of the division of one complex number by another. It is a fact of life that one cannot write down a similar difference quotient for functions from $\mathbb{R}^3$ into $\mathbb{R}^3$ because one cannot divide by elements (vectors) of $\mathbb{R}^3$ in any suitable way.

Keep your eyes open for the remarkable consequences that follow from this difference quotient's having a limit.

Exercises to Paragraph 3.2.2

1. Compute the first derivative of the following functions:
   (a) $f(z) = 2z^3 + iz$;
   (b) $g(z) = 1/z, (z \neq 0)$;
   (c) $f(\zeta) = (\zeta + 1)/(\zeta - 1), (\zeta \neq 1)$;
   (d) $G(z) = f(g(z))$ with f, g as in (c) and (b);
   (e) $H(z) = g(f(z))$.

2. Let $f(z) = \bar{z}$. Prove that $f'(z_0)$ exists for no z_0 in $\mathbb{C}$. *Hint:* Write $z_0 = x_0 + iy_0$ and let z approach z_0 along both vertical and horizontal lines. Both approaches must yield the same limit in the difference quotient, by definition of f'.

3. Argue that if $f'(z_0)$ exists and z is close to z_0, then $|f(z) - f(z_0)|$ is approximately $|f'(z_0)| \, |z - z_0|$. Classic calculus!

4. Prove that a function $f(z)$ differentiable at z_0 is continuous at z_0. *Hint:* Use the difference quotient.

5. Recall the Mean Value Theorem of real differential calculus for functions $y = f(x)$ on a closed interval $a \leq x \leq b$. Does this theorem have a complex analog? Discuss.

6. Let $g(z) = z\bar{z}$. Prove that $g'(z_0)$ exists if and only if $z_0 = 0$. Contrast Exercise 2.

3.2.3 Definition of Analytic Function

This will be very important to us. Let $f: \Omega \to \mathbb{C}$. If $f'(z_0)$ exists at all points $z_0 \in \Omega$, then a function $f': \Omega \to \mathbb{C}$ is defined via $z \to f'(z)$. This new function, it would appear, may or may not be continuous, have a derivative, etc. We say that f is *analytic at* z_0 if and only if

    (i) $f'(z)$ exists for all points z in some open set containing z_0 (in particular, $f'(z_0)$ exists);
    (ii) $f' = f'(z)$ is a continuous function of z in some open set containing z_0.

In other words, f is analytic at z_0, provided it is continuously (complex) differentiable in an open neighborhood of z_0.

Here are some synonyms:

    "analytic" = "holomorphic" = "regular" = "regular analytic"

Thus, analyticity is a "local" property: From the definition we see that f is analytic at z_0 if and only if f is analytic at *all* points in some neighborhood of z_0. We want this.

We define f *analytic on* Ω (open!) in the obvious way.

Warning: Some authors define f to be analytic if it merely has a complex derivative; they do not require continuity of the derivative. By a remarkable theorem of Goursat, however, a complex derivative *is* continuous. Hence, both definitions of analyticity are equivalent. Authors who use the seemingly less restrictive definition must then prove Goursat's Theorem in order to have the continuity of f' at their disposal. We avoid the work of proving Goursat's Theorem but, on the other hand, we must verify the continuity (as well as existence) of f' for each f we wish to regard as analytic.

Some Examples of Analytic Functions

1. The constant functions, of course.

2. More generally, any polynomial in z, for a polynomial is differentiable and its derivative is again a polynomial, hence continuous. Thus, polynomials satisfy the requirement for analyticity.

3. Any rational function $(f(z)/g(z))$ where f and g are polynomials and we avoid points where the function is not defined (say, $g(z) \neq 0$).

One immediate aim: Find more analytic functions. To do this, we must use the Cauchy–Riemann equations, which we now derive.

We close this section by suggesting a good exercise: Prove that the conjugation function $f(z) = \bar{z}$ is nowhere analytic, even though it is everywhere continuous.

Exercises to Paragraph 3.2.3

1. At what points of the z-plane are the following functions defined and analytic?
    (a) $f(z) = z^2 + 1$,
    (b) $g(z) = 1/(z^2 + 1)$,
    (c) $F(z) = 1/z$.

2. Why is $g(z) = z\bar{z}$ not analytic at $z = 0$? For $g'(0)$ exists!

3. Is $f(z) = |z|$ analytic anywhere in the z-plane?

4. Prove from the definition of $f'(z_0)$ that if f is analytic at z_0, then we may write

$$f(z) = f(z_0) + f'(z_0)(z - z_0) + \varepsilon(z; z_0)(z - z_0),$$

where $\varepsilon(z; z_0) \to 0$ as $z \to z_0$. *Hint:* Get ε from

$$\frac{f(z) - f(z_0)}{z - z_0} = f'(z_0) + \varepsilon(z; z_0).$$

Compare Theorem 3 of Chapter 1.

5. *Some real calculus.* Observe that $f(x) = |x|$ is not differentiable at $x = 0$. Verify that $F(x) = \int_{-1}^{x} |t|\, dt$ is differentiable for *all* x. What is $F'(x)$? Conclude $F''(x)$ doesn't exist at $x = 0$ even though $F'(x)$ does.

Now, by integrating twice, construct a real function $G(x)$ such that $G'(x)$, $G''(x)$ exist for all x but $G'''(x)$ doesn't exist at $x = 0$.

Generalize to get real functions with k but not $k + 1$ derivatives. Compare Exercise 2 for Paragraph 1.3.5.

6. If $f(z)$ is analytic in Ω, is $f'(z)$ also analytic in Ω (so that $f''(z)$ exists, etc.)? *Comment:* This is quite deep.

7. Is a function analytic in Ω continuous in Ω?

8. If f and g are analytic and $f(g(\zeta))$ is defined on some domain, is $f(g(\zeta))$ analytic there?

9. Given that $f(z) = u(z) + iv(z)$ is analytic at z_0, prove that $f'(z_0) = u_x(z_0) + iv_x(z_0)$ by letting z approach z_0 on a horizontal line through z_0. Prove also that $f'(z_0) = v_y(z_0) - iu_y(z_0)$. Conclude the *Cauchy–Riemann equations* $u_x = v_y$, $u_y = -v_x$.

10. Prove that if $f'(z) \equiv 0$ for all z in Ω, then f is constant in Ω. By the way, how do you prove this basic result in real differential calculus?

11. *An important issue.* We will eventually show that if f is analytic and $f(z_0) = 0$, then $f(z) = (z - z_0)^k g(z)$, where k is a positive integer (*not* a fraction), $g(z)$ is analytic at z_0, and $g(z_0) \neq 0$. Compare the factorization of a polynomial with a root of order k at z_0. In this spirit, given that $f(z_0) = 0$, define $f_1(z) = f(z)/(z - z_0)$ for $z \neq z_0$.
(a) Why is $f_1(z)$ analytic if $z \neq z_0$?
(b) How should we define $f_1(z_0)$ so that f_1 is *continuous* at z_0 (what is $\lim_{z \to z_0} f_1(z)$?)?
(c) With this definition of $f_1(z_0)$, is f_1 analytic at z_0 (so that $f(z)$ equals $(z - z_0)f_1(z)$, a product of *analytic* functions)?
(d) If so, we may now apply the same reasoning to f_1. Does the process of factoring out $z - z_0$ stop after k repetition? These last two questions are nontrivial. See Chapter 5.

12. *A long exercise on inverse functions.* Suppose the function f maps a set S onto a set S' in a one-to-one manner, that is, $z_1 \neq z_2$ in S implies $f(z_1) \neq f(z_2)$ in S'. Then f has an inverse function g mapping S' onto S, which will satisfy both $g(f(z)) = z$, $f(g(w)) = w$. This definition prompts two general questions: (i) Under what conditions is a function (analytic, say) one-to-one on S? (ii) If f has a nice property, does its inverse g also enjoy that property?

We remark that in the situation of interest to us, the case $S = \Omega$ is a domain and f is analytic, there is no graceful general answer to (i). We will discuss instances of (ii) below.
(a) An example: let Ω be the open first quadrant,

$$\Omega = \{z = x + iy \mid x > 0, y > 0\},$$

and define $f(z) = z^2$ on Ω. Verify that f is one-to-one on Ω. Then describe the image set $f(\Omega)$ in the w-plane and describe the inverse function $z = g(w)$. (Polar coordinates might help in specifying z.)
(b) Observe that $w = f(z) = z^2$ does not have an inverse function on the entire plane $S = \mathbb{C}$. Compare question (i) above. This is the old quandary of *two* square roots $\sqrt{w}$.
(c) Now we turn to question (ii) and continuity. Consider the following nontrivial theorem, Brouwer's "Invariance of Domain": *If f is a one-to-one continuous function, defined on the open set S, then the image $S' = f(S)$ is also open (f is an "open mapping").*

Use Brouwer's result to prove that if f is a one-to-one continuous mapping of the domain Ω onto the domain Ω', then the inverse function $g: \Omega' \to \Omega$ is also continuous. (This is simply a matter of definition, thanks to Brouwer.)

Preview: We'll see in Chapter 8 that a nonconstant analytic function is an open mapping, even if it is not one-to-one!

(d) *For the topologist:* Prove that if S is a compact (closed and bounded) set and f maps S one-to-one onto S' continuously, then the inverse function g is continuous on S'. No mention of analyticity here.

(e) Now we ask about analyticity of the inverse. Prove the following (this is independent of (a) through (d) above):

Let the analytic function f map the domain Ω one-to-one onto the domain Ω', and suppose further that $f'(z) \neq 0$ and that the inverse function g is continuous on Ω'. Then

(i) $g'(w)$ *exists for each w in Ω'*;

(ii) *in fact, $g'(w) = 1/f'(z)$, where $w = f(z)$*;

(iii) *thus g' is continuous and so g is analytic.*

Hint: Use the difference quotient to prove (i).

We are not too happy with this statement. As (c) above indicates, the same result follows from even weaker hypotheses. Namely, it is true that $\Omega' = f(\Omega)$ is always a domain, if f is analytic, and that $f' \neq 0$ and g is continuous if f is one-to-one. The truth is better conveyed by the (imprecise) dictum: *The inverse of an analytic function is analytic.*

13. *A proof of the Chain Rule.* We proved in the text that $f(g(\zeta))' = f'(g(\zeta))g'(\zeta)$, provided $g(\zeta) = a\zeta + b$. Now let $f(z), g(\zeta)$ be arbitrary analytic functions such that the composite $f(g(\zeta))$ is defined in some domain. Proceed as follows: Write, as in Exercise 4,

$$f(z) = f(z_0) + f'(z_0)(z - z_0) + \varepsilon_1(z; z_0)(z - z_0),$$

$$g(\zeta) = g(\zeta_0) + g'(\zeta_0)(\zeta - \zeta_0) + \varepsilon_2(\zeta; \zeta_0)(\zeta - \zeta_0),$$

and let $\zeta \to \zeta_0$ in the difference quotient

$$\frac{f(g(\zeta)) - f(g(\zeta_0))}{\zeta - \zeta_0}$$

with $z_0 = g(\zeta_0)$.

14. True or false?

(a) The product and quotient (where defined) of two complex analytic functions are analytic.

(b) The product and quotient (where defined) of two real harmonic functions are harmonic.

Moral: Analytic function theory has an even richer structure than harmonic function theory.

Section 3.3 THE CAUCHY-RIEMANN EQUATIONS

Let $f: \Omega \to \mathbb{C}$, with $f(z) = u(z) + iv(z)$ as usual. Since $z = x + iy$ is identified with the point (x, y), we also write $u = u(z) = u(x, y)$, $v = v(z) = v(x, y)$.

It is natural to inquire whether we can study the complex derivative $f'(z)$ in terms of something we already know, namely, the real partial derivatives of u and v. We might ask for a criterion for the analyticity of f in terms of the differentiability (with respect to x and y) of its real and imaginary parts u and v. This might enable us to manufacture new analytic functions by putting together suitable u's and v's. The criterion is this.

CAUCHY–RIEMANN EQUATIONS

If f is analytic at $z \in \Omega$, then

$$f'(z) = u_x(z) + iv_x(z) = v_y(z) - iu_y(z)$$

so that u, v satisfy the system of first-order differential equations

$$u_x(z) = v_y(z), \qquad u_y(z) = -v_x(z).$$

Conversely, if u and v are in $\mathscr{C}^1(\Omega)$ and together satisfy these differential equations for all $z \in \Omega$, then $f = u + iv$ is analytic in Ω.

Proof: Given $f'(z_0)$ exists, we let z approach z_0 along a horizontal line $y = y_0$ and a vertical line $x = x_0$, obtaining two expressions for the same limit. For $z = x + iy_0$, we have

$$f'(z_0) = \lim_{z \to z_0} \frac{f(z) - f(z_0)}{z - z_0}$$

$$= \lim_{x \to x_0} \left\{ \frac{u(x, y_0) - u(x_0, y_0)}{x - x_0} + i \, \frac{v(x, y_0) - v(x_0, y_0)}{x - x_0} \right\}$$

$$= u_x(x_0, y_0) + iv_x(x_0, y_0).$$

On the other hand, suppose $z = x_0 + iy$. Then we also have

$$f'(z_0) = \lim_{y \to y_0} \left\{ \frac{u(x_0, y) - u(x_0, y_0)}{i(y - y_0)} + i \, \frac{v(x_0, y) - v(x_0, y_0)}{i(y - y_0)} \right\}$$

$$= -iu_y(x_0, y_0) + v_y(x_0, y_0).$$

The Cauchy–Riemann equations follow from the uniqueness of $f'(z_0)$; we've shown its real part to be u_x and also v_y, whence $u_x = v_y$, etc.

For the converse, we are given that u and v are in $\mathscr{C}^1(\Omega)$ and $u_x = v_y$, $u_y = -v_x$. Hence,

$$f(z) - f(z_0) = \{u(z) - u(z_0)\} + i\{v(z) - v(z_0)\}$$

$$= \{u_x(z_0) \cdot (x - x_0) + u_y(z_0) \cdot (y - y_0)$$

$$+ \varepsilon_1 \cdot (x - x_0) + \varepsilon_2 \cdot (y - y_0)\}$$

$$+ i\{v_x(z_0) \cdot (x - x_0) + v_y(z_0) \cdot (y - y_0)$$

$$+ \varepsilon_3 \cdot (x - x_0) + \varepsilon_4 \cdot (y - y_0)\}$$

by Theorem 3 of Chapter 1. Using the Cauchy–Riemann equations, we rewrite the last expression thus:

$$f(z) - f(z_0) = (u_x(z_0) + iv_x(z_0)) \cdot (z - z_0)$$
$$+ (\varepsilon_1 + i\varepsilon_3) \cdot (x - x_0) + (\varepsilon_2 + i\varepsilon_4) \cdot (y - y_0)$$

Hence, the difference quotient for the derivative becomes

$$\frac{f(z) - f(z_0)}{z - z_0} = u_x(z_0) + iv_x(z_0)$$
$$+ (\varepsilon_1 + i\varepsilon_3)\frac{x - x_0}{z - z_0} + (\varepsilon_2 + i\varepsilon_4)\frac{y - y_0}{z - z_0}.$$

Since $|x - x_0| \le |z - z_0|$, $|y - y_0| \le |z - z_0|$ and $\varepsilon_1, \varepsilon_2, \varepsilon_3, \varepsilon_4$ tend to zero as $z = (x, y)$ approaches $z_0 = (x_0, y_0)$, we conclude that $f'(z_0)$ exists and, in fact, equals $u_x(z_0) + iv_x(z_0)$. Also, f' is continuous because the partial derivatives are. Thus, f is analytic and the converse is proved. Done.

The Cauchy-Riemann Equations in Polar Coordinates

Certain prominent analytic functions are usually defined in terms of the polar coordinates (r, θ) of a point rather than Euclidean coordinates (x, y). For example (see Section 3.4),

$$f(z) = \ln r + i\theta,$$

where $z = x + iy = r(\cos \theta + i \sin \theta), r > 0$, will define the very important complex logarithm; $\log z = f(z)$. We will discuss this in the next section.

If f is any complex function, then it may be written

$$f(z) = u(r, \theta) + iv(r, \theta).$$

If f is analytic, then its real and imaginary parts u, v will satisfy the polar form of the Cauchy–Riemann equations,

$$\frac{\partial u}{\partial r}(r, \theta) = \frac{1}{r}\frac{\partial v}{\partial \theta}(r, \theta), \qquad \frac{\partial u}{\partial \theta}(r, \theta) = -r\frac{\partial v}{\partial r}(r, \theta);$$

that is,

$$u_r = \frac{1}{r}v_\theta, \qquad u_\theta = -rv_r.$$

Conversely, if u, v satisfy these equations and are continuously differentiable, then f is analytic. The proof of this amounts to translating the Cauchy–Riemann equations in x, y into r, θ, that is, the two-variable Chain Rule. We leave the details as an exercise to be done by everyone once in his life.

Preview: The Relation of Harmonic to Analytic Functions

Let $f(z) = u(z) + iv(z)$ be analytic in Ω so that $u_x = v_y, u_y = -v_x$. It follows that $u_{xx} = v_{yx}$ and $u_{yy} = -v_{xy}$, provided these second partial derivatives exist. If these partials exist and are continuous, then $v_{yx} = v_{xy}$, whence we conclude that $u_{xx} + u_{yy} = 0$; that is, u is harmonic. Similarly, we have that v is harmonic (verify this!). This fact is usually stated thus: *The real and imaginary parts of an analytic function are harmonic.*

What does this mean?

1. We will see that many of our results on harmonic functions (e.g., Mean-Value Properties, Maximum Principle, Liouville's Theorem) have immediate extensions to the case of complex analytic functions.

2. One way to construct analytic functions $f(z)$ (as we will see) will be to start with harmonic $u(x, y)$ and find a harmonic $v(x, y)$ such that $u_x = v_y, u_y = -v_x$. Then $f = u + iv$ will be analytic (why?).

Caution: In deriving the harmonicity of the real and imaginary parts of an analytic function, we had to assume that the second derivatives u_{xx}, u_{xy}, u_{yx}, etc., exist. It is a remarkable fact (proof later) that these second partials *do* exist, provided $f = u + iv$ is analytic. This follows from the fact (at least as remarkable) that if f' exists, then f'' exists (whence by the same reasoning f''' exists, and so on). An analytic function has *all* complex derivatives $f', f'', \ldots, f^{(n)}, \ldots$.

Why is this? The reason seems to be that if $f = u + iv$ is analytic, then (as we have seen)

$$u_x - v_y = 0, \qquad u_y + v_x = 0.$$

That is, the real and imaginary parts of f give solutions to a simultaneous system of first-order partial differential equations: the Cauchy–Riemann equations of course. This is much stronger than the mere *existence* of the partial derivatives.

Notice, therefore, that the property of being analytic for a complex function $w = f(z)$ is much stronger than that of being continuously differentiable in the superficially analogous case of a real function $y = f(x)$. Moreover, this is in spite of the fact that the definition of f' via the difference quotient looks the same in each case. The existence of the two-dimensional complex limit of the difference quotient as z approaches z_0 is a powerful hypothesis.

At this point you might reread the first paragraphs of Chapter 1 with profit.

Exercises to Section 3.3

1. Verify the Cauchy–Riemann equations for $f(z) = z^2$.

2. Use the Cauchy–Riemann equations to show that $f(z) = \bar{z}$ is not analytic. This question was asked in Section 3.2, also.

3. Prove in two ways that if $f'(z) \equiv 0$ identically, then $f(z)$ is constant. See the set of exercises for Paragraph 3.2.3.

4. Prove that an analytic function which assumes only real values on a domain Ω is constant.

5. *The Chain Rule under coordinate change.* Convince yourself that if $x = x(r, \theta)$, $y = y(r, \theta)$ and $r = r(x, y)$, $\theta = \theta(x, y)$, then $u_x = u_r r_x + u_\theta \theta_x$. Here, the subscripts r, x denote partial derivatives and we can write u for both $u(x, y)$ and $u(x(r, \theta), y(r, \theta))$ so that u_r makes sense.

6. Verify the Cauchy–Riemann equations in polar coordinates as given in the text. Use Exercise 5.

7. Verify that $f(z) = \ln r + i\theta$, $-\pi < \theta < \pi$, satisfies the Cauchy–Riemann equations in polar coordinates. Here, $z \neq 0$ has polar coordinates (r, θ). (This $f(z)$ is the complex logarithm.)

8. Verify that $f(z) = e^x(\cos y + i \sin y)$ is analytic for all $z = x + iy$. This is the complex exponential function (refer to Section 3.4).

9. For $z = x + iy$, verify that $f(z) = (x^2 - y^2) + \lambda xyi$ (λ real) is analytic if and only if $\lambda = 2$.

Section 3.4 THE EXPONENTIAL AND RELATED FUNCTIONS

3.4.0 Introduction

Our gallery of analytic functions contains only polynomials and quotients of polynomials thus far. In determining analyticity for these functions, we used the fact that we could differentiate z^n explicitly. Now we construct some more functions and prove them to be analytic. The chief tool in this will be the Cauchy–Riemann equations. The functions we construct will all depend somehow on the exponential function, with which we now begin.

3.4.1 The Exponential Function

For $z = x + iy \in \mathbb{C}$ we define the *exponential function* exp: $\mathbb{C} \to \mathbb{C}$ by

$$\exp z = e^x(\cos y + i \sin y).$$

This generalizes the real exponential function, for when $y = 0$ we have $z = x$ and so $\exp z = \exp x = e^x$. We will note other similarities with e^x. See Exercise 12 for a full-scale motivation.

Analyticity: The real and imaginary parts of $\exp z$ are

$$u(z) = u(x, y) = e^x \cos y, \qquad v(z) = v(x, y) = e^x \sin y.$$

We check that the partials satisfy

$$u_x(x, y) = e^x \cos y = v_y(x, y),$$
$$u_y(x, y) = -e^x \sin y = -v_x(x, y).$$

Thus, u and v comprise a set of solutions to the Cauchy–Riemann equations. Since the partials are continuous throughout $\mathbb{C}$, we conclude that *exp z is analytic in all of* $\mathbb{C}$; that is, it is an *entire* analytic function, $\Omega = \mathbb{C}$.

The Derivative: Knowing that $f'(z) = u_x(z) + iv_x(z)$ (refer to Section 3.3), we see immediately that *exp z is its own derivative*:

$$\frac{d}{dz} \exp z = \exp z.$$

We recall that the real exponential function satisfies $(e^x)' = e^x$. This is further evidence that we have chosen the correct generalization of e^x.

Periodicity: In this respect exp z will be unlike the real exponential function, which is not periodic. Let $z = x + iy$. We observe that

$$\exp(z + 2\pi i) = \exp(x + i(y + 2\pi))$$
$$= e^x(\cos(y + 2\pi) + i \sin(y + 2\pi))$$
$$= e^x(\cos y + i \sin y)$$
$$= \exp z.$$

By a repetition of this argument we may show (see Figure 3.4) that, for any fixed z and every integer k,

$$\exp(z + 2k\pi i) = \exp z.$$

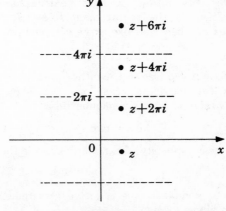

Figure   3.4

$$\exp z = \exp(z + 2\pi i) = \exp(z + 4\pi i) = \cdots$$

It follows from the properties of sine and cosine that if the complex number b is such that

$$\exp(z + b) = \exp z$$

for all $z \in \mathbb{C}$, then b *must* be of the form $2k\pi i$ for some integer k (proof?). We summarize this by saying that *exp z is periodic with period $2\pi i$.*

Addition Formula: This formula says that

$$\exp(z_1 + z_2) = (\exp z_1)(\exp z_2);$$

that is, *the exponential of a sum is the product of the exponentials.* This is a generalization of the calculus result $e^{a+b} = e^a e^b$, which works for real a, b.

To prove it, let $z_1 = x_1 + iy_1$, $z_2 = x_2 + iy_2$. Then

$$\begin{aligned}
\exp(z_1 + z_2) &= \exp((x_1 + x_2) + i(y_1 + y_2)) \\
&= e^{x_1+x_2}(\cos(y_1 + y_2) + i\sin(y_1 + y_2)) \\
&= e^{x_1}e^{x_2}\{(\cos y_1 \cos y_2 - \sin y_1 \sin y_2) \\
&\qquad + i(\sin y_1 \cos y_2 + \cos y_1 \sin y_2)\} \\
&= e^{x_1}(\cos y_1 + i\sin y_1)\, e^{x_2}(\cos y_2 + i\sin y_2) \\
&= (\exp z_1)(\exp z_2).
\end{aligned}$$

The key here was, of course, the addition formulas for sine and cosine, familiar from trigonometry.

What fact about exp z can we deduce anew from the addition formula by letting $z_1 = z$, $z_2 = 2\pi i$?

Further Properties of the Exponential Function

Complex Numbers as Exponents: Later we will give a meaning to a^z, where a and z are complex. The most important operation of this type, however, is the following: If e is the usual base for the natural logarithm, determined by the requirement $\ln e = 1$, and if $z = x + iy$ is any complex number, we raise e to a complex power by writing

$$e^z = \exp z.$$

We will feel free to use either notation.

In the particular case $x = 0$, $z = iy$, our convention becomes *Euler's relation* (worth remembering):

$$e^{iy} = \exp iy = \cos y + i\sin y.$$

A basic property of exponentiation is the *addition formula*

$$e^{z_1+z_2} = e^{z_1}e^{z_2}$$

for z_1, z_2 complex. This is just a restatement of the addition formula proved above for exp.

Of course we have

$$(e^z)' = \frac{d}{dz}e^z = e^z,$$

since $e^z = \exp z$.

The Polar Form of a Complex Number: This form is a useful combination of the complex exponential function and polar coordinates. For us, it will often be the "right" way to use polar coordinates.

Let $z = x + iy$ have polar coordinates (r, θ), where $r = |z|$ and θ is the usual angle (in radians). We call θ the *argument* of z and write $\theta = \arg z$. Now we note that

$$x = r \cos \theta, \qquad y = r \sin \theta.$$

It follows, therefore, that

$$z = x + iy = r(\cos \theta + i \sin \theta).$$

Now we make the crucial observation that, since θ is a real number, Euler's relation gives

$$e^{i\theta} = \exp i\theta = \cos \theta + i \sin \theta.$$

This is essentially our definition of $e^{i\theta}$. We conclude, therefore, that $z = x + iy$ may be written in polar form as

$$z = re^{i\theta}.$$

For example, if $z = 1 + i$, then $r = \sqrt{2}$, $\theta = \pi/4$, so that

$$z = 1 + i = \sqrt{2}e^{\pi i/4},$$

while if $z = -1$, then we have

$$-1 = e^{\pi i}$$

a real number.

Note that $|e^{i\theta}| = 1$ (Pythagoras' Theorem), so that $|z| = |r|\,|e^{i\theta}| = r$, as expected.

The use of the exponential $re^{i\theta}$ rather than the ordinary pair (r, θ) of polar coordinates is justified by the following arithmetic observation: If z_1, z_2 are complex numbers with

$$r_1 = |z_1|, \qquad \theta_1 = \arg z_1, \qquad r_2 = |z_2|, \qquad \theta_2 = \arg z_2,$$

then we have, thanks to the addition law,

$$z_1 z_2 = (r_1 e^{i\theta_1})(r_2 e^{i\theta_2}) = r_1 r_2 e^{i\theta_1} e^{i\theta_2} = r_1 r_2 e^{i(\theta_1 + \theta_2)}.$$

Thus, we may conclude that the product $z_1 z_2$ satisfies (see Figure 3.5)

   (i) $|z_1 z_2| = r_1 r_2 = |z_1|\,|z_2|$ (known already);
   (ii) $\arg(z_1 z_2) = \theta_1 + \theta_2 = \arg z_1 + \arg z_2$.

Now we can multiply two complex numbers geometrically! The product $z_1 z_2$ has absolute value equal to the product of the absolute values of the two factors and has argument equal to the *sum* of the factors (see Figure 3.5). We will apply this to powers and roots in a moment.

Note that in writing $z = re^{i\theta}$ it is not the variable z that appears in the exponent.

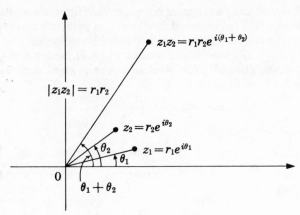

Figure   3.5

Afterthought

You may object that in writing z as $re^{i\theta}$ we overlooked that chronic difficulty with polar coordinates, the fact that the angle arg z is not a uniquely defined real number θ, but rather that any number $\theta + 2k\pi$, $k = 0 \pm 1, \pm 2,\ldots$, determines the same angle. However, this objection evaporates when we recall that the exponential function has period $2\pi i$ or, in the same vein,

$$e^{i(\theta + 2k\pi)} = e^{i\theta}e^{2\pi ik} = e^{i\theta}(e^{2\pi i})^k = e^{i\theta}.$$

All is well.

Polar Form Applied to nth Powers and nth Roots: If $z = re^{i\theta}$, then $z^2 = r^2e^{2i\theta}$, and by repeated applications of this idea we conclude

$$z^n = (re^{i\theta})^n = r^n e^{ni\theta}$$

for any nonnegative integer n. The formula is also true for $n < 0$ (check!). Now let's consider the converse problem from algebra.

Problem: the nth Roots of a Complex Number

Given a nonzero complex number z, compute all ζ such that $\zeta^n = z$.

Such a ζ is, of course, an nth root of z, and we might write $\zeta = z^{1/n}$, keeping in mind, however, that ζ is not unique when $|n| > 1$. The equation $\zeta = z^{1/n}$ determines a set of ζ's. We seek an explicit formula for them.

The key to the problem of nth roots is to write z in polar form:

$$z = re^{i\theta} (= re^{i(\theta + 2k\pi)}, k = 0, \pm 1,\ldots).$$

Now we define n complex numbers $\zeta_0, \zeta_1,\ldots, \zeta_{n-1}$ by

$$\zeta_k = r^{1/n}e^{i(\theta + 2k\pi)/n} \qquad (k = 0, 1,\ldots, n - 1).$$

We claim that these n complex numbers are distinct and give all solutions to $\zeta^n = z$. This will solve our problem.

First, however, by $r^{1/n}$ we mean the unique *positive* real number whose nth power equals r. You should convince yourself that this $r^{1/n}$ actually exists (completeness of the real numbers) and is unique (if $0 < a < b$, then $a^n < b^n$).

Now we check that $(\zeta_k)^n = z$. Observe that

$$(\zeta_k)^n = (r^{1/n}e^{i(\theta + 2k\pi)/n})^n = re^{i(\theta + 2k\pi)} = re^{i\theta} = z.$$

We used the formula for nth powers and the periodicity of the complex exponential, $e^{2k\pi i} = 1$ for integers k.

To prove that $\zeta_0, \zeta_1, \ldots, \zeta_{n-1}$ are distinct, assume $\zeta_h = \zeta_k$ with $0 \le h < k \le n - 1$. Thus, $\zeta_k \zeta_h^{-1} = 1$. On the other hand,

$$\zeta_k \zeta_h^{-1} = (r^{1/n}e^{i(\theta + 2k\pi)/n})(r^{-1/n}e^{-i(\theta + 2h\pi)/n})$$

$$= e^{2(k-h)\pi i/n}.$$

But this last number cannot equal unity because $e^{2m\pi i} = 1$ if and only if m is an integer (periodicity!) and $m = (k - h)/n$ is not an integer. Done.

Finally, we must verify that if $\zeta^n = z$, then ζ is one of $\zeta_0, \ldots, \zeta_{n-1}$. Let t be a variable, and consider the equation $t^n = z$; that is, $t^n - z = 0$. Since $t = \zeta_k$ is a root of this equation for $k = 0, 1, \ldots, n - 1$, we have the factorization

$$t^n - z = (t - \zeta_0)(t - \zeta_1)\cdots(t - \zeta_{n-1}).$$

This follows from successive division of $t^n - z$ by the factors $t - \zeta_0$, $t - \zeta_1$, and so on. Thus, $\zeta^n - z = 0$ implies

$$(\zeta - \zeta_0)(\zeta - \zeta_1)\cdots(\zeta - \zeta_{n-1}) = 0,$$

whence ζ must be one of $\zeta_0, \ldots, \zeta_{n-1}$ (why?).

This is a special instance of the fact that a polynomial of degree n has at most n distinct roots.

A Remark on the Roots of Unity

We defined $\zeta_k = r^{1/n}e^{i(\theta + 2k\pi)/n}$. See Figure 3.6. Thus, we have

$$\zeta_k = \zeta_0 e^{2k\pi i/n}.$$

Now note that $\omega_k = e^{2k\pi i/n}$ is an *n*th root of unity, $(\omega_k)^n = 1$. This insight leads to the following conclusion: *Let ζ satisfy $\zeta^n = z$. Then all nth roots of z (nonzero) are given by $\zeta\omega_0, \zeta\omega_1, \ldots, \zeta\omega_{n-1}$, where the ω_k's are the nth roots of unity.*

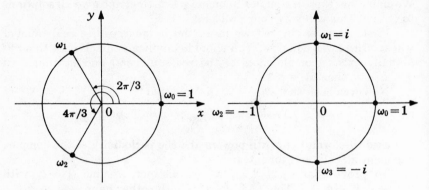

The cube roots of unity   The fourth roots of unity

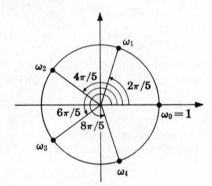

The fifth roots of unity

Figure   3.6

Exercises to Paragraph 3.4.1

1. Compute $\exp z$ for the following z:
   (a) $1 + \pi i$,
   (b) $\pi i/2$,
   (c) $-14\pi i$,
   (d) $3\pi i$,
   (e) 0.

2. Prove, using properties of the real sine and cosine, that if $\exp z = \exp(z + b)$ for all z, then b is an integer multiple of $2\pi i$.

3. Verify that $f(z) = e^{2\pi i z}$ is periodic with period 1.

4. Prove that $\exp z \neq 0$ for all z. Use the definition, or the fact that $e^z e^{-z} = 1$.

5. If $w \neq 0$, does there exist z such that $w = e^z$? How many such z?

6. Describe the set of points w of the form $w = \exp z$ with $z = x + iy$, x arbitrary, $-\pi < y \leq \pi$.

7. Write the following in polar form $re^{i\theta}$:
   (a) $(1 + i)/\sqrt{2}$,
   (b) $-i$,
   (c) -1.

8. Verify $e^{\pi i} + 1 = 0$. Note all the famous numbers that appear here! This is worth remembering.

9. Concoct a formula similar to that in Exercise 8, but which involves the number 2 as well as $e, \pi, i, 1, 0$.

10. Sketch in the plane the three points ω satisfying $\omega^3 = 1$. Now locate all six sixth roots of unity ($z^6 = 1$). Where would the twelfth roots of unity lie?

11. (a) Write in polar form both square roots of $(1 + i)/\sqrt{2}$.
    (b) Likewise for $-(1 + i)/\sqrt{2}$.

12. *Why we define* $e^{x+iy} = e^x(\cos y + i \sin y)$.
    (a) Surely we desire $e^{x+iy} = e^x e^{iy}$. Thus, if $e^{iy} = U(y) + iV(y)$, we must motivate $U(y) = \cos y$, $V(y) = \sin y$. We do this as follows:
    (b) Surely we desire $(d/dy) e^{iy} = ie^{iy}$, whence $(d^2/dy^2) e^{iy} = -e^{iy}$. Prove we must have, therefore,

$$U''(y) = -U(y), \qquad V''(y) = -V(y).$$

   (c) From elementary differential equations,

$$U(y) = C_1 \cos y + C_2 \sin y, \qquad V(y) = C_3 \cos y + C_4 \sin y.$$

   (d) But surely we desire $e^{iy} = 1$ when $y = 0$. Prove we must have, therefore, $U(0) = 1$, $U'(0) = 0$, $V(0) = 0$, $V'(0) = 1$.
   (e) Conclude:

$$U(y) = \cos y, \qquad V(y) = \sin y.$$

13. Prove $e^z = e^\zeta$ implies $z - \zeta = 2k\pi i$ for some integer k. *Hint:* By the addition law, this is equivalent to proving $e^b = 1$ implies $b = 2k\pi i$ for some integer k.

14. *The formula of de Moivre.*
    (a) Prove, for any integer n,

$$(\cos \theta + i \sin \theta)^n = \cos n\theta + i \sin n\theta.$$

   Hint: Consider the nth power of $e^{i\theta}$.
   (b) Deduce:

$$\cos 3\theta = \cos^3 \theta - 3 \cos \theta \sin^2 \theta.$$

   Thus, the angle 3θ may be treated in terms of θ.
   (c) Obtain a similar formula for $\sin 3\theta$.

15. *Local one-oneness.* Prove that the exponential mapping $w = e^z$ is locally one-to-one; that is, given a point z_0, there is an open neighborhood of z_0 in which $e^z = e^\zeta$ only if $z = \zeta$. *Hint:* Exercise 13.

3.4.2 The Logarithm

We would like to define the logarithm $\log z$ as the inverse function of the exponential so that, for all complex a, b (with $b \neq 0$),

$$\log \exp a = a, \qquad \exp \log b = b.$$

We recall that these formulas hold for real numbers a, b, provided $b > 0$, where $\log b$ means the natural logarithm $\ln b$ and $\exp a = e^a$.

However, we cannot define an inverse for the complex exponential because this function is not one-to-one. In fact, we have seen that it is periodic,

$$\exp(a + 2\pi k i) = \exp a \qquad (k = 0, \pm1, \pm2, \ldots)$$

for all complex a. Thus, for example, $\exp 0 = \exp 2\pi i = 1$, so it is not clear whether we should define $\log 1 = 0$ or $\log 1 = 2\pi i$ (or $\log 1 = 4\pi i$, etc.).

To resolve this quandary, we must examine the mapping properties of the exponential function more closely. The following result is basic. It provides a geometric picture of the exponential function. See Figure 3.7.

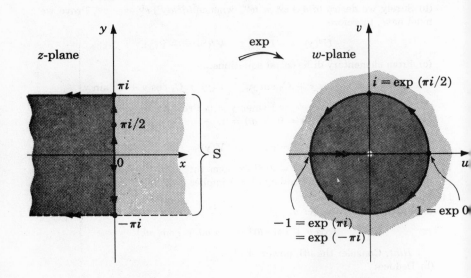

Note: Left half ($x < 0$) of strip S is mapped one-to-one onto punctured disc $0 < |w| < 1$.

Figure   3.7

THEOREM 1

The exponential function maps the horizontal strip

$$S = \{z = x + iy \mid -\pi < y \le \pi\}$$

in the z-plane onto the punctured w-plane $\mathbb{C} - \{0\}$ *in a one-to-one fashion. In particular, the horizontal line* $y = \pi$ *is mapped onto the negative real axis in the w-plane.*

Proof: It is straightforward to prove that $\exp z \neq 0$ and that $\exp(x + i\pi)$ lies on the negative real u-axis. We leave this to you.

Let $w \neq 0$. We want to prove $w = \exp z$ for some $z \in S$. Now w has polar coordinates (ρ, φ) with $\rho > 0$ and $-\pi < \varphi \leq \pi$. Define $z = x + iy$, where $x = \ln \rho$, $y = \varphi$. Note $z \in S$. To see that $w = \exp z$, simply note $\exp z = e^{\ln \rho}(\cos \varphi + i \sin \varphi) = \rho(\cos \varphi + i \sin \varphi) = w$. This proves that $\exp$ maps S onto $\mathbb{C} - \{0\}$.

To prove that $\exp$ is one-to-one when restricted to S, we will suppose $z_1, z_2 \in S$ and $\exp z_1 = \exp z_2$, and show that $z_1 = z_2$.

Now $\exp z_1 = \exp z_2$ implies $(\exp z_1)/(\exp z_2) = 1$, whence $\exp(z_1 - z_2) = 1$. But the periodicity of $\exp$ implies that $z_1 - z_2$ must be of the form $2k\pi i$, with k an integer (check this!). We claim now that $k = 0$; that is, $z_1 = z_2$. This is because z_1, z_2 are both in the strip S, whence $|z_1 - z_2| < 2\pi$. This proves $k = 0$, and so $\exp$ is one-to-one as a function from S to $\mathbb{C} - \{0\}$. Done.

A Generalization

Since $\exp z = \exp(z \pm 2\pi i) = \exp(z \pm 4\pi i) = \cdots$, we may extend Theorem 1 to other strips in the z-plane. Thus, for each real b, let

$$S_b = \{z = x + iy \mid b - 2\pi < y \leq b\}$$

be the horizontal strip of height 2π whose upper edge is the horizontal line $y = b$. We state

COROLLARY 2

The exponential function maps each horizontal strip S_b in the z-plane onto the punctured w-plane $\mathbb{C} - \{0\}$ in a one-to-one fashion. In particular, the horizontal line $y = b$ is mapped onto the ray extending from the origin $w = 0$ determined by the angle $\varphi = b$.

See Figure 3.8.

The Definition of the Logarithm

Having the foregoing information on the exponential, we proceed to define a kind of inverse for it. Let $z \neq 0$, $z = re^{i\theta}$ as usual. We define the expression $\log z$ by the formula

$$\log z = \ln|z| + i \arg z = \ln r + i\theta.$$

This is *not* a function, because $z = re^{i\theta} = re^{i(\theta \pm 2\pi)} = \cdots$, but $\ln r + i\theta$, $\ln r + i(\theta \pm 2\pi)$, $\ln r + i(\theta \pm 4\pi), \cdots$ are all different complex numbers. Thus, the expression $\log z$ is not "single valued" because $\arg z$ is not.

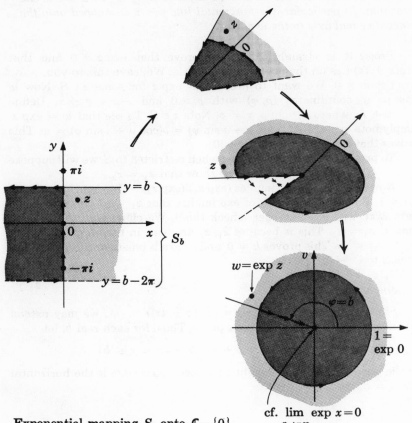

Exponential mapping S_b onto $\mathbb{C} - \{0\}$

Figure 3.8

To obtain an honest function, we proceed as follows: Let S_π be the horizontal strip in the w-plane given by

$$S_\pi = \{w = u + iv \mid -\pi < v \leq \pi\}.$$

Now let us agree to write z in $\mathbb{C} - \{0\}$ as

$$z = re^{i\theta} \qquad (r > 0, -\pi < \theta \leq \pi).$$

Having restricted $\theta = \arg z$ in this way, we once again define

$$\log z = \log re^{i\theta} = \ln r + i\theta.$$

You should check that $\log z$ is now a function defined for all $z \neq 0$ *with values in the horizontal strip* S_π. Hence, we may write

$$\log: \mathbb{C} - \{0\} \rightarrow S_\pi.$$

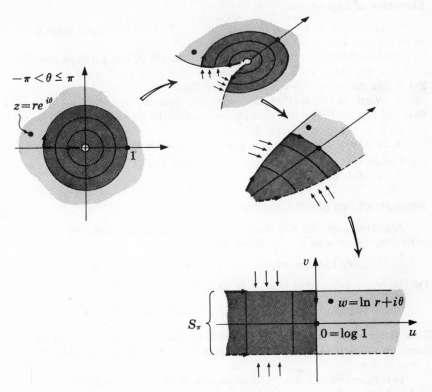

$$w = \log z, \text{ principal branch}$$

Figure   3.9

This function, with the restriction $-\pi < \theta \leq \pi$, is called the *principal branch of the logarithm*. See Figure 3.9. Usually, for purposes of continuity, we make the slightly stronger restriction $-\pi < \theta < \pi$; more on this point in a moment. We will discuss the properties of this function below.

More generally, however, a *branch* of the logarithm is defined by selecting a horizontal ("target") strip S_b in the w-plane, then agreeing that $\theta = \arg z$ satisfies

$$b - 2\pi < \theta \leq b,$$

and defining $\log z = \ln r + i\theta$ as usual. This gives a function onto a strip in the w-plane:

$$\log: \mathbb{C} - \{0\} \to S_b.$$

Important: Different strips S_b give different (though related) functions or "branches" $\log z$. The key to many problems is choice of an appropriate branch of the logarithm.

Examples of Logarithms

1. Let $z = 1 = e^0 = e^{2\pi i} = e^{4\pi i} = \cdots$. For the principal branch we have $\log 1 = \ln 1 + i\theta = 0 + i0 = 0$.

On the other hand, if we choose the branch of $\log z$ determined by $\pi < \theta \le 3\pi$, then we agree to write $1 = e^{2\pi i}$ so that $\log 1 = \ln 1 + 2\pi i = 2\pi i$. Note for this branch $\log 1 \in S_{3\pi}$.

2. What is $\log i$? For the principal branch, $\theta = \arg i = \pi/2$, so that $\log i = \pi i/2$. For the branch determined by $3\pi < \theta \le 5\pi$, we have $i = 9\pi/2$, so that $\log i = 9\pi i/2$.

3. If $z = 1 + i$, then the principal branch of $\log z$ is given by $\log z = \ln r + i\theta = \ln \sqrt{2} + \pi i/4 = (\ln 2)/2 + \pi i/4$.

4. Log 0 is not defined.

Some Properties of the Logarithm

Log and Exp: To see that these are mutual inverses, let $z = re^{i\theta}$ with $-\pi < \theta \le \pi$ as above. Then we compute

$$\exp \log z = \exp(\ln r + i\theta) = e^{\ln r} e^{i\theta} = re^{i\theta} = z.$$

On the other hand, if $w = u + iv$ with $-\pi < v \le \pi$, then

$$\log \exp w = \log e^u e^{iv} = \ln e^u + iv = u + iv = w.$$

This shows that the principal branch of the logarithm is the inverse function of exp when exp is restricted to the strip determined by $-\pi < v \le \pi$.

The moral of all this is that *exp, defined on a horizontal strip S_b of height 2π, is the inverse of some appropriately chosen branch of the logarithm.* In particular, the mapping

$$\log: \mathbb{C} - \{0\} \to S_b$$

is one-to-one and onto.

Continuity: The principal branch of $\log z$ is continuous on the z-plane with the nonpositive x-axis removed, that is, on the domain of points $z = re^{i\theta}$ with $r > 0$, $-\pi < \theta < \pi$ (note $\theta \ne \pi$). This is true because its real part, $\ln r$, and imaginary part, θ, are continuous functions (think of them in polar coordinates) on this domain.

Now we see why we might wish to delete the entire nonpositive x-axis in defining the principal branch $\log z$. For the function θ is not defined at the origin and, moreover, cannot be extended to a *continuous* function on the negative x-axis as well. For as $z = re^{i\theta}$ approaches any point ζ on the negative x-axis *from above* (see Figure 3.10), we have

$$\lim_{z \to \zeta} \theta = \pi,$$

while if z approaches ζ from below, we get

$$\lim_{z \to \zeta} \theta = -\pi.$$

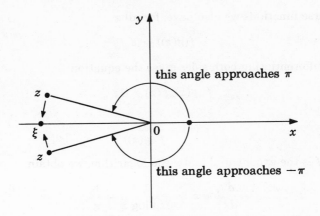

Showing θ, as a function of z,
is not continuous at $z = \xi$

Figure   3.10

Hence, θ (and log z) will not be continuous at ξ. Thus, it is convenient to delete these troublesome ξ's.

Analyticity: There are several ways of verifying that every branch of log z is analytic. In the spirit of this section we will employ the Cauchy–Riemann equations in polar coordinates.

The real and imaginary parts of log $z = u + iv$ are

$$u(r, \theta) = \ln r, \qquad v(r, \theta) = \theta.$$

The partial derivatives are readily computed:

$$u_r = \frac{1}{r}, \qquad u_\theta = 0, \qquad v_r = 0, \qquad v_\theta = 1.$$

It follows immediately that u, v satisfy the Cauchy–Riemann equations (see Section 3.3),

$$u_r = \frac{1}{r}v_\theta, \qquad u_\theta = -rv_r,$$

and since all requirements of continuity are met, we may conclude that log z is analytic.

The Derivative: What is $(d/dz)\log z$? We could build it using the partials of u, v computed above and the standard trig relations between (x, y) and (r, θ). But here is a second way, useful when our function is the inverse of something known. It is standard calculus.

The Chain Rule for complex functions tells us that

$$(f(g(z)))' = f'(g(z))g'(z).$$

For inverse functions we also have, for all z,

$$f(g(z)) = z$$

so that differentiating both sides gives the equation

$$f'(g(z))g'(z) = 1;$$

that is,

$$g'(z) = \frac{1}{f'(g(z))}.$$

Thus, if f is the exponential and g the logarithm, we obtain

$$\frac{d}{dz}\log z = \frac{1}{\exp\log z} = \frac{1}{z}.$$

This answers our questions. Compare the answer with the well-known $(d/dx)\ln x = 1/x$.

The Addition Law Revisited: For $a, b > 0$, the natural logarithm satisfies:

$$\ln ab = \ln a + \ln b.$$

This generalizes in the complex case to the addition law

$$\log z_1 z_2 = \log z_1 + \log z_2 + 2k\pi i,$$

where k is some integer. The $2k\pi i$ term reflects the fact that we do not specify the branches of log that occur.

To establish the formula, write $\log z_1 z_2 = \log z_1 + \log z_2 + \zeta$ and prove by exponentiating both sides that the error term ζ satisfies $e^\zeta = 1$, whence ζ must have the form $2k\pi i$.

A special case of this law is given by

$$\log z^m = m\log z + 2k\pi i.$$

Here m is a given integer and k depends on the branch.

Another Expression for the nth Roots of z: We saw earlier (end of Paragraph 3.4.1) that if $z = re^{i\theta}$, then the n complex numbers

$$\zeta_k = r^{1/n}e^{i(\theta+2k\pi)/n} \qquad (k = 0, 1, \ldots, n-1)$$

each satisfy $\zeta^n = z$. When we write $z^{1/n}$ we mean the set $\zeta_0, \zeta_1, \ldots, \zeta^{n-1}$ of nth roots of z.

There is another quite useful way of writing $z^{1/n}$. It is

$$z^{1/n} = \exp\left(\frac{1}{n}\log z\right).$$

Let us examine the right-hand side here to verify equality. Suppose that

$$z = re^{i\theta} = re^{i(\theta\pm2\pi)} = re^{i(\theta\pm4\pi)} = \cdots$$

as usual. Then, depending on our choice of a branch, log z may be any of the numbers

$$\ln r + i(\theta + 2k\pi) \qquad (k = 0, \pm 1, \pm 2, \ldots).$$

Thus, $(1/n) \log z$ might be any one of the numbers

$$\frac{\ln r^{1/n} + i(\theta + 2k\pi)}{n} \qquad (k = 0, \pm 1, 2, \ldots).$$

Finally, using the fact that $\exp \tau = \exp(\tau + 2\pi i)$, you should convince yourself that

$$\exp\left(\frac{1}{n} \log z\right) = r^{1/n} e^{i(\theta + 2k\pi)/n} \qquad (k = 0, 1, \ldots, n - 1).$$

But these last n numbers are the nth roots $\zeta_0, \zeta_1, \ldots, \zeta_{n-1}$ of z found earlier. This justifies the new formula for $z^{1/n}$.

We will see more of this when we discuss z^c with c complex.

Exercises to Paragraph 3.4.2

1. Compute $\log z$ (principal branch) for the following z:
   (a) $-1 + i$,
   (b) $2 + 2i$,
   (c) $4e^{i\pi/7}$,
   (d) $-i$.

2. Likewise for the branch determined by $\pi < \theta \le 3\pi$.

3. If x is real and positive, can $\log x$ be complex? Explain.

4. *Mapping by the exponential $w = \exp z$.* Sketch the following sets in the z-plane and their images under the exponential function in the w-plane. Indicate where the boundaries are mapped. Here, $z = x + iy$.

   $$\Omega_1 = \{z \mid x < 0, -\pi < y \le \pi\}.$$
   $$\Omega_2 = \{z \mid 0 < x < \ln 2, -\pi < y \le \pi\}.$$
   $$\Omega_3 = \{z \mid x < 0, 0 < y < \pi\}.$$
   $$\Omega_4 = \{z \mid x \ge 0, -\pi < y \le \pi\}.$$

5. Let $w = \log z$ denote the principal branch of the logarithm. Where does $\log z$ map the following sets? Sketch as in Exercise 4. *Important:* $-\pi < \arg z \le \pi$.

   The punctured disc $0 < |z| < 1$; the upper half-disc $|z| < 1, y > 0$; the annulus $1 < |z| < 2$; the set $|z| \ge 1$ complementary to the open unit disc.

6. Where does $w = f(z) = e^{2\pi z}$ map the horizontal strip $-1 < y \le 1$? Write down an inverse function for $f(z)$ whose image is the strip.

7. To remember the defining formula for $\log z$, write $z = |z| e^{i \arg z}$ and apply $\log$ to this product, using the addition law and properties of the natural logarithm.

8. Let $h(z) = e^{f(z)}$, where $f(z)$ is a continuous function defined on a domain Ω.
   (a) Prove $h(z)$ is analytic on Ω if $f(z)$ is.

(b) Prove $f(z)$ is constant on Ω if $f(z)$ is analytic on Ω and $h(z)$ is constant on Ω.

(c) Now work (b) without assuming f analytic.

9. Suppose f is analytic on a domain Ω in the z-plane, and that the image set $f(\Omega)$ does not intersect some ray $\varphi = $ constant in the w-plane.

(a) Show that one can select a branch of $\log w$ so that the composite $h(z) = \log f(z)$ is analytic in Ω.

(b) Suppose now that $h(z)$ is constant in Ω. Prove that $f(z)$ must be constant in Ω. *Hint:* There is a one-line proof.

(c) Is (b) true if f is not analytic?

3.4.3 Complex Trigonometric Functions

Now we use the exponential function to construct more elementary analytic functions.

For real y, we have the Euler relations (Paragraph 3.4.1)

$$e^{iy} = \cos y + i \sin y, \qquad e^{-iy} = \cos y - i \sin y.$$

Solving for $\cos y$ and $\sin y$ in these two equations yields

$$\cos y = \frac{e^{iy} + e^{-iy}}{2}, \qquad \sin y = \frac{e^{iy} - e^{-iy}}{2i}.$$

These expressions are sometimes useful in ordinary calculus. But now we extend them to the case of a complex variable, defining

$$\cos z = \frac{e^{iz} + e^{-iz}}{2}, \qquad \sin z = \frac{e^{iz} - e^{-iz}}{2i}.$$

These functions are analytic because they are built from the analytic e^{iz} and e^{-iz}. It is easy to check that their derivatives are given by familiar formulas:

$$\frac{d}{dz} \sin z = \cos z, \qquad \frac{d}{dz} \cos z = -\sin z.$$

Just as in trigonometry, the other basic functions are defined in terms of sine and cosine. Thus,

$$\tan z = \frac{\sin z}{\cos z}, \qquad \cot z = \frac{\cos z}{\sin z}$$

$$\sec z = \frac{1}{\cos z}, \qquad \csc z = \frac{1}{\sin z}.$$

These functions are analytic in any domain where the denominators do not vanish. See the exercises for this subsection.

We leave it to you to verify the formulas (reminiscent of ordinary calculus)

$$\frac{d}{dz} \tan z = \sec^2 z, \qquad \frac{d}{dz} \cot z = -\csc^2 z,$$

$$\frac{d}{dz} \sec z = \sec z \tan z, \qquad \frac{d}{dz} \csc z = -\csc z \cot z.$$

The complex trig functions satisfy many of the formulas of their real counterparts. For example, you might verify the Pythagorean relation

$$\sin^2 z + \cos^2 z = 1$$

and the periodicity relations

$$\sin z = \sin(z + 2k\pi), \qquad \cos z = \cos(z + 2k\pi),$$

with $k = 0, \pm 1, \pm 2, \ldots$. Note $2k\pi$ here, not $2k\pi i$. See Figure 3.11.

Figure 3.11

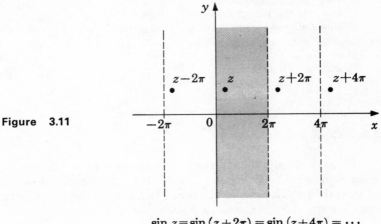

$$\sin z = \sin(z \pm 2\pi) = \sin(z \pm 4\pi) = \cdots$$
$$\cos z = \sin(z \pm 2\pi) = \cos(z \pm 4\pi) = \cdots$$

Exercises to Paragraph 3.4.3

1. Prove that the derivatives of $\sin z$, $\cos z$ are $\cos z$, $-\sin z$, respectively.
2. Prove that $\sin z = 0$ only if $z = x = k\pi$ for some integer k.
3. Find all z such that $\cos z = 0$.
4. What is the domain of definition of $\tan z$?
5. Prove $(d/dz)(\tan z) = \sec^2 z$.
6. (a) Prove $\sin^2 z + \cos^2 z \equiv 1$ from the definition of $\sin z$, $\cos z$.
   (b) State and prove addition formulas for $\sin(z_1 + z_2)$ and $\cos(z_1 + z_2)$.
7. True or false? $\cos z = \sin(z + \pi/2)$ for all z. Reason?

8. Prove that $\sin(z + \tau) = \sin z$ for all z if and only if $\tau = 2k\pi$ for some integer k.

9. (a) Given any complex w_0, prove that there exists z_0 satisfying $w_0 = \sin z_0$. *Hint:* First solve for e^{iz_0} in terms of w_0 via the quadratic equation $(e^{iz_0})^2 - 2iw_0 e^{iz_0} - 1 = 0$. (Where did we get this?)
(b) Solve $\sin z = -5$.

10. Prove that the z_0 found above is unique, provided it is required to lie in the vertical strip $-\pi/2 < x \le \pi/2$ or any such strip of width π.

11. What is the image in the w-plane of the vertical strip $-\pi/2 < x \le \pi/2$ under the mapping $w = \sin z$? See the preceding exercise.

12. In the spirit of Exercises 9 through 11, construct an inverse function (arcsine) for the sine function as we did for the exponential function (strips, branches). Is the arcsine analytic?

13. *Complex hyperbolic functions.*
(a) Generalize the real hyperbolic functions

$$\sinh x = \frac{(e^x - e^{-x})}{2}, \qquad \cosh x = \frac{(e^x + e^{-x})}{2}$$

by replacing x with z in these formulas.
(b) Compute the derivatives of $\sinh z$, $\cosh z$.
(c) How are the squares of $\sinh z$, $\cosh z$ related?

14. Solve the following complex differential equations for $w = f(z)$:
(a) $w' - w = 0$,
(b) $w'' + w = 0$,
(c) $w'' - w = 0$.

3.4.4 Complex Exponents

We have already defined z^m when m is an integer and, more generally, when m is a rational number. For if $m = a/b$ is rational but not an integer, then $z^{a/b}$ is a *set* of numbers (for example, the two square roots of z). We recall for $z \ne 0$ and m rational that

$$z^m = \exp(m \log z).$$

Now we define z^c for c complex by a similar formula:

$$z^c = \exp(c \log z) \qquad (z \ne 0).$$

Of course z^c stands for a set of numbers, since $\log z$ is involved here. Selection of a branch of the logarithm (restricting $-\pi < \arg z \le \pi$, say) leads to a "single-valued" function $f(z) = z^c$.
An example: We have in the case $z = i$, $c = i$,

$$i^i = \exp(i \log i)$$

$$= \exp\left(i\left(i\left(\frac{\pi}{2} + 2k\pi\right)\right)\right)$$

$$= \exp\left(-\frac{\pi}{2} - 2k\pi\right)$$

$$= e^{-\pi/2}e^{-2k\pi}$$

with $k = 0, \pm 1, \ldots$

The Derivative of z^c: We suppose that a branch of the logarithm has been chosen so that $f(z) = z^c$ is a single-valued function, analytic (why?) in some domain (which?). We have

$$\frac{d}{dz}z^c = \frac{d}{dz}\exp(c\log z)$$

$$= \exp(c\log z)cz^{-1} \qquad (Chain\ Rule!)$$

$$= c\exp(c\log z)\exp(-\log z) \qquad (trick!)$$

$$= c\exp((c-1)\log z) \qquad (why?)$$

$$= cz^{c-1}.$$

That is,

$$\frac{d}{dz}z^c = cz^{c-1} \qquad (z \neq 0).$$

This compares nicely with $(d/dx)x^c = cx^{c-1}$ from calculus.

Exercises to Paragraph 3.4.4

1. Write in polar form:
   (a) π^i,
   (b) π^{2i},
   (c) $(1 + i)^i$.

2. State and prove an addition law relating z^{b+c} and $z^b z^c$ for z, b, c complex. Discuss any ambiguities due to the multivaluedness of z^c. Of course the law you state will have its origin in the addition law for the logarithm (Paragraph 3.4.2).

Section 3.5 THE HARMONIC CONJUGATE

We are answering the question: "Do analytic functions exist?" Our work thus far has yielded two methods of verifying analyticity, namely, direct differentiation (polynomials) and the Cauchy–Riemann equations. Now we present a third method for obtaining analytic functions. A fourth method, power series, will be discussed in Chapter 5.

Suppose $u = u(z)$ is a real-valued harmonic function defined for all

z in some domain Ω. We wish to construct a *harmonic conjugate* $v = v(z)$ for u, that is, a function v such that, for all z in Ω, the sum

$$f(z) = u(z) + iv(z)$$

is complex analytic. It is necessary and sufficient for the analyticity of f that the partial derivatives of v be continuous and satisfy $v_x = -u_y$, $v_y = u_x$ throughout Ω.

We will construct $v(z)$ by using a certain line integral. For motivation, suppose v existed. Then its "total (exact) differential" is defined as

$$dv = v_x\,dx + v_y\,dy,$$

and we have, from Section 1.4, a formula for $v(z)$:

$$v(z) - v(z_0) = \int_{z_0}^z dv,$$

where the integral is independent of the curve from z_0 (any selected point) and z.

We will play a variation on this theme. By the Cauchy–Riemann equations, dv should equal

$$-u_y\,dx + u_x\,dy.$$

Now we claim that if Ω is simply connected (no holes) and if z, z_0 are in Ω, then the integral

$$\int_{z_0}^z [-u_y\,dx + u_x\,dy]$$

is independent of the path in Ω from z_0 to z. This is true essentially because the Jordan domain between any two nonintersecting curves from z_0 to z lies inside Ω, and so the line integral from z_0 to z and back to z_0 (over one curve and back by the reverse of the other) is equal to a certain double integral over this Jordan domain (Green's Theorem). But this double integral vanishes because u is harmonic in Ω. You'll enjoy verifying these assertions.

Hence, the integral displayed above is independent of the path and defines a function $v(z)$:

$$v(z) - v(z_0) = \int_{z_0}^z dv = \int_{z_0}^z [-u_y\,dx + u_x\,dy].$$

We may choose the real value $v(z_0)$ at will, of course. See Figure 3.12.

Here is an example. We are given $u(x, y) = x^2 - y^2$, harmonic in $\mathbb{C}$. This gives us a candidate for dv, namely,

$$-u_y\,dx + u_x\,dy = 2y\,dx + 2x\,dy.$$

Line integrals of this differential are certainly independent of the path because (as we observe) it is the differential of any function of the form $2xy + c$ with c real. To conclude (no need to carry out the integral), we

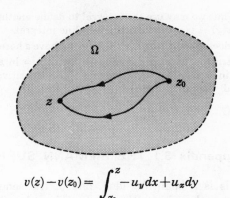

Figure   3.12

$$v(z) - v(z_0) = \int_{z_0}^{z} -u_y\,dx + u_x\,dy$$

may define $v(x, y) = 2xy + c$ for any real c and obtain a harmonic function that is conjugate to the original $u(x, y)$. This leads to

$$f(z) = u(z) + iv(z) = (x^2 - y^2) + i(2xy + c)$$
$$= z^2 + ci,$$

which is surely analytic in $\mathbb{C}$.

We summarize our findings in the following statement.

THEOREM 3

Given a real harmonic function $u(x, y)$ on a simply connected domain Ω, then there exist infinitely many harmonic conjugates $v(x, y)$ for u on Ω, each differing by an additive constant. Hence, the given u is the real part of infinitely many analytic functions

$$f(z) = u(z) + iv(z)$$

on Ω, each differing by a purely imaginary additive constant.

You should convince yourself that our argument used the data that u was harmonic and Ω simply connected. In this regard, you might observe that the function $u(z) = \ln|z|$, harmonic in the punctured plane $\mathbb{C} - \{0\}$, does *not* have a harmonic conjugate in the full domain $\mathbb{C} - \{0\}$, though it does have harmonic conjugates on simply connected subdomains of $\mathbb{C} - \{0\}$ (such as $\mathbb{C}$ with the nonpositive x-axis deleted). Branches again!

Exercises to Section 3.5

1. Given $u(x, y) = 2xy$, compute all $v(x, y)$ such that $f = u + iv$ is analytic.
2. Prove that if $u = u(x, y)$ is harmonic in a simply connected domain Ω (a disc, say), then the line integral $\int_{z_0}^{z} [-u_y\,dx + u_x\,dy]$ is independent of the path in Ω from z_0 to z.

Thus we may use the integral to define another function of z by agreeing that $v(z) - v(z_0)$ equals the given line integral.

3. Why does the function $u(z) = \ln|z|$ not have a harmonic conjugate in $\mathbb{C} - \{0\}$?

4. Construct a harmonic conjugate for $u(z) = \ln|z|$ on the simply connected domain Ω obtained by removing the nonpositive x-axis from the xy-plane. (Recall the "slit plane.")

Appendix 3.1 THE RIEMANN SURFACE FOR log z

This is a beautiful idea, one of the seminal ideas of nineteenth century mathematics. We will describe it briefly.

The formula $\log z = \ln r + i\theta$ is ambiguous because $\theta = \arg z$ has infinitely many values for each $z \neq 0$. We dealt with this ambiguity in an *ad hoc* fashion by defining "branches" of the logarithm. Unfortunately, we had to delete the nonpositive x-axis to maintain continuity of $\log z$. This was somewhat unnatural.

Riemann's idea runs in the opposite direction. Rather than delete points, he added new points, building a huge so-called Riemann surface on which all branches of $\log z$ could coexist at once, and freeing complex function theory forever from its confinement within the flat z-plane.

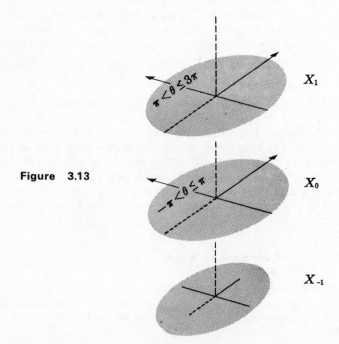

Figure 3.13

We construct the Riemann surface for log z as follows:

(i) Let X_k ($k = 0, \pm 1, \pm 2, \ldots$) denote the punctured plane $\mathbb{C} - \{0\}$ with polar coordinates r, θ and agree $(2k - 1)\pi < \theta \leq (2k + 1)\pi$ on X_k.

(ii) Imagine the punctured planes X_k stacked one above the other (in 3-space) with X_{k+1} hovering above X_k. See Figure 3.13.

(iii) Slit each plane X_k along its negative x-axis, $0 > x > -\infty$.

(iv) Join X_k to X_{k+1} along the slit negative x-axes by gluing the negative x-axis ($\theta = (2k + 1)\pi$) of X_k, the lower plane, along the top edge of the third quadrant, $((2k + 1)\pi < \theta \leq (2k + \frac{3}{2})\pi)$ of the plane X_{k+1} just above. See Figure 3.14.

(v) Let X denote the union of all these planes X_k joined as just described. We call X the *Riemann surface* for log z.

Note that X deserves the title "surface." Locally it looks like the z-plane in that every point of X is situated inside an open disc on X, which is essentially the same as a standard open disc in the z-plane. Now, however, the distinct coordinate pairs (r, θ), $(r, \theta \pm 2\pi)$, $(r, \theta \pm 4\pi)$, ... correspond to *distinct* points on the surface X. We define the logarithm of the point (r, θ) by the familiar formula $\ln r + i\theta$. This gives a nice single-valued function log: $X \to \mathbb{C}$.

Figure   3.14

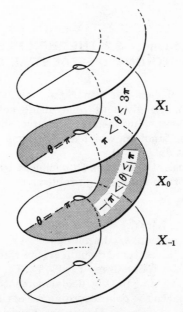

A portion of the Riemann
surface for log z

You may check that this function is single-valued (we've just argued this), one-to-one, and onto $\mathbb{C}$. Moreover, it is possible to define $\exp: \mathbb{C} \to X$ (rather than $\mathbb{C} \to \mathbb{C} - \{0\}$) and thereby obtain an inverse function for log.

Finally, it is reasonable to call the function log "analytic on X." For each point (r, θ) of X has an open neighborhood on which log agrees with some branch of the usual complex logarithm (or two branches glued together, in the case that (r, θ) lies on a seam joining X_k with X_{k+1}). We may therefore do complex calculus on the surface X.

It is standard to study certain other "functions" (such as $z^{1/2}$) by constructing *their* natural Riemann surfaces. You might ask someone to show you the surface for the square root function. Or you may construct it yourself: Take *two* copies of the complex plane and start cutting and pasting.

Geometry, as well as function theory, was greatly enriched by Riemann's insight. The following sort of question now becomes meaningful: Given some surface, specified by geometric data only, is it perhaps the Riemann surface of some mysterious complex analytic function? Are there analytic functions whose natural domains of definition are the hollow globe, the torus (surface of a doughnut), the torus with many holes, the infinite cylinder, some other favorite surface of yours? But this is another story.

Appendix 3.2 THE DIRICHLET PROBLEM WITHOUT DIFFERENTIAL EQUATIONS

A friend of ours has remarked, "Analytic functions are precious because each one gives us ready-made solutions to boundary-value problems." These solutions are, of course, the real and imaginary parts of the analytic function, which are harmonic and hence solutions to the Laplace equation $\Delta u = 0$.

We'll illustrate our friend's comment by starting with an analytic function and then concocting problems that are solved with the aid of that function.

Example 1: Let $f(z) = \log z = \ln|z| + i \arg z$. Thus, $\ln|z|$ and $\arg z$ (that is, each branch) are harmonic.

Consider the pictured domain Ω (the right-hand half of an annulus) bounded by the four curves $\Gamma_1, \Gamma_2, \Gamma_3, \Gamma_4$. See Figure 3.15. Now we introduce a heat flow in Ω by declaring that the temperature at a point ζ on $\partial\Omega$ is equal to the polar coordinate θ of the point ζ. (Say $-\pi < \theta \leq \pi$ for definiteness.) Thus, if $B(\zeta)$ is the boundary temperature, it is constantly equal to $\pi/2$ on Γ_2 and $-\pi/2$ on Γ_4, while it varies with the angle on Γ_1 and Γ_3.

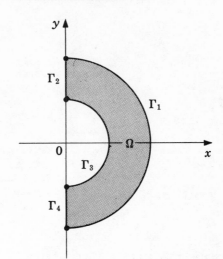

Figure 3.15

It should be clear now that the Dirichlet problem

$$\Delta U = 0 \quad \text{on } \Omega, \qquad U = B \quad \text{on } \partial\Omega,$$

is solved by the suitable branch of the steady-state temperature $U(z) = \arg z$.

Example 2: Again using $\log z$, we can solve the following variant of the Dirichlet problem: Ω is the punctured unit disc. $B = 0$ identically on the unit circle. We seek a function $U(z)$ harmonic in Ω such that $U = B = 0$ for $|z| = 1$ and also $\lim_{z \to 0} U(z) = +\infty$. Thus, the origin is an infinite source of heat.

A moment's thought shows that $U(z) = -\ln|z|$ is a solution. Is it unique?

Example 3: Now we examine $-\exp(-z) = -e^{-x}(\cos y - i \sin y)$. We observe that the real part of this function equals e^{-x} on the horizontal lines $y = \pm\pi$, while it equals $-\cos y$ on the y-axis.

Thus, we let Ω be the semi-infinite strip

$$\Omega = \{(x, y) \mid x < 0, -\pi < y < \pi\}$$

and define a boundary temperature $B(x, y)$ on $\partial\Omega$ by

$$B(x, \pm\pi) = e^{-x}, \qquad B(0, y) = -\cos y.$$

See Figure 3.16. Thus, the boundary is very hot as $x \to -\infty$, while it is relatively cool, $|B(0, y)| \leq 1$, along the right side.

Of course a steady-state temperature on the strip Ω, whose boundary behavior is that of B, is given by the real part $U(x, y) = -e^{-x}\cos y$ of $f(z) = -\exp(-z)$.

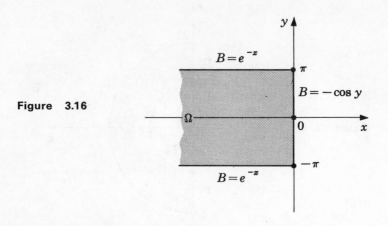

Figure   3.16

Example 4: We use the same function $-e^{-x}\cos y$. This time let Ω be a well-chosen rectangle,

$$\Omega = \left\{ (x, y) \mid 0 < x < 1, \ -\frac{\pi}{2} < y < \frac{\pi}{2} \right\}.$$

We impose a boundary temperature along the sides of Ω by defining (see Figure 3.17)

$$B(0, y) = -\cos y, \qquad B\left(x, \frac{\pm \pi}{2}\right) = 0,$$

$$B(1, y) = -\frac{1}{e}\cos y.$$

Where is this relatively warm or cool?

The steady-state temperature on the rectangle with boundary behavior equal to that of B is again given by $U(x, y) = -e^{-x}\cos y$.

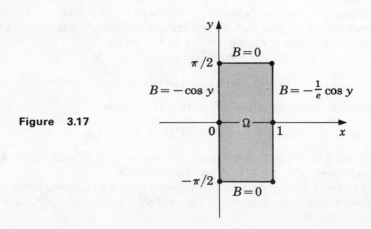

Figure   3.17

Example 5: Now we see what cos z affords us. First we calculate its real and imaginary parts (from the definition), obtaining

$$\cos z = \cos x \cosh y - i \sin x \sinh y.$$

Keeping the real part in mind, we define Ω to be the infinite strip bounded by the vertical lines $x = 0$, $x = \pi/2$. See Figure 3.18. Now we impose a boundary temperature on Ω by defining

$$B(y) = \begin{cases} \cosh y, & x = 0 \\ 0, & x = \pi/2 \end{cases}$$

Thus the left-hand edge gets hot as $|y| \to \infty$.

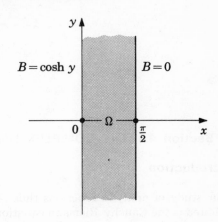

Figure 3.18

A harmonic function on Ω, which agrees with B on $\partial\Omega$, is given by the real part of cos z, namely,

$$U(x, y) = \cos x \cosh y.$$

Summary

Our gallery of analytic functions furnishes an opportunity for studying many completed boundary-value problems involving harmonic functions.

4

Integrals of Analytic Functions

Section 4.1 THE COMPLEX LINE INTEGRAL

4.1.0 Introduction

In our study of analytic functions thus far, the only property we have exploited is the Cauchy–Riemann equations. For further investigations we must have at our disposal a means of integrating complex functions and a means of expanding such functions in power series. We turn first to integration.

This chapter divides logically into three parts: first, the complex line integral is defined in Section 4.1; second, the basic theorems *about* integration are obtained in Sections 4.2 and 4.3. These are the Cauchy Integral Theorem and Cauchy Integral Formula. Third, the remainder of the chapter (Sections 4.4 through 4.11) is devoted to results that follow from the two theorems named. These results are the complex analytic versions of the qualitative theorems about harmonic functions obtained in Chapter 2.

4.1.1 Basics

Throughout this chapter Ω is a plane domain, not necessarily bounded, and Γ is a piecewise-smooth curve lying entirely inside Ω and not necessarily closed. For most of our theorems the function $f(z)$ will be supposed analytic.

We ask, "What is meant by the complex line integral

$$\int_\Gamma f(z)\, dz$$

and how do we compute it?" Let us deal with the definition now; the computation will be treated later in various stages.

Here is our philosophy: *The key to treating complex line integrals of the above form is to break everything up into real and imaginary parts and so reduce to the familiar case of line integrals of real-valued functions which we know how to handle.*

Thus, we write

$$f(z) = u(z) + iv(z)$$

as usual, and since $z = x + iy$, we have

$$dz = dx + i\, dy.$$

Therefore, breaking things up, we obtain

$$\int_\Gamma f(z)\, dz = \int_\Gamma (u(z) + iv(z))(dx + i\, dy)$$

$$= \int_\Gamma (u(z)\, dx - v(z)\, dy) + i \int_\Gamma (v(z)\, dx + u(z)\, dy).$$

Now note that we already have a meaning for both of these last two line integrals. Recall that if Γ is parametrized by the function $\gamma(t) = (\gamma_1(t), \gamma_2(t))$ with $a \le t \le b$, then the real part above is interpreted thus:

$$\int_\Gamma (u(z)\, dx - v(z)\, dy) = \int_a^b [u(\gamma(t))\gamma'_1(t) - v(\gamma(t))\gamma'_2(t)]\, dt.$$

In the right-hand integral here, everything has been "pulled back" onto the t-axis. Likewise, of course, the imaginary part of $\int_\Gamma f(z)\, dz$ becomes

$$\int_\Gamma (v(z)\, dx + u(z)\, dy) = \int_a^b [v(\gamma(t))\gamma'_1(t) + u(\gamma(t))\gamma'_2(t)]\, dt.$$

An Example of a Complex Line Integral

We interpret $\int_\Gamma z\, dz$, where Γ is the quarter-circle from $z_0 = 1$ to $z_1 = i$ (see Figure 4.1) parametrized by $\gamma(t) = (x, y) = (\cos t, \sin t)$, $0 \le t \le \pi/2$.

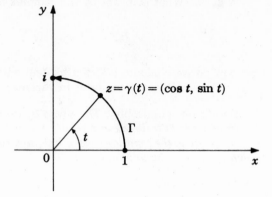

Figure 4.1

$$z = \gamma(t) = (\cos t,\ \sin t)$$

From our definition, we have

$$\int_\Gamma z\,dz = \int_\Gamma (x + iy)(dx + i\,dy)$$

$$= \int_\Gamma (x\,dx - y\,dy) + i \int_\Gamma (y\,dx + x\,dy)$$

$$= \int_0^{\pi/2} (-2 \cos t \sin t)\,dt$$

$$+ i \int_0^{\pi/2} (-\sin^2 t + \cos^2 t)\,dt$$

$$= -\int_0^{\pi/2} \sin 2t\,dt + i \int_0^{\pi/2} \cos 2t\,dt.$$

Thus, the complex integral is reduced to two real integrals. You may compute these as usual to obtain

$$\int_\Gamma z\,dz = -1.$$

Good news: We'll soon see much faster ways to do this.

Exercises to Paragraph 4.1.1

1. Compute $\int_\Gamma z^2\,dz$ for various curves Γ, as follows:
   (a) Let Γ be the straight-line segment from $z_0 = 0$ to $z_1 = 1 + i$ parametrized by increasing x. Thus, $z = x + ix$ on Γ, and the curve is traced out as x increases from $x = 0$ to $x = 1$; no need to introduce a superfluous parameter t. Rewrite the integral above as two (real and imaginary) integrals in x and compute each in the good old way.
   (b) This time let Γ be the same segment traversed in the opposite direction, from z_1 to z_0. Again, $z = x + ix$, but now x decreases from $x = 1$ (lower

limit of integration) to $x = 0$ (upper limit), that is, $\int_1^0$. Or, if you are fastidious, let $z = (1 - t) + i(1 - t)$ as t varies from $t = 0$ to $t = 1$. In either case your result should be the same, namely, the negative of your result in (a).

(c) This time let Γ consist of two line segments, the first Γ_1 from $z_0 = 0$ to $z_2 = 1$, the second Γ_2 from z_2 to $z_1 = 1 + i$. Parametrize Γ_1 by x, Γ_2 by y, and use the fact $\int_\Gamma = \int_{\Gamma_1} + \int_{\Gamma_2}$. Does your value here agree with your result in (a)?

2. Compute the very important integral $\int_C z^{-1} dz$, where $C = C(0; 1)$ is the unit circle traversed, as usual, in the positive direction and parametrized by θ, $z = e^{i\theta}$, $0 \le \theta \le 2\pi$. Check first that $dz = de^{i\theta} = ie^{i\theta} d\theta$.

3. *A simplified notation.* Rather than invoke $\gamma(t)$, we suppose Γ is given by $z(t) = x(t) + iy(t)$ with $a \le t \le b$, and that, on Γ, $dz = z'(t) dt = (x'(t) + iy'(t)) dt$. Verify that $\int_\Gamma f(z) dz$ is the same thing as $\int_a^b f(z(t)) z'(t) dt$, where this latter integral is computed by first multiplying out the integrand. *Useful:* complex function f, real parameter t.

4. Prove the following by reducing everything to real and imaginary parts. For (c), refer to the Exercises to Paragraph 1.4.1.
   (a) $\int_\Gamma cf(z) dz = c \int_\Gamma f(z) dz$, for $c \in \mathbb{C}$.
   (b) $\int_\Gamma (f(z) + g(z)) dz = \int_\Gamma f(z) dz + \int_\Gamma g(z) dz$.
   (c) $\int_{-\Gamma} f(z) dz = -\int_\Gamma f(z) dz$.

4.1.2 On Evaluating $\int_\Gamma f(z)\, dz$

This evaluation will be a major theme of the remaining chapters.

The Fundamental Theorem of Calculus was useful in evaluating real integrals. It contains a great deal of information; here is one version.

THEOREM

Let $g(x)$ be a continuous real-valued function, $a \le x \le b$. Define $G(x)$ for $a \le x \le b$ by the integral

$$G(x) = \int_a^x g(t) \, dt.$$

Then

   (i) *$G(x)$ is continuously differentiable in the interval $a < x < b$;*

   (ii) *$G'(x) = g(x)$;*

   (iii) *if also $F'(x) = g(x)$, then $F(x)$ and $G(x)$ differ by a constant and*

$$\int_a^b g(t) \, dt = F(b) - F(a).$$

Statement (iii) here outlines the familiar process for evaluating definite integrals from calculus: First find $F(x)$ whose derivative is the integrand $g(x)$ (so-called antidifferentiation or indefinite integration); then compute the number $F(b) - F(a)$.

We wish now to demonstrate that this theorem has a complex version. It is sometimes useful in evaluating complex line integrals

$$\int_{\Gamma} f(z) \, dz.$$

First we must mention that it does not make good sense to imitate the real-variable situation and thereby *define G* by

$$G(\zeta) = \int_{z_0}^{\zeta} f(z) \, dz,$$

where z_0 and ζ are joined by a curve Γ and the integral is the line integral along Γ defined above. This is because we have no guarantee that $G(\zeta)$ depends only on the point ζ and not on the arbitrary choice of curve Γ from z_0 to ζ. Contrast the real variable case, with only one curve (a piece of the x-axis) from a to x. We will see in Section 4.2 that in the important case where $f(z)$ is analytic in certain nice domains Ω (details later), then it *is* possible to define an antiderivative $G(\zeta)$ as above.

However, we may state the following partial result now. It enables us to integrate functions that have a known antiderivative (or "primitive," as complex analysts often term it).

THEOREM 1

Let $f(z)$ be continuous in a domain Ω and suppose that $F(z)$ is a function defined, analytic, and satisfying $F'(z) = f(z)$ at all points z of Ω. If Γ is a curve in Ω from z_0 to z_1, then

$$\int_{\Gamma} f(z) \, dz = \int_{\Gamma} F'(z) \, dz = F(z_1) - F(z_0).$$

Hence, the value of the integral depends only on F and on the end points of Γ, but not on Γ itself.

Note: This means that if Γ_1, Γ_2 are curves from z_0 to z_1, and f is as above, then

$$\int_{\Gamma_1} f(z) \, dz = F(z_1) - F(z_0) = \int_{\Gamma_2} f(z) \, dz.$$

Proof: We write, as usual, $f = u + iv$, $F = U + iV$. By the Cauchy–Riemann equations for U, V and the fact that $u = U_x$, $v = V_x$, we have

$$\int_{\Gamma} f(z) \, dz = \int_{\Gamma} \left[u \, dx - v \, dy \right] + i \int_{\Gamma} \left[v \, dx + u \, dy \right]$$

$$= \int_{\Gamma} \left[U_x \, dx + U_y \, dy \right] + i \int_{\Gamma} \left[V_x \, dx + V_y \, dy \right]$$

$$= \int_{\Gamma} dU + i \int_{\Gamma} dV.$$

Here we have used the exact differential notation: $dU = U_x\,dx + U_y\,dy$ and likewise for dV.

Now Theorem 8 of Section 1.4, on the line integral of an exact differential, allows us to conclude that

$$\int_\Gamma dU = U(z_1) - U(z_0), \quad \int_\Gamma dV = V(z_1) - V(z_0).$$

Our result follows immediately. Done.

Comments

1. If we write dF for $F'(z)\,dz$, then our result becomes

$$\int_\Gamma dF = F(z_1) - F(z_0).$$

Thus, our theorem is the "complexification" of Theorem 8, Chapter 1, and ultimately of one version of the Fundamental Theorem of Calculus.

2. To apply this result to the general line integral $\int_\Gamma f(z)\,dz$, we must, of course, be able to recognize $f(z)$ as $F'(z)$ for some $F(z)$ defined (and analytic!) throughout a domain Ω containing Γ.

3. The Cauchy Integral Theorem will give us a better way of looking at the problem of

$$\int_\Gamma f(z)\,dz$$

for analytic f. It avoids the troublesome $f = F'$.

Some Complex Line Integrals Evaluated

Example 1: Compute $\int_\Gamma z^2\,dz$, where Γ is the straight-line segment from the origin $z_0 = 0$ to the point $z_1 = (1, 1) = 1 + i$.

First approach: We parametrize Γ by the variable x (a very natural parameter, you will agree). Then, if z is on Γ, z has the form $z = x + ix = (1 + i)x$. See Figure 4.2. Thus we obtain, by brute force,

$$\int_\Gamma z^2\,dz = \int_{x=0}^{x=1} (1 + i)^2 x^2\,d((1 + i)x) = (1 + i)^3 \int_0^1 x^2\,dx$$

$$= (-2 + 2i)\frac{x^3}{3}\bigg]_0^1 = \frac{2}{3}(-1 + i).$$

Second approach: We observe that z^2 is the derivative of $F(z) = z^3/3$ because $F'(z) = z^2$. Since $F(z)$ is analytic throughout $\Omega = \mathbb{C}$, it is certainly defined and analytic on Γ. Thus, Theorem 1 may be applied, giving

$$\int_\Gamma z^2\,dz = \frac{z_1{}^3}{3} - \frac{z_0{}^3}{3} = \frac{(1 + i)^3}{3}$$

$$= \frac{2}{3}(-1 + i).$$

as above.

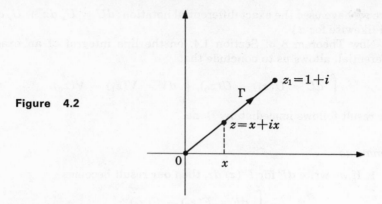

Figure 4.2

Example 2: Compute $\int_\Gamma (dz/z)$, where Γ is the unit circle parametrized in the counterclockwise direction. *This is a most important integral.*

At the outset it is necessary to realize that even though the integrand $1/z$ is the derivative of $\log z$, the logarithm is *not* defined and analytic on the full unit circle Γ. See Figure 4.3. Reason: $\log z$ is a continuous function only if we consider a *branch*, and to select a branch of $\log z$ we must omit some ray leading from the origin in the z-plane (e.g., $-\pi < \theta < \pi$, omitting the ray $\theta = -\pi$). Hence, it is incorrect to appeal to Theorem 1 and argue that, since $z_0 = z_1$,

$$\int_\Gamma \frac{dz}{z} = \log z_1 - \log z_0 = 0.$$

This is false! In fact the correct value of the integral will be $2\pi i$.

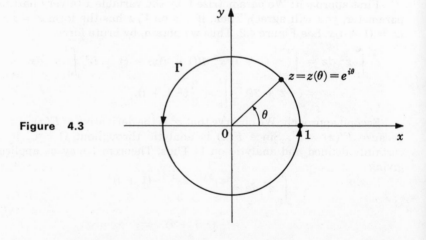

Figure 4.3

To compute this value, we note $z = e^{i\theta}$ on the circle Γ, so that $dz = ie^{i\theta}\, d\theta$. This yields

$$\int_\Gamma \frac{dz}{z} = \int_{\theta=0}^{\theta=2\pi} e^{-i\theta}(ie^{i\theta}\, d\theta) = i \int_0^{2\pi} d\theta = 2\pi i.$$

We will meet this integral again and again. Remember that it's not zero!

Exercises to Paragraph 4.1.2

1. Compute $\int_\Gamma z^2\, dz$ for Γ given below. Use Theorem 1.
   (a) Let Γ be the straight-line segment from $z_0 = 0$ to $z_1 = 1 + i$. This was worked out in Example 1 above and (using the parametrization of Γ) in Exercises for Paragraph 4.1.1.
   (b) Now let Γ be the straight line traversed from z_1 to z_0. Note that you obtain the negative of the result in (a).
   (c) Let Γ be *any* curve from z_0 to z_1. Does it matter?

2. Show that, for any point z_0, $\int_\Gamma (z - z_0)^{-1}\, dz = 2\pi i$ if $\Gamma = C(z_0; r)$. Note that the value is independent of $r > 0$ as well as of the center z_0. This is a slight generalization of the important computation in Example 2. *Hint:* $z - z_0 = re^{i\theta}$ on Γ.

3. (a) Compute $\int_\Gamma z^{-2}\, dz$ where Γ is the unit circle $C(0; 1)$ with its usual positive (counterclockwise) parametrization.
   (b) Generalize to $\int_\Gamma z^{-k}\, dz$, k an integer ≥ 2.

4. Prove that if f is a continuous complex function in Ω such that $f = F'$ for some (analytic) function in Ω and if Γ is a closed curve (loop) contained in Ω, then $\int_\Gamma f(z)\, dz = 0$. *Hint:* Use Theorem 1.
   Note that this result gives an immediate solution to Exercise 3.

5. Prove that $\int_\Gamma z^{-1}\, dz = 0$ if Γ is any loop that does not enclose the origin. *Hint:* Construct an analytic branch of $\log z$ by deleting a ray extending from the origin that does not intersect Γ. Then apply Theorem 1 (or Exercise 4 above).

6. Calculate $\int_\Gamma e^z\, dz$, where Γ is the graph of $y = \sin x$ parametrized by x from $x = 0$ to $x = \pi$. *Warning:* This is easy. Think first.

7. *Example 2 the hard way*
   (a) Verify $\int_\Gamma z^{-1}\, dz = 2\pi i$, where Γ is the positively oriented unit circle, by using the parametrization $\alpha(\theta) = (\cos\theta, \sin\theta)$, $0 \leq \theta \leq 2\pi$, and breaking things up into real and imaginary parts, obtaining two ordinary integrals in θ. Note how much faster it is to use $z = e^{i\theta}$, $dz = ie^{i\theta}\, d\theta$ as we did in the text. See Exercise 3 of Paragraph 4.1.1 in this regard.
   (b) Let $z = z(\theta) = e^{i\theta}$, $0 \leq \theta \leq 2\pi$, parametrize the unit circle. Now we will see that the tangent vector can be obtained from this "complex" parametrization. Prove in fact that $z'(\theta) = dz/d\theta = ie^{i\theta}$ is the tangent to the circle at $e^{i\theta}$, provided we interpret $z'(\theta)$ as a vector with tail end on the curve at $e^{i\theta}$ and arrowhead at $e^{i\theta} + ie^{i\theta}$. Picture!

8. Give a domain Ω and function $f(z)$ analytic throughout Ω such that $f(z)$ is not the derivative $F'(z)$ of another function $F(z)$ analytic throughout Ω. *Hint:* We'll see later that Ω must fail to be simply connected.

9. *Local one-to-oneness.* Let $f(z)$ be analytic in Ω and $f'(z_0) \neq 0$ at z_0 in Ω. Prove that f is locally one-to-one in a neighborhood of z_0; that is, there is a disc $D(z_0; R)$ in Ω on which the function f is one-to-one. (Compare real calculus: Nonzero derivative implies the graph is rising or falling, whence the function is locally one-to-one.) Prove the complex result as follows:

(a) Prove or assume: If $g(z)$ is continuous in Ω and $g(z_0) \neq 0$, then there is an open disc $D(z_0; R)$ such that

$$\int_{[z_1, z_2]} g(z)\,dz \neq 0$$

provided the integral is taken over any line segment $[z_1, z_2]$ with distinct end points z_1, z_2 in the disc. This lemma is not trivial.

(b) Apply (a) to integrals $\int_{[z_1, z_2]} f'(z)\,dz$, which by Theorem 1 equal $f(z_2) - f(z_1)$, to conclude one-to-oneness.

Advice: Look for other instances in which we prove by integration certain theorems that make no mention of integration. This is different from real calculus.

4.1.3 The *ML*-inequality

Soon it will be crucial to estimate the absolute value

$$\left| \int_\Gamma f(z)\,dz \right| .$$

This estimation should be in terms of the "sizes" of the function $f(z)$ and the curve Γ. It will be accomplished, again and again, by the following important theorem.

THEOREM 2

Let $f(z)$ be a continuous function defined on the curve Γ. Suppose that $|f(z)| \leq M$ for all z on Γ and that $L = $ length Γ. Then

$$\left| \int_\Gamma f(z)\,dz \right| \leq \int_\Gamma |f(z)|\,|dz| \leq ML.$$

Note: By $|dz|$ we mean ds, where s is the arc length parameter on Γ.

Proof: Write $I = \int_\Gamma f(z)\,dz$. For some angle ω, we have $I = |I|e^{i\omega}$. Thus,

$$|I| = \int_\Gamma e^{-i\omega} f(z)\,dz = \int_a^b e^{-i\omega} f(z(t)) z'(t)\,dt,$$

where $z(t)$ is a parametrization of Γ, $a \leq t \leq b$. As usual,

$$e^{-i\omega} f(z(t)) z'(t) = U(t) + iV(t).$$

Since $|I|$, U, V are real, it must be the case that

$$|I| = \int_a^b U(t)\, dt, \qquad \int_a^b V(t)\, dt = 0.$$

Since everything is real, we have surely

$$|I| \le \int_a^b |U(t)|\, dt.$$

But also

$$|U(t)| \le |e^{-i\omega} f(z(t)) z'(t)| = |f(z(t))|\, |z'(t)|,$$

since U is the real part of the complex number $U + iV$. Thus,

$$|I| \le \int_a^b |f(z(t))|\, |z'(t)|\, dt = \int_\Gamma |f(z)|\, |dz|.$$

This is the first inequality of the statement. The second inequality is clear from elementary considerations of the real integral of positive functions and the fact that $\int_\Gamma |dz| = L$. Done.

Comments

1. Our proof depended on a trick, multiplying through by $e^{-i\omega}$. This helped reduce things to the real case. A more direct proof, which depends on the interpretation of the integral as a limit of finite sums, is offered in the exercises for this subsection.

2. Exercise 2 will show you how the ML-inequality is generally applied. We'll encounter this situation very frequently.

3. Recall that f, if continuous, assumes its maximum modulus $\max |f(z)|$ on Γ, so surely $|f(z| \le M$ for some M and all z on Γ.

An Example

We estimate $|\int_\Gamma z^2\, dz|$, where Γ is the straight-line segment from $z_0 = 0$ to $z_1 = 1 + i$. See Figure 4.2.

First observe that for z on Γ, $|z^2| \le |1 + i|^2 = 2$. In fact, the best possible estimate here is $M = 2.$

Likewise, $L = \text{length } \Gamma = |z_1 - z_0| = \sqrt{2}$. Thus, the ML-inequality gives $|\int_\Gamma z^2\, dz| \le 2\sqrt{2}$.

Of course we were able to integrate this function exactly. We know that $|\int_\Gamma z^2\, dz| = |2(-1 + i)/3| = 2\sqrt{2}/3$. In cases such as this, it is pointless to estimate. The true purpose of the ML-inequality is to prove theorems.

Exercises to Paragraph 4.1.3

1. (a) Use the ML-inequality to obtain the estimate

$$\left| \int_\Gamma (z - z_0)^{-1} \, dz \right| \le 2\pi,$$

where $\Gamma = C(z_0 ; r)$, $r > 0$. *Hint:* Compute $M = \max |z - z_0|^{-1}$ on Γ.
   (b) Compare this with the known absolute value of $\int_\Gamma (z - z_0)^{-1} \, dz$.

2. *The standard application of the ML-inequality.* Suppose that two hypotheses are satisfied by a continuous function f, namely, (i) $|f(z)| \le M$ for all z in the punctured disc $D(z_0 ; R) - \{z_0\}$, and (ii) the value $I_r = \int_{C(z_0;r)} f(z) \, dz$ is the same for all r, $0 < r < R$. Prove $I_r = 0$ for $0 < r < R$.

   We will soon see that (ii) holds if f is analytic in the punctured disc. Thus, the ML-inequality is often used to show that a certain integral vanishes.

3. *Another approach*
   (a) Convince yourself that the integral I above equals a certain limit of finite complex sums of the form $\sum_{k=1}^m f(z(t_k)) \, (z(t_k) - z(t_{k-1}))$, where $z(t)$ is the parametrization, $a \le t \le b$, and $a = t_0 \le t_1 \le \cdots \le t_m = b$ is a sampling of values from the parameter interval. Note the similarity with the Riemann integral of real calculus.
   (b) Prove the first inequality in the statement of Theorem 2 by applying the triangle inequality to the sum in (a) and basic observations about limits and absolute values.

Section 4.2 THE CAUCHY INTEGRAL THEOREM

This is the most-used complex version of the Fundamental Theorem of Calculus (one real variable) or Green's Theorem (real-valued functions of x, y). It will enable us to compute certain integrals very easily and also to prove certain other basic theorems, most notably the Cauchy Integral Formula of Section 4.3.

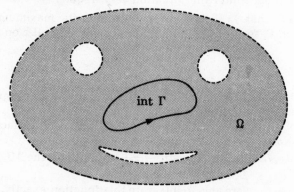

Note: int Γ is a subset of Ω;
$f(z)$ is analytic throughout Ω.

Figure 4.4

CAUCHY INTEGRAL THEOREM

Let $f(z)$ be analytic in a domain Ω and let Γ be a closed Jordan curve inside Ω whose interior is contained in Ω so that $f(z)$ is analytic on and inside Γ. Then

$$\int_\Gamma f(z)\, dz = 0.$$

Note: This theorem does *not* apply to the situation $f(z) = 1/z$, $\Omega = \mathbb{C} - \{0\}$, $\Gamma =$ the unit circle. Reason: The full interior of Γ is not contained in Ω; that is, $f(z)$ is not analytic at all points inside Γ. Example 2 of Section 4.1 showed that this integral equals $2\pi i$, not zero.

Proof: We have, writing int Γ for the interior of Γ,

$$\int_\Gamma f(z)\, dz = \int_\Gamma (u + iv)(dx + i\, dy)$$

$$= \int_\Gamma (u\, dx - v\, dy) + i \int_\Gamma (v\, dx + u\, dy)$$

$$= \iint_{\text{int } \Gamma} (-v_x - u_y)\, dx\, dy + i \iint_{\text{int } \Gamma} (u_x - v_y)\, dx\, dy$$

$$= 0 + i0 = 0$$

by Green's Theorem and the Cauchy–Riemann equations. Done.

As a first application, we may now compute such recalcitrant integrals as

$$\int_\Gamma \cos(\sin z)dz, \qquad \int_\Gamma e^{z^2}\, dz,$$

where Γ is the unit circle $C(0; 1)$ with its usual counterclockwise orientation. The Cauchy Integral Theorem assures us that each of these integrals equals zero. Note that we obtain these results *without* finding antiderivatives for the integrands $\cos(\sin z)$ and e^{z^2}. Theorem 1 would be of little help in these cases.

Generalizations

Extending the Cauchy Integral Theorem to curves Γ less special than Jordan curves (which do not intersect themselves, simple loops) is a study in itself. We will now discuss three such generalizations.

The Case Γ Is a Figure-Eight: Let $f(z)$ be analytic in a domain Ω that contains Γ and all points inside each "lobe" (see Figure 4.5). In this case we break Γ into two closed curves, Γ_1 and Γ_2, each of which is

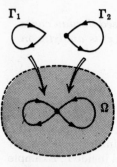

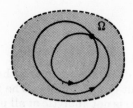

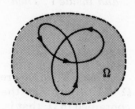

$\Gamma = \Gamma_1 + \Gamma_2$,
two Jordan curves

$\Gamma = $ (inner loop) $+$
(outer loop)

Γ breaks up into
four Jordan curves
(or two, if you are
observant).

Figure   4.5

now nonself-intersecting, and apply the preceding version of Cauchy's Integral Theorem to each curve. We obtain

$$\int_\Gamma f(z)\, dz = \int_{\Gamma_1} f(z)\, dz + \int_{\Gamma_2} f(z)\, dz$$

$$= 0 + 0 = 0.$$

Moral

Even if Γ intersects itself and so is not a Jordan curve, it may be possible to decompose Γ into finitely many Jordan curves. If $f(z)$ is analytic in the interior of each of these new Jordan curves (one must check this carefully!), then $\int_\Gamma f(z)\, dz = 0$.

A General Statement: We will state, discuss, and occasionally apply this version of the theorem despite the fact that a rigorous proof would require too much topology and will not be given here. This version is geometrically appealing and worthy of meditation because it gets close to the essence of what is happening.

STRONG CAUCHY INTEGRAL THEOREM

Let $f(z)$ be analytic in a domain Ω and let Γ be a (not necessarily simple) closed curve in Ω which can be shrunk to a point within Ω. Then

$$\int_\Gamma f(z)\, dz = 0.$$

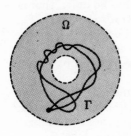  

Shrinking Γ to a point within Ω.

Figure   4.6

Explanation: Imagine that we obtain the complicated curve Γ by taking a simple closed circle made of thread and dropping it into the domain Ω. See Figure 4.6. Now suppose that the thread begins to shrink in time so that the length of the curve gets smaller and smaller until, after a finite time interval, the curve of thread has shrunk to a simple point inside Ω. Suppose further that the shrinking curve lay within the domain Ω at every instant, and never touched a point outside Ω. In this case, we say that Γ has been shrunk (deformed) to a point within Ω.

The proof of the strong version (we do not give it here) depends on the following nontrivial fact: The desired value

$$\int_\Gamma f(z)\, dz$$

does not change as we replace the original curve Γ by the ever shorter shrunken versions, provided the shrinking is carried out in a sufficiently smooth manner and, of course, occurs inside Ω. Since the curve Γ eventually shrinks to one point, and since the integral of $f(z)$ *over one point* in Ω is zero (proof?), the original integral must have been zero. Done!

At this point it is instructive to consider once more the situation $f(z) = 1/z$, $\Omega = \mathbb{C} - \{0\}$, Γ = the unit circle. (See the note following our original statement of the Cauchy Integral Theorem.) In our present language the circle Γ cannot be shrunk to a point inside Ω because it is looped around the "bad point" 0, at which f is not analytic. Thus, the Strong Cauchy Integral Theorem does not apply (and, in fact, the value of the integral is different from zero, as we know).

Note also that the first situation we discussed, in which Γ was a figure-eight, may be handled by the Strong Cauchy Integral stated here. For the figure-eight, Γ may surely be shrunk to a point within Ω, the domain of analyticity for $f(z)$.

The Case Ω Is Simply Connected: We now mention an important special case of the strong version of the theorem given above. Recall that the domain Ω is said to be *simply connected* if it has no holes, that is, if all points inside every closed loop that lies in Ω also lie in Ω. For example, the open disc is simply connected, as is the xy-plane and the xy-plane with the negative x-axis deleted. On the other hand, the punctured disc and punctured plane (one point removed) are not simply connected.

Consider the following fact: *If Ω is a simply connected domain, then every closed curve Γ in Ω may be shrunk to a point within Ω.* This is true because the curve Γ cannot enclose any holes and so cannot be snagged on any points not in Ω as it is shrinking. Because of this fact, we may state this corollary of the strong version.

CAUCHY INTEGRAL THEOREM (SIMPLY CONNECTED CASE)

Let $f(z)$ be analytic in a simply connected domain Ω. Let Γ be any closed curve in Ω. Then

$$\int_\Gamma f(z)\, dz = 0.$$

Further Results

The next result is of the "independence-of-path" variety and is essentially a restatement of the Cauchy Integral Theorem. Note that we require $f(z)$ to be analytic at all points between the two curves (which may intersect). See Figure 4.7.

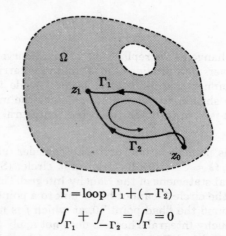

Figure 4.7

$$\Gamma = \text{loop } \Gamma_1 + (-\Gamma_2)$$

$$\int_{\Gamma_1} + \int_{-\Gamma_2} = \int_\Gamma = 0$$

COROLLARY 3

Let $f(z)$ be analytic in Ω, and let Γ_1, Γ_2 be piecewise-smooth curves from the point z_0 to the point z_1, all lying inside Ω and such that all points between the two curves are also in Ω. Then

$$\int_{\Gamma_1} f(z)\, dz = \int_{\Gamma_2} f(z)\, dz.$$

Proof:

$$\int_{\Gamma_1} f(z)\, dz - \int_{\Gamma_2} f(z)\, dz = \int_{\Gamma_1} f(z)\, dz + \int_{-\Gamma_2} f(z)\, dz$$

$$= \int_{\Gamma} f(z)\, dz = 0$$

by the Cauchy Integral Theorem. Here, $-\Gamma_2$ is the curve Γ_2 traversed in the opposite direction, and Γ is the curve formed by Γ_1 followed by $-\Gamma_2$. The first equality depends on the fact that reversing the path of integration changes the sign of the integral. Done.

Comment

In the proof above we did not specify *which* version of the Cauchy Theorem was being used. If Γ_1 touches Γ_2 only at the end points, then $\Gamma = \Gamma_1 + (-\Gamma_2)$ is a Jordan curve, and our original version suffices. But the crucial hypothesis is the analyticity of $f(z)$ between the curves. If this holds, *some* version of Cauchy's Theorem can be found which guarantees $\int_{\Gamma_1} = \int_{\Gamma_2}$. Here is a standard application.

COROLLARY 4

If $f(z)$ is analytic in the simply connected domain Ω and if z, z_0 are points of Ω, then

$$F(z) = \int_{z_0}^{z} f(\zeta)\, d\zeta$$

is a value independent of the path from z_0 to z in Ω and hence defines a continuous function $F = F(z)$ in Ω.

Proof: Use Corollary 3.

Remarks

1. In the Appendix to this section we will see in fact that $F(z)$ is analytic and $F'(z) = f(z)$. This is another version of the Fundamental Theorem of Calculus.

2. A different choice of z_0 in Corollary 4 leads to a different function. How different?

3. The result above is false if Ω is the punctured plane $\mathbb{C} - \{0\}$, which is not simply connected. For consider $f(z) = 1/z$. We have referred to this phenomenon already (see Exercises to Paragraph 1.4.3 following Green's Theorem, log z). It may be stated thus: Differentials that are "exact" on simply connected subdomains may fail to be exact on the full, nonsimply connected domain. Advanced theories make much of this relationship between geometry and calculus: more "holes," more inexact differentials.

The Case of Several Curves

Now we deal with a different situation. Thus far we have discussed integration over a single, possibly quite complicated, closed curve Γ. The idea has been that if $f(z)$ is analytic inside the curve Γ, then its integral around Γ vanishes. Now we show that this conclusion may also be true if Γ is replaced by finitely many curves of a certain type, namely, the positively oriented curves that comprise the full boundary of a Jordan domain.

We recall that a bounded domain Ω is a *k-connected Jordan domain* if its boundary $\partial\Omega$ consists of precisely k Jordan curves (simple closed curves) that are positively oriented with respect to Ω (that is, Ω lies on our left as we traverse the curve in the direction of parametrization).

CAUCHY INTEGRAL THEOREM FOR JORDAN DOMAINS

Let Ω be a k-connected Jordan domain, and let $f(z)$ be analytic on some domain Ω^+ containing Ω and $\partial\Omega$. Then

$$\int_{\partial\Omega} f(z)\, dz = 0.$$

Proof: Actually, we have done all the work for this already, in proving our first version of the Cauchy Theorem at the beginning of this section. The proof given there works here, word for word, provided we replace Γ, int Γ, and Ω there by $\partial\Omega$, Ω, and Ω^+ here. The reason that the same proof works here is that we already have Green's Theorem (Chapter 1) for arbitrary Jordan domains. We leave details to you. Done.

This version has a pleasing and useful corollary (see Figure 4.8). The loops Γ_1, Γ_2 are similarly parametrized (both counterclockwise, say) and the function $f(z)$ is analytic in the domain (call it Ω^+) that contains both curves Γ_1, Γ_2 and the region between them (though $f(z)$ may fail to be analytic at some points encircled by the inner curve).

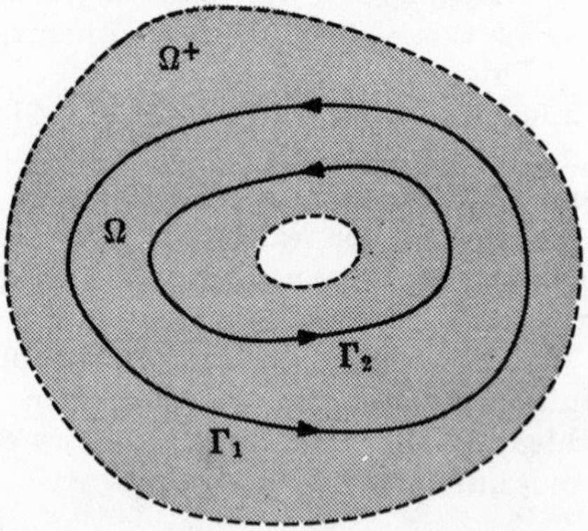

Figure   4.8

Now let the region between the two curves be denoted by Ω, and observe that Ω becomes a 2-connected Jordan domain if we agree that its boundary consists of the curves Γ_1 and $-\Gamma_2$. That is, we reverse the orientation of Γ_2 so that Ω is now to its left. The preceding theorem says that $f(z)\,dz$ integrates as follows:

$$0 = \int_{\partial\Omega} = \int_{\Gamma_1 - \Gamma_2} = \int_{\Gamma_1} + \int_{-\Gamma_2} = \int_{\Gamma_1} - \int_{\Gamma_2}.$$

In other words, the integral of $f(z)$ around Γ_1 equals the integral of $f(z)$ around (the original) Γ_2. Let us state this officially.

COROLLARY 5

Let Γ_1, Γ_2 be similarly oriented piecewise-smooth simple closed curves, and suppose $f(z)$ is analytic in a domain containing the curves and all points between them. Then

$$\int_{\Gamma_1} f(z)\,dz = \int_{\Gamma_2} f(z)\,dz.$$

Comments

1. The two integrals in Figure 4.8 may each fail to be zero because f might not be analytic everywhere inside the inner curve.

2. A topologist might say that the integral of $f(z)\,dz$ around a loop is "invariant" under those deformations of the loop that take place entirely inside the set of points where f is analytic.

3. Corollary 5 allows us to conclude that

$$\int_{\Gamma} \frac{dz}{z} = 2\pi i$$

when Γ is *any* piecewise-smooth simple loop around the origin, not only the unit circle. Keep your eyes open for opportunities to replace unpalatable parametrized loops by much simpler circles!

Culture: The Cauchy–Goursat Integral Theorem

We mentioned earlier that many authors do not require that $f'(z)$ be continuous in defining the analyticity of $f(z)$. These people cannot use Green's Theorem (as we did) to prove the Integral Theorem, since Green's Theorem requires continuous partials. The French mathematician E. Goursat (*ca.* 1900) gave a remarkable proof of the Integral Theorem, which requires only that $f'(z)$ *exist*. Thus, this result is sometimes called the Cauchy–Goursat Theorem. But, as we have already remarked, the Cauchy Integral Formula (Section 4.3) will imply that all these definitions of analyticity are equivalent.

Exercises to Section 4.2

1. Compute $\int_{|z|=1} f(z)\, dz$, where the integral is taken over the positively oriented unit circle and $f(z)$ is given below. How many can you do without the Cauchy Integral Theorem?
   (a) z^2.
   (b) $\sin(\cos z)$.
   (c) $z + (1/z)$.
   (d) $1/(z - \frac{1}{2})$.
   (e) $1/z^2$.
   (f) $z^7/(z^2 + 4)$.

2. Consider $I = \int_{\Gamma} f(z)\, dz$, where f is continuous in a domain Ω containing the closed loop Γ. True or false?
   (a) If f is analytic in Ω and Ω is a disc, then $I = 0$.
   (b) If $I = 0$, then f is analytic at all points of the plane on and inside Γ.
   (c) If f is analytic in Ω and $I \neq 0$, then not all points enclosed by Γ are contained in Ω.
   (d) If f is analytic in Ω, then $I = 0$.
   (e) If Ω is the punctured unit disc $0 < |z| < 1$ and if $I = 0$ for all Γ in Ω, then $\lim_{z \to 0} f(z)$ exists as a finite complex number.
   (f) If f is analytic in the punctured unit disc Ω and if $\lim_{z \to 0} f(z)$ exists as a finite complex number, then $I = 0$ for all loops Γ in Ω.
   (g) If Ω has holes, the Cauchy Theorem may fail to apply.

3. (a) Verify by direct computation that $\int_{\Gamma_1} z^{-1}\, dz = \pi i = -\int_{\Gamma_2} z^{-1}\, dz$, where Γ_1 is the upper half-circle traversed from $z_0 = 1$ to $z_1 = -1$ and given by $z = e^{i\theta}, 0 \leq \theta \leq \pi$, while Γ_2 is the lower half-circle from z_0 to z_1 (same end points) given by $z = e^{-i\theta}, 0 \leq \theta \leq \pi$.
   (b) Explain why the different values obtained for $\int_{\Gamma_1}$ and $\int_{\Gamma_2}$ do not contradict the result of Corollary 3.

4. Let Ω be the simply connected domain obtained by deleting the nonpositive x-axis from the plane $\mathbb{C}$. Define $F(z) = \int_1^z \zeta^{-1}\, d\zeta$, where the integral is taken over any curve from 1 to z lying inside Ω.
   (a) What is $F(1)$?
   (b) What is $F(e)$?
   (c) Give the real and imaginary parts of $F(z)$ in terms of familiar functions.
   (d) What is the derivative of $F(z)$?

5. Suppose $f(z) = c_{-k}z^{-k} + \cdots + c_{-1}z^{-1} + g(z)$ with $c_{-k} \neq 0$, $k > 0$, and $g(z)$ analytic in the entire z-plane. Thus, f is analytic except at $z = 0$. Prove $\int_{\Gamma} f(z)\, dz = 2\pi i c_{-1}$, where Γ is the positively oriented unit circle. Meditate upon the fact that the integral fails to vanish only because of the term in $1/z$. What "function" is the integral of $1/z$?

   Preview: The theory of Laurent series will tell us how to express certain nonanalytic $f(z)$ in the form [polynomial in $1/z$] + [analytic] as above. In this case, integration reduces to computing the coefficient ("residue") c_{-1}.

6. Given that f is analytic in the punctured disc $0 < |z - z_0| < R$ and that $|f(z)| \leq M$ there, prove that $\int_{|z-z_0|=r} f(z)\, dz = 0$ for $0 < r < R$.

7. Given that $g(z)$ is analytic in the open disc $|z - z_0| < R$, prove that

$$\int_{|z-z_0|=r} \frac{g(z) - g(z_0)}{z - z_0}\, dz = 0 \qquad \text{for } 0 < r < R.$$

Hint: Apply Exercise 6 to the integrand here.

8. Given that $g(z)$ is analytic in the disc $|z - z_0| < R$, prove that

$$g(z_0) = \frac{1}{2\pi i} \int_{|z-z_0|=r} \frac{g(z)}{z - z_0} \, dz,$$

provided $0 < r < R$. This is the *Cauchy Integral Formula.* Hint: $g(z) = (g(z) - g(z_0)) + g(z_0)$. Use the preceding exercises.

9. In the open disc $|z| < 1$, sketch a closed curve Γ that intersects itself four times. Then break up your curve into a finite number of simple closed (Jordan) curves. By repeated use of the Jordan curve version of Cauchy's Theorem, argue that $\int_\Gamma f(z) \, dz$ vanishes if f is analytic in $|z| < 1$.

10. Verify that the function $S(t, z) = (1 - t)z$ defined for $0 \le t \le 1$, $z \in D(0; r)$, shrinks the disc down to the origin $z = 0$ as t ("time") varies from $t = 0$ to $t = 1$. Conclude that any curve Γ lying in the disc is also shrunk to the point $z = 0$. This is a special case of our assertion that a closed curve inside a simply connected domain may be shrunk to a point within that domain.

11. *The complex differential $f(z) \, dz$.* Let $f(z) = u(x, y) + iv(x, y)$ as usual, with $u, v \in \mathscr{C}^1(\Omega)$. We write $f(z) \, dz$ as $f \, dx + if \, dy$, reminiscent of the real differential $p \, dx + q \, dy$. Prove that if Ω is simply connected, then the following four statements are equivalent. Use results from the text when applicable.
    (a) $\int_\Gamma [f \, dx + if \, dy] = 0$ for all closed curves Γ in Ω.
    (b) $f \, dx + if \, dy$ is *exact* in Ω (that is, there exists a complex function $F = U + iV$ with $U, V \in \mathscr{C}^1(\Omega)$ such that

$$f \, dx + if \, dy = dF \, (\equiv F_x \, dx + F_y \, dy).$$

(c) The partial derivatives f_y (that is, $u_y + iv_y$) and $(if)_x$ are equal throughout Ω.
    (d) $f(z)$ is analytic in Ω.
        The point of this exercise is that (a), (b), (c) are already known to be equivalent for real differentials $p \, dx + q \, dy$ in the situation $p, q \in \mathscr{C}^1(\Omega)$ (so that (c) makes sense) and Ω simply connected. See Section 1.4.

12. *How have we used Green's Theorem?* Answer: We converted a given line integral into an integral of $p \, dx + q \, dy$ and then invoked Green's Theorem to obtain a double integral. Retrace our steps in each of the following situations:
    (a) *The Inside-Outside Theorem.* Given

$$\int_\Gamma \frac{\partial u}{\partial n} \, ds,$$

that is, $\int_\Gamma \nabla u \cdot N \, ds$.
    (b) *Curl and circulation.* Given $\int_\Gamma V \cdot T \, ds$, where V is a "flow vector." See the Exercises to Paragraph 1.4.3.
    (c) *The Divergence Theorem.* Given $\int_\Gamma V \cdot N \, ds$. Compare (a).
    (d) *The Cauchy Integral Theorem.* Given $\int_\Gamma f(z) \, dz$.
    (e) Can you think of any others?

13. Suppose $f(z)$ is analytic in a domain Ω, but $f(z)$ is not the derivative of a function defined and analytic everywhere in Ω. What can you conclude about Ω? What is your favorite example of this phenomenon?

Appendix to Section 4.2 THE DERIVATIVE OF AN INTEGRAL

We give a rather general theorem on path-independent integrals. This will be crucial in the characterization of analyticity known as Morera's Theorem (Section 4.10).

THEOREM

Let $f(z)$ be continuous in a domain Ω and moreover let the integral $F(z) = \int_{z_0}^{z} f(\zeta)\, d\zeta$ be the same for all polygonal paths from z_0 to z with z_0, z in Ω. Then $F(z)$ is analytic in Ω and, in fact, $F'(z) = f(z)$.

Note: This says that the derivative of the integral (taken with respect to the upper limit of integration) is the original integrand. Compare the Fundamental Theorem of Calculus:

$$\frac{d}{dx} \int_a^x f(t)\, dt = f(x)$$

Proof: It should be clear that $F(z)$ is well defined as a function, thanks to the path-independence hypothesis. See Corollary 4.

Now we show that $F'(z_1) = f(z_1)$ for all z_1 in Ω. This will yield in particular the analyticity of F and will complete the proof of the theorem.

We will work directly with the difference quotient of F at z_1. Observe that, by the independence-of-path hypothesis,

$$\frac{F(z) - F(z_1)}{z - z_1} = \frac{1}{z - z_1} \left\{ \int_{z_0}^{z} f(\zeta)\, d\zeta - \int_{z_0}^{z_1} f(\zeta)\, d\zeta \right\}$$

$$= \frac{1}{z - z_1} \int_{z_1}^{z} f(\zeta)\, d\zeta,$$

where all integrals are taken over polygonal paths (from z_0 to z, etc.). We wish to take the limit here as z approaches z_1 and obtain $f(z_1)$. Hence, let us introduce $f(z_1)$ by rewriting $f(\zeta)$,

$$f(\zeta) = f(z_1) + (f(\zeta) - f(z_1)).$$

The last integral above becomes two integrals (note z is fixed):

$$\frac{f(z_1)}{z - z_1} \int_{z_1}^{z} d\zeta + \int_{z_1}^{z} \frac{f(\zeta) - f(z_1)}{z - z_1}\, d\zeta.$$

The first term here is easily seen to be $f(z_1)$. Hence, it will suffice to show that

$$\lim_{z \to z_1} \int_{z_1}^{z} \frac{f(\zeta) - f(z_1)}{z - z_1}\, d\zeta = 0, \qquad (z \text{ in } \Omega, z \neq z_1).$$

Some meditation should convince you that the *ML*-inequality of Paragraph 4.1.3 is our only hope here.

To do this, we first let I_z denote the integral in the preceding equation. Since z will ultimately approach z_1, there is no loss in assuming that I_z is computed on a straight segment from z_1 to z lying inside the domain Ω. Hence, we must have, by the *ML*-inequality (Theorem 2), that

$$|I_z| \leq \max\left\{\left|\frac{f(\zeta) - f(z_1)}{z - z_1}\right|\right\} |z - z_1|,$$

where ζ varies on the straight line from z_1 to z. Thus, for fixed z close to z_1, we cancel to get

$$|I_z| \leq \max|f(\zeta) - f(z_1)|.$$

Now let z approach z_1. Then ζ also approaches z_1 and so $f(\zeta)$ approaches $f(z_1)$. We conclude that

$$\lim_{z \to z_1} |I_z| = 0,$$

as desired. Done.

Comments

1. The process discussed in the theorem is *not* what is meant by differentiating under the integral sign. In fact, we did not even suppose that *f* was differentiable. (It is! See Morera's Theorem, Section 4.10).

2. You should check that everything here happened on polygonal paths (that is, on finite sequences of line segments linked end-to-end).

Section 4.3 THE CAUCHY INTEGRAL FORMULA

This is very important.

CAUCHY INTEGRAL FORMULA

Let $f(z)$ be analytic in a domain Ω, and let Γ be a simple closed (Jordan) curve inside Ω whose interior is also contained in Ω. If z_0 is any point interior to Γ, then the value $f(z_0)$ is given by

$$f(z_0) = \frac{1}{2\pi i} \int_\Gamma \frac{f(z)}{z - z_0} \, dz.$$

Comments

1. The variable z is "integrated out," leaving a number that depends on z_0. Remarkably, this number is $f(z_0)$.

2. The point z_0 must lie inside Γ, never *on* Γ, so that the denominator $z - z_0$ never vanishes on Γ.

3. This result is in the great tradition of integral representation theorems. Recall Green's III and the Poisson Integral Formula from the first two chapters.

4. In the case $f(z) \equiv 1$ and $z_0 = 0$, then the formula yields $\int_\Gamma (dz/z) = 2\pi i$ if Γ winds once around the origin. This agrees with the value we computed in Example 2 of Paragraph 4.1.1.

Proof of the Formula: Let $C(r) = C(z_0; r)$ be the circle of radius $r > 0$ centered at z_0, and let r be chosen small enough that $C(r)$ lies inside Γ as in Figure 4.9. By Corollary 5 of Section 4.2, we may as well be integrating around $C(r)$; that is,

$$\frac{1}{2\pi i} \int_\Gamma \frac{f(z)}{z - z_0}\, dz = \frac{1}{2\pi i} \int_{C(r)} \frac{f(z)}{z - z_0}\, dz.$$

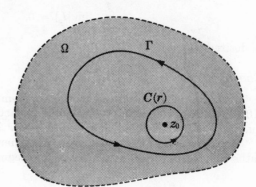

Figure 4.9

Now we try the useful trick of writing

$$f(z) = (f(z) - f(z_0)) + f(z_0)$$

so that the value $f(z_0)$ appears in our computations:

$$\frac{1}{2\pi i} \int_\Gamma \frac{f(z)}{z - z_0}\, dz = \frac{f(z_0)}{2\pi i} \int_{C(r)} \frac{dz}{z - z_0} + \frac{1}{2\pi i} \int_{C(r)} \frac{f(z) - f(z_0)}{z - z_0}\, dz.$$

We deal now with each of the right-hand terms. The first term on the right equals $f(z_0)$ because the value of the integral is $2\pi i$, as we saw in Section 4.1.

Now we claim that the second integral on the right is zero. This, of course, will complete the proof. We'll show that if $r > 0$ is chosen sufficiently small, then this second integral can be made arbitrarily small in absolute value. But the value of this integral is independent of r (why?). Hence, this value must be zero. This style of argument is standard.

Since $f'(z_0)$ exists, the integrand $(f(z) - f(z_0))/(z - z_0)$ is bounded in absolute value on some disc centered at z_0 (why is this so?). Thus, $|f(z) - f(z_0)|/|z - z_0| \leq M$ for all z near z_0. The *ML*-inequality takes the form

$$\left| \frac{1}{2\pi i} \int_{C(r)} \frac{f(z) - f(z_0)}{z - z_0} \, dz \right| \leq \frac{1}{2\pi} M \cdot \text{length } C(r) = Mr,$$

since $C(r)$ has length $2\pi r$.

Since r may be taken very close to zero, the value of the integral about each $C(r)$ must be zero. Done.

Exercises to Section 4.3

1. Compute the following integrals, applying theorems whenever possible. In each case, Γ is the unit circle $|z| = 1$ with positive orientation. *Moral:* The Integral Formula has practical uses.
   (a) $\int_\Gamma z^{-1} \cos z \, dz$.
   (b) $\int_\Gamma (z - (\pi/4))^{-1} \cos z \, dz$.
   (c) $\int_\Gamma (z - 2)^{-1} e^z \, dz$.
   (d) $\int_\Gamma z(z - (\pi/4))^{-1} \, dz$.
   (e) $\int_\Gamma (e^z - 1)z^{-1} \, dz$.
   (f) $\int_\Gamma (z - \frac{1}{2})^{-1} \cos 2\pi z \, dz$.

2. In the proof of the Integral Formula, we claimed that if f is analytic at z_0, then the value $\int_{C(z_0;r)} (f(z) - f(z_0))(z - z_0)^{-1} \, dz$ is the same for all (sufficiently small) r. Cite a previous result that justifies our claim.

3. *Prove the Circumferential Mean Value Theorem:* If f is analytic at z_0, then

$$f(z_0) = \frac{1}{2\pi} \int_0^{2\pi} f(z_0 + re^{i\theta}) \, d\theta$$

   for sufficiently small $r > 0$. *Hint:* $z = z_0 + re^{i\theta}$ on $C(z_0; r)$.

4. Verify

$$\int_{|z|=2} e^z(z(z - 1))^{-1} \, dz = 2\pi i(1 + e)$$

   by rewriting the integrand as a sum of two functions and applying the Cauchy Integral Formula to each term.

5. *Further calisthenics with the ML-inequality.* Suppose we know that $g(z)$ is analytic in the punctured disc $0 < |z - z_0| < R$. Which of these additional hypotheses (concerning g and z_0) allow us to conclude that

$$\int_{C(z_0;r)} g(z) \, dz = 0 \qquad \text{for } 0 < r < R?$$

   Note that the value of this integral is independent of r, by Corollary 5.
   (a) g is defined and analytic at z_0 also.
   (b) $g(z) = 0$ indentically for $0 < |z - z_0| < R$.
   (c) g is bounded (that is, $|g(z)| \leq M$) for $0 < |z - z_0| < R$.
   (d) g is defined and continuous at z_0 also.

(e) $\lim_{z \to z_0} g(z)$ exists as a finite complex number.

(f) $g(z) = 2\pi i$ identically for $0 < |z - z_0| < R$.

(g) $g(z) = 1/(z - z_0)$.

(h) $g(z) = 1/(z - z_0)^2$.

(i) $\lim_{z \to z_0} g(z) = \infty$.

(j) $g(z) = (f(z) - f(z_0))/(z - z_0)$ with f analytic in $|z - z_0| < R$.

(k) $\lim_{z \to z_0} g(z)$ does not exist as a finite complex number nor as ∞.

You might observe that each of (a), (b), (c), (d), (e), (f), (j) implies (c).

6. Discuss our progress in computing the complex line integral, as follows:

(a) The Cauchy Integral Theorem enables us to integrate $\int_\Gamma f(z)\, dz$, where f is analytic at all points on and within the closed curve Γ.

(b) The Integral Formula enables us to integrate functions $f(z)$ which fail to be analytic at certain points z_0 inside Γ, namely, functions of the form $f(z) = g(z)/(z - z_0)$ with $g(z)$ analytic (and fairly well known). What is $\int_\Gamma f(z)\, dz$?

(c) The procedure in (b) may be extended to functions of the form $f(z) = g(z)/(z - z_0)(z - z_1)$ and so on, provided $z_0 \neq z_1$, by writing $f(z)$ as a sum of two quotients with denominators $z - z_0, z - z_1$.

(d) Further questions: What if $f(z) = g(z)/(z - z_0)^n$ with $n \geq 2$? What is $\int_\Gamma f(z)\, dz$? Which functions $f(z)$ can be put in this form?

7. A question asked during the 1970 Berkeley Ph.D. oral examination: "What is the most important function in complex function theory"? The expected answer: $1/(z - z_0)$. Do you agree? What reasons can you give? Keep this question in mind as you read on.

8. (a) Prove that an analytic function is uniquely determined by its boundary values; that is, if f and g are analytic in a domain Ω, and if $f(z) = g(z)$ for all z on a loop Γ which is contained, together with its interior, in Ω, then $f(z) = g(z)$ for all z inside Γ as well.

(b) What about existence? If $f(z)$ is a complex-valued function defined on Γ, as in (a), does there exist a function analytic in Ω which agrees with the given f on Γ? Note that f is defined only on Γ at the start. Part (a) implies that there is at most one analytic extension of f into the interior of the loop Γ.

9. *The Poisson Integral Formula via Cauchy's.* We obtain this again (see Chapter 2), using only the Cauchy Integral Theorem and Formula, avoiding Green's III and the Green's function. Given $f = u + iv$ analytic in a domain containing the disc $|\zeta| < R$ and its (positively oriented) boundary $|z| = R$.

(a) For $|\zeta| < R$, verify that the "inverse point" $\zeta^* = R^2/\bar{\zeta}$ is outside the circle $|z| = R$ and lies on the ray from the origin through ζ.

(b) Show that

$$\int_{|z|=R} f(z)(z - \zeta^*)^{-1}\, dz = \int_{|z|=R} \bar{\zeta} f(z)(z\bar{\zeta} - R^2)^{-1}\, dz = 0.$$

(c) Apply (b) to prove that

$$f(\zeta) = \frac{1}{2\pi i} \int_{|z|=R} \left[\frac{1}{z - \zeta} + \frac{\bar{\zeta}}{R^2 - z\bar{\zeta}} \right] f(z)\, dz$$

$$= \frac{1}{2\pi i} \int_{|z|=R} \frac{(R^2 - \zeta\bar{\zeta})f(z)\, dz}{(z - \zeta)(R^2 - z\bar{\zeta})}.$$

(d) Using polar coordinates $z = Re^{i\theta}$, $\zeta = \rho e^{i\varphi}$, check that

$$\frac{dz}{(z - \zeta)(R^2 - z\bar{\zeta})} = \frac{z^{-1}\,dz}{(z - \zeta)(\bar{z} - \bar{\zeta})} = \frac{i\,d\theta}{R^2 - 2R\rho\cos(\theta - \varphi) + \rho^2}.$$

(e) Combine (c) and (d) to obtain

$$f(\zeta) = \frac{1}{2\pi}\int_0^{2\pi}\frac{(R^2 - \rho^2)f(Re^{i\theta})\,d\theta}{R^2 - 2R\rho\cos(\theta - \varphi) + \rho^2}.$$

(f) Note that $f(Re^{i\theta})$ is the only complex number in (e). Hence, we may take real parts to obtain

$$u(\zeta) = u(\rho\cos\varphi, \rho\sin\varphi) = \frac{1}{2\pi}\int_0^{2\pi}\frac{(R^2 - \rho^2)U(\theta)\,d\theta}{R^2 - 2R\rho\cos(\theta - \varphi) + \rho^2}$$

where $U(\theta) = u(Re^{i\theta})$ is the value of u on the boundary $|z| = R$.

This is the Poisson Integral Formula for the harmonic function u. This line of proof requires that we know that u is the real part of an analytic function.

10. *The Cauchy Integral Formula via Poisson's.* Now observe that Cauchy's Formula follows from Poisson's plus the Cauchy Integral Theorem plus the harmonicity of u, v if $f = u + iv$ is analytic on a domain containing $|\zeta| \le R$, as follows:

(a) Defining $U(\theta) = u(Re^{i\theta})$, $V(\theta) = v(Re^{i\theta})$ and adding the Poisson representations for $u(\zeta)$ and $iv(\zeta)$, obtain the first equality in part (c) of Exercise 9. *Hint:* Work backwards in Exercise 9.

(b) This is the same as part (b) of Exercise 9.

(c) Conclude the Cauchy Formula for the circle $|z| = R$.

(d) Obtain the Cauchy Formula for an arbitrary loop Γ enclosing ζ by the usual "deformation of path" argument (see Corollary 5).

11. *The Cauchy Integral Formula for k-connected Jordan Domains.* Verify that the proof given in this section essentially gives a more general result, namely,

THEOREM

Let Ω be a k-connected Jordan domain. Suppose f is analytic on a domain Ω^+ containing Ω and $\partial\Omega$. Then for each point z_0 in Ω,

$$f(z_0) = \frac{1}{2\pi i}\int_{\partial\Omega}\frac{f(z)}{z - z_0}\,dz,$$

where the standard positive orientation is taken on the Jordan curves comprising $\partial\Omega$.

12. Use Exercise 11 to conclude that if C_1, C_2 are concentric circles of radii r_1, r_2, respectively, with $r_1 < r_2$, each with its standard counterclockwise parametrization, and if f is analytic on and between the two circles, then

$$f(z_0) = \frac{1}{2\pi i}\left[\int_{C_2} - \int_{C_1}\right]\left(\frac{f(z)}{z - z_0}\,dz\right)$$

for every point z_0 between the two circles.

Section 4.4 HIGHER DERIVATIVES OF ANALYTIC FUNCTIONS

If f is analytic in a domain Ω and ζ is a point of Ω, then we have just seen that

$$f(\zeta) = \frac{1}{2\pi i} \int_\Gamma \frac{f(z)}{z - \zeta}\, dz,$$

provided Γ is a simple closed curve in Ω whose interior includes the point ζ and is contained entirely in Ω. Since f is analytic, the derivative $f'(\zeta)$ also exists. How do we compute it in terms of the integral formula?

Just as in the case of Green's III and the Poisson Integral Formula, our approach is straightforward calculus: We differentiate under the integral sign. We have

$$f'(\zeta) = \frac{d}{d\zeta} f(\zeta) = \frac{1}{2\pi i} \frac{d}{d\zeta} \int_\Gamma \frac{f(z)}{z - \zeta}\, dz$$

$$= \frac{1}{2\pi i} \int_\Gamma \frac{\partial}{\partial\zeta} \frac{f(z)}{z - \zeta}\, dz$$

$$= \frac{1}{2\pi i} \int_\Gamma \frac{f(z)}{(z - \zeta)^2}\, dz.$$

Note that this last integral is finite because ζ is not on the curve Γ and the integrand is continuous in z on Γ. To justify the crucial third equality in the chain here, we use the following lemma. It is the complex version of the result for real variables in Chapter 2.

LEMMA 6

Let Ω be a domain and Γ a (not necessarily closed) curve that is disjoint from Ω. Let $F(z, \zeta)$ and the partial derivative $F_\zeta(z, \zeta)$ be continuous complex-valued functions of the pair (z, ζ) with z in Γ, ζ in Ω. Then the integral

$$\int_\Gamma F(z, \zeta)\, dz$$

is an analytic function of ζ in Ω and, moreover,

$$\frac{d}{d\zeta} \int_\Gamma F(z, \zeta)\, dz = \int_\Gamma F_\zeta(z, \zeta)\, dz.$$

Note: Since $z = z(s)$ on Γ and also $\zeta = \xi + i\eta$ in Ω, we may write $F(z, \zeta) = U(s, \xi, \eta) + iV(s, \xi, \eta)$. The continuity hypothesis of the lemma is equivalent to the continuity of U, V and certain **partial** derivatives as real-valued functions of three real variables.

Proof: This may be reduced to the real (Chapter 2) by using the functions $U(s, \zeta, \eta)$ and $V(s, \zeta, \eta)$, breaking all integrals into real and imaginary parts, and (as we establish in Exercise 7 below) using the fact that complex differentiation decomposes into real differentiation; thus,

$$\frac{\partial}{\partial \zeta} = \frac{1}{2}\left(\frac{\partial}{\partial \xi} - i\,\frac{\partial}{\partial \eta}\right).$$

We leave the details to you.

Having obtained an integral representation for $f'(\zeta)$ by differentiating under the integral sign, there is nothing to prevent us from differentiating once again to obtain

$$f''(\zeta) = \frac{2}{2\pi i}\int_{\Gamma} \frac{f(z)}{(z - \zeta)^3}\,dz.$$

Clearly, we may repeat this process as often as we wish! There is always a simple function of ζ waiting to be differentiated (note that the function may be complicated in z, the variable of integration). We sum up this phenomenon as follows:

Theorem 7

Let f be analytic in Ω. Then all derivatives $f'(\zeta), f''(\zeta), \ldots, f^{(n)}(\zeta), \ldots$ exist and are analytic at all ζ in Ω. Moreover, we have the formula

$$f^{(n)}(\zeta) = \frac{n!}{2\pi i}\int_{\Gamma} \frac{f(z)}{(z - \zeta)^{n+1}}\,dz \qquad (n = 0, 1, \ldots),$$

where Γ is any simple closed curve inside Ω whose interior contains ζ and lies entirely inside Ω.

Comments

1. Is it clear that $f^{(n)}(\zeta)$ has a continuous derivative?

2. Keep in mind that this result depends on a routine process (differentiating under the integral sign: calculus!) and a deeper result (the Cauchy Integral Formula).

3. In differentiating the integrand $f(z)/(z - \zeta)^n$, we never touch $f(z)$ itself, but only $1/(z - \zeta)^n$.

4. Later we will see that an analytic function enjoys an even stronger property with regard to derivatives: It may be expanded in a power series whose coefficients involve all derivatives of f. That is, we will see that

$$f(z) = \sum_{n=0}^{\infty} a_n(z - z_0)^n,$$

where the coefficients satisfy

$$a_n = \frac{f^{(n)}(z_0)}{n!} \qquad (n = 0, 1, \ldots).$$

An Extension: Integrals of Cauchy Type

Let $\phi(z)$ be *any* function continuous on a curve Γ (not necessarily closed). For ζ *not* on Γ, we define a new function $f(\zeta)$ by the integral formula

$$f(\zeta) = \frac{1}{2\pi i} \int_\Gamma \frac{\phi(z)}{z - \zeta}\, dz.$$

Note that the integral yields a finite value because the integrand is defined and continuous in the variable z. Moreover, $f(\zeta)$ is continuous in ζ, essentially because the integrand is continuous in ζ. Finally, we may differentiate

$$f'(\zeta) = \frac{1}{2\pi i} \int_\Gamma \frac{\phi(z)}{(z - \zeta)^2}\, dz.$$

The differentiation process may be repeated. Thus, we have constructed an analytic function $f(\zeta)$ in the domain $\Omega = \mathbb{C} - \Gamma$. The function $f(\zeta)$ is called a *Cauchy integral* or *integral of Cauchy type*. We emphasize that $f(\zeta)$ is analytic even though $\phi(z)$ was not; in fact, $\phi(z)$ wasn't even defined on an open set!

What other methods of constructing analytic functions can you recall?

Caution: We remark that, in the situation described here, it is not generally true that

$$\lim_{\zeta \to z} f(\zeta) = \phi(z) \qquad (\zeta \in \Omega, z \in \Gamma).$$

That is, the original function $\phi(z)$ does not give the "boundary values" (note that $\Gamma = \partial\Omega$) of the analytic function $f(\zeta)$, $\zeta \in \Omega$. In fact, not every continuous $\phi(z)$ on a closed curve Γ can be realized as the boundary value function of an analytic function $f(\zeta)$, no matter what method of obtaining $f(\zeta)$ is used. Contrast the cases of harmonic and analytic functions: If Γ is the unit circle, then every continuous *real-valued* $\phi(z)$, $z \in \Gamma$, is the boundary value of a real harmonic function defined in the interior of the circle (Poisson Integral Formula), but no such real-valued $\phi(z)$ is the boundary value of an analytic function, unless, of course, $\phi(z)$ is constant (proof?).

Exercises to Section 4.4

1. (a) What property of the integrand in $(1/2\pi i) \int_\Gamma f(z)(z - \zeta)^{-2}\, dz$ guarantees the existence of the integral? Here, Γ, ζ, f are as in the text.
   (b) What is the key to proving that the value of this integral is $f'(\zeta)$?

(c) The function $f^{(n)}(\zeta) = (n!/2\pi i) \int_\Gamma f(z)(z - \zeta)^{-n-1} \, dz$ is seen to be continuous, either because it is differentiable or because it is an integral. Give the details of each argument.

2. (a) Verify $2\pi i \cos \zeta = \int_\Gamma (z - \zeta)^{-2} \sin z \, dz$, where Γ is the usual unit circle $|z| = 1$ and also $|\zeta| < 1$.
   (b) Is this equation true for $|\zeta| > 1$?

3. Compute:
   (a) $\int_{|z|=1} z^{-6} \sin z \, dz$,
   (b) $\int_{|z|=2} e^z z^{-2} (z - 1)^{-1} \, dz$.

4. Verify $\lim_{n \to \infty} \int_{|z|=1} e^z z^{-n} \, dz = 0$.

5. Compute a formula for $f'(\zeta)$, where $f(\zeta) = \int_{|z|=1} (z - \zeta)^{-1} \sin xy \, dz$ with $z = x + iy$ and $|\zeta| \neq 1$. (Here, f is a Cauchy integral.)

6. Write down a formula that enables us to compute the line integral $\int_\Gamma f(z) \, dz$, where $f(z) = g(z)/(z - z_0)^m$ with m a positive integer, g analytic on and inside the closed loop Γ, and z_0 a point inside Γ. Note that several of the exercises above will be handled by this formula (which is a special case of the Residue Theorem of Chapter 7).

7. Verify that if $f(\zeta) = u(\zeta) + iv(\zeta)$ is analytic, then

$$\frac{d}{d\zeta} f(\zeta) = \frac{1}{2} \left(\frac{\partial}{\partial \xi} - i \frac{\partial}{\partial \eta} \right) (u(\xi, \eta) + iv(\xi, \eta)).$$

This relates real and complex derivatives. It was used in the proof of Lemma 6.

Section 4.5 HARMONICITY OF u(z) AND v(z)

The foregoing technical facts about existence of derivatives now yield a cloudburst of results on the behavior of analytic functions.

We have already mentioned the following fact.

THEOREM 8

Let $f = u + iv$ be analytic in Ω. Then $u = u(z)$, $v = v(z)$ are harmonic in Ω.

Proof: Since the higher derivatives $f''(z)$, $f'''(z)$ exist, you may verify that the second partials u_{xx}, u_{yy} and also v_{xx}, v_{yy} exist and are continuous for each z in Ω. But, as we saw in Chapter 3, these functions therefore satisfy $u_{xx} + u_{yy} = 0$, $v_{xx} + v_{yy} = 0$ as a consequence of the Cauchy–Riemann equations. Done.

Moral

1. An analytic function $f(z)$ determines two real harmonic functions $u(z)$, $v(z)$, related by the Cauchy–Riemann equations.

2. A real harmonic function $u(z)$ in a simply connected domain and a single value $f(z_0)$ with real part $u(z_0)$ determine an analytic

function $f(z) = u(z) + iv(z)$ (since we may solve for $v(z)$, the harmonic conjugate of $u(z)$, as in Chapter 3).

These two points emphasize the close connection between the studies of harmonic and analytic functions in the plane.

We may draw a further conclusion on boundary values of an analytic function $f(z) = u(z) + iv(z)$. Since $u(z)$ is determined (inside the unit disc, say) by its boundary values on the circle, it follows from the second point above that $f(z)$ is determined up to an additive constant by the boundary values of its *real part* $u(z)$. It is naive to hope, therefore, that we may prescribe arbitrary *real and imaginary* boundary values and then expect to construct an analytic function with those boundary values. The Cauchy–Riemann equations are looming here.

Exercises to Section 4.5

1. True or false?
   (a) If u is harmonic in Ω, then it is the real part of an analytic function f in Ω.
   (b) If u is harmonic in Ω, then it is the imaginary part of an analytic function g in Ω.
   (c) The real and imaginary parts of a complex analytic function are harmonic.
   (d) If two analytic functions f and g have the same real part u, then $f = g$ identically.
   (e) If $f(z) = u(z) + iv(z)$ with u, v harmonic, then f is analytic.
   (f) There exists an analytic $f(z)$ with real part $x^2 - y^2$ which satisfies $f(0) = 2\pi i$.

2. Why could we not conclude the harmonicity of the real and imaginary parts of every analytic function immediately after deriving the Cauchy–Riemann equations in Chapter 3?

3. Let $H(w)$ be real harmonic and $w = f(z)$ complex analytic. Prove that $h(z) = H(f(z))$ is real harmonic (assuming it is defined on some domain the z-plane), as follows: Construct $g(w) = H(w) + iG(w)$, where G is a harmonic conjugate for the given H. Prove that $h(z)$ is the real part of the analytic (why?) function $g(f(z))$.

4. A direct approach to $\Delta h(z) = 0$ as in Exercise 3. Writing $z = x + iy$, $w = f(z) = u(x, y) + iv(x, y)$, and applying the Chain Rule for partial derivatives (that is, $h_x = H(u(x, y), v(x, y))_x = H_u u_x + H_v v_x$, etc.), the Cauchy–Riemann equations for u, v, and the fact that $H_{uu} + H_{vv} = 0$, verify that $h_{xx} + h_{yy} = 0$. There are some nice cancellations to be seen here.

5. Let $H(w) = \ln|w|$ for $w \neq 0$ and let $f(z) = e^z$. Verify that $h(z) = H(f(z))$ is harmonic by computing it exactly. (It's well known!)

6. In Exercise 1 above, statements (a) and (b) are false. What reasonable restriction on the domain Ω would make them true?

7. Write down a harmonic function $u(z)$ in the punctured plane $\Omega = \mathbb{C} - \{0\}$ which is not the real part of an analytic function in the full domain Ω.

8. *Some Dirichlet problems solved by inspection.* Let Ω be the upper half-plane $y > 0$, $\partial\Omega$ the x-axis. Suppose $B(x)$ is a polynomial in x with real coefficients. We will find $H(z) = H(x, y)$ continuous in $\overline{\Omega}$ and harmonic in Ω such that $H(x, 0) = B(x)$. (Dirichlet problem for upper half-plane.)
   (a) Show that the polynomial $B(z)$ in the complex variable is analytic.

(b) Argue that $H(z) = \operatorname{Re} B(z)$, the real part of $B(z)$, is a solution to the Dirichlet problem.

(c) Show how this method generalizes to any real $B(x)$ that is the restriction (to the x-axis) of a function which is analytic in Ω. (That is, $B(x) = \sin x$, e^x, etc.)

(d) Observe that this method fails for $B(x) = 1/(1 + x^2)$, because $B(z) = 1/(1 + z)^2$ is not analytic at $z = i$ in Ω.

Appendix to Section 4.5 ANALYTIC FUNCTIONS AND THE DIRICHLET PROBLEM

We pause now to discuss one reason why analytic functions are so important: They may be used to solve boundary value problems. Here is the idea.

As usual, we are given a Jordan domain Ω, which is the interior of a smooth loop $\Gamma = \partial\Omega$. See Figure 4.10. We are also given a continuous real-valued function $B(z)$ for z on Γ (the "boundary values"). To solve the Dirichlet problem, we must find $H(z)$ harmonic in Ω, continuous in $\overline{\Omega}$, and satisfying $H(z) = B(z)$ for z on $\partial\Omega$.

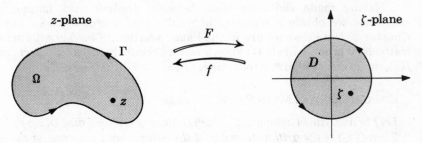

Figure 4.10

A classical method of solution is this: Translate this problem into a closely related Dirichlet problem on the ordinary disc (a much simpler domain) and then use the Poisson Integral Formula to solve the problem on the disc. Moreover, we assert that this "translation" is done by an analytic function from the disc onto Ω (and its inverse).

Here are the basic details. Let Ω be as above (in the z-plane) and let D be the disc $|\zeta| < 1$ in the ζ-plane. Suppose we can find a one-to-one analytic function $f: \overline{D} \to \overline{\Omega}$, $z = f(\zeta)$, with analytic inverse $F: \overline{\Omega} \to \overline{D}$, $\zeta = F(z)$. Then, if $H(z)$ is harmonic in Ω, it's not hard to see that

$$h(\zeta) = H(f(\zeta))$$

is harmonic in the disc D. This follows from the analyticity of f. See the exercises. Likewise, $b(\zeta) = B(f(\zeta))$ is continuous on ∂D, $|\zeta| = 1$.

Using the Poisson Integral Formula and the function $b(\zeta)$, we can compute a harmonic $h(\zeta)$ on the disc satisfying $h(\zeta) = b(\zeta)$ on the boundary ∂D. To solve the original Dirichlet problem on Ω, we translate $h(\zeta)$ back to Ω. Just define, for z in Ω,

$$H(z) = h(F(z)).$$

Again, $H(z)$ is harmonic (because F is analytic) and $H(z) = B(z)$ on $\Gamma = \partial\Omega$.

Moral

The solution of the Dirichlet problem on certain domains Ω may be reduced to the search for analytic functions that map Ω in a nice way ("conformally") onto a standard disc. This is the stuff of Chapter 8. See Paragraph 8.2.3 for a solved Dirichlet problem.

Section 4.6 CIRCUMFERENTIAL AND SOLID MEANS

Having made the connection between analytic and harmonic functions, we obtain a sequence of results reminiscent of those from Chapter 2. Note that we are in the happy position of having alternate methods of proof, namely (1) known results about the harmonic parts of $f(z)$, or (2) direct analytic tools such as the Cauchy Integral Formula.

Circumferential Mean-Value Theorem

Let f be analytic in a domain Ω which contains the closed disc $\bar{D}(z_0; r)$. Then $f(z_0)$ is the arithmetic mean of the values that f assumes on the circumference of the disc. That is,

$$f(z_0) = \frac{1}{2\pi} \int_{\theta=0}^{\theta=2\pi} f(z_0 + re^{i\theta})\, d\theta.$$

Note: As θ varies from 0 to 2π, the complex number $z_0 + re^{i\theta}$ traverses the circle of radius r centered at the point z_0. See Figure 4.11. Note the similarity with polar coordinates (r, θ) centered at z_0.

First Proof: Break up f into real and imaginary parts and apply the analogous result for harmonic functions. Done.

Second Proof: We apply the Cauchy Integral Formula, noting that for z on $\Gamma = C(z_0; r)$ we have, since r is constant,

$$dz = d(z_0 + re^{i\theta}) = r\, de^{i\theta} = ire^{i\theta}\, d\theta.$$

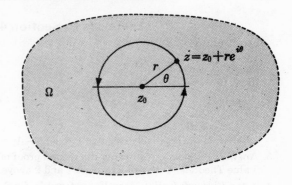

Figure 4.11

Thus,

$$f(z_0) = \frac{1}{2\pi i} \int_\Gamma \frac{f(z)}{z - z_0}\, dz$$

$$= \frac{1}{2\pi i} \int_{\theta=0}^{\theta=2\pi} \frac{f(z_0 + re^{i\theta})}{re^{i\theta}}\, ire^{i\theta}\, d\theta$$

$$= \frac{1}{2\pi} \int_{\theta=0}^{\theta=2\pi} f(z_0 + re^{i\theta})\, d\theta,$$

as claimed. Done.

Question

Why is the integral in the theorem not always zero by the Cauchy Integral Theorem? (Look closely!)

Comment

The first proof shows that the theorem is true for an even wider class of function than the analytic, namely, the *complex harmonic functions* $f(z) = u(z) + iv(z)$ with u, v harmonic (but no relations necessarily between u and v).

SOLID MEAN-VALUE THEOREM

Let f, Ω, $\overline{D}(z_0; r)$ be as in the preceding theorem. Then

$$f(z_0) = \frac{1}{\pi r^2} \iint_D f(z)\, dx\, dy.$$

Note: The double integral may be broken into real and imaginary parts with f.

Proof: As in the first proof of the preceding theorem.

Exercises to Section 4.6

1. Compute:
   (a) $\int_0^{2\pi} \cos(7e^{i\theta})\, d\theta$,
   (b) $\int_0^{2\pi} \log(1 + (1/\pi)e^{i\theta})\, d\theta$.

2. Compute:
   (a) $\int_{|z|=7} \cos z\, dz$,
   (b) $\int_{\Gamma} \log z\, dz$,
   where Γ is the positively oriented circle $|z - 1| = 1/\pi$.

3. Answer the question following the second proof of the Circumferential Mean-Value Theorem. Compare Exercises 1 and 2 above.

4. In the Circumferential Mean-Value formula for $f(z_0)$, the radius r appears in the integrand on the right-hand side, but does not appear on the left (that is, $f(z_0)$). What corollary of the Cauchy Integral Theorem explains this phenomenon?

5. Compute $(1/\pi) \int_D \int \cos z\, dx\, dy$, where D is given by $|z| < 1$.

6. Let f be analytic at z_0. Prove that f cannot have a strict local maximum at z_0; that is, we cannot have $|f(z)| < |f(z_0)|$ for all z in some punctured disc $0 < |z - z_0| < R$. *Hint:* Use the fact that the upper bound $M(r) = \max_{|z-z_0|=r} |f(z)|$ is assumed and then apply the ML-inequality to a well-chosen integral.

Section 4.7 THE MAXIMUM MODULUS PRINCIPLE

This is the complex analog of the Maximum Principle for harmonic functions. Recall that the latter principle says that a real nonconstant harmonic function cannot assume a maximum value in a domain Ω. For complex-valued functions $f(z)$, we cannot speak of the "largest" or "maximum" value attained, since the complex numbers are not ordered. Use of $|f(z)|$ enables us to create a similar theory, however.

Let Ω be a domain as usual and let f be a complex-valued function defined in Ω. We say that f *attains its maximum modulus at z_0 in Ω* if and only if $|f(z)| \leq |f(z_0)|$ for all z in Ω. Geometrically, this means that $f(z_0)$ is at least as far from the origin of the w-plane (note $w = f(z)$) as is any other point of the set $f(\Omega)$ of values.

Strong Maximum Modulus Principle

Let f be a nonconstant analytic function defined in a plane domain Ω. Then f does not assume its maximum modulus at any point of Ω.

Proof: If the set $f(\Omega)$ is unbounded, then the theorem is clear.

Now suppose the set $f(\Omega)$ is bounded. Assume f attains its maximum modulus at z_0 in Ω. We will transfer this to the harmonic case and obtain a contradiction. First, however, we must translate the bounded set $f(\Omega)$ away from the origin while retaining our hypothesis about maximum modulus. This is done for purely technical reasons, and our motivation should become clear in a moment.

Thus, let $g(z) = f(z) + cf(z_0)$, where c is a positive real number large enough that $g(z) \neq 0$ for z in Ω and that a branch of the logarithm may be defined at all points of the set $g(\Omega)$. See Figure 4.12. Note that $g(z)$ is analytic and also attains its maximum modulus at z_0.

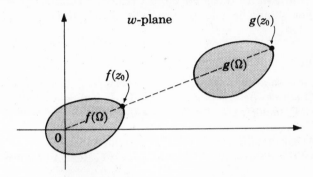

Translating $f(\Omega)$ away from $w=0$

Figure   4.12

Now we define $h(z) = \ln|g(z)|$. This is harmonic because it is the real part of $\log g(z)$. Since the natural logarithm is an increasing function of its variable, in this case $|g(z)|$, we conclude that $h(z)$ assumes its maximum at z_0 in Ω. But $h(z)$ is nonconstant because $f(z)$ is nonconstant (exercise!). This contradicts the maximum principle for harmonic functions. Done.

Still paralleling the harmonic theory, we have

Weak Maximum Modulus Principle

Let Ω be a bounded domain and f a nonconstant function analytic on Ω and continuous on its closure $\overline{\Omega}$. Then f attains its maximum modulus $|f(z)|$ at certain points z of the boundary $\partial\Omega$ only.

Proof: The function $|f(z)|$ attains a maximum somewhere on $\overline{\Omega}$ because it is continuous and $\overline{\Omega}$ is compact (that is, closed and bounded). This is a standard fact of analysis. But by the Strong Principle asserted previously, the maximum cannot be attained in the open set Ω, and so must occur somewhere on the boundary. Done.

It is also possible to prove

Minimum Modulus Principle

Let f be nonconstant analytic on a domain Ω with $f(z) \neq 0$ for z in Ω. Then $f(z)$ does not attain its minimum modulus $|f(z)|$ at any point of Ω.

We will leave the definition of minimum modulus and the proof of this result to you. Note that $f(z) \neq 0$ implies that $1/f(z)$ is defined and analytic.

Application to Boundary-Value Questions

This is a favorite theme of ours. Using the results of the last few sections, you should be able to prove quite simply the following uniqueness assertion.

ASSERTION

Let f and g be analytic in a domain Ω and let Γ be a Jordan curve inside Ω whose interior is also inside Ω. If $f(z) = g(z)$ for all z on the curve Γ, then $f(z) = g(z)$ for all z in the interior of Γ as well.

Note that it suffices to prove that the function $f - g$ vanishes throughout the interior of Γ if it vanishes on Γ. Refer to Figure 4.13.

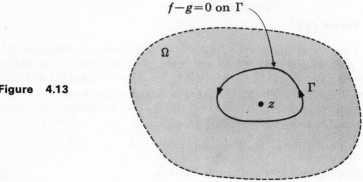

$f-g=0$ on Γ

Ω

Γ

$\bullet z$

Figure   4.13

Question

Is it true that $f = g$ throughout Ω and not merely inside Γ?

Exercises to Section 4.7

1. Let f be analytic on the domain Ω. Prove that if the image $f(\Omega)$ is compact (closed and bounded) in the w-plane, then it must be a single point; that is, f is constant. *Hint:* $|w|$ assumes its maximum on a compact set. (Can you prove this latter fact?)

    We comment that the Open Mapping Theorem (Paragraph 8.1.4) will give us much better information about $f(\Omega)$.

2. (a) Prove that there exists z satisfying $|z| = 1$, $|\cos z| > 1$.
    (b) Using continuity and (a), prove z exists satisfying $|z| < 1$, $|\cos z| > 1$.
    (c) Contrast $\cos x$, x real.

3. Proof or counterexample: If f is a nonconstant analytic function defined in $\Omega = \mathbb{C}$ (that is, f is *entire*), then $\lim_{z \to \infty} f(z) = \infty$.

4. Let f be nonconstant analytic in $\mathbb{C}$. Define $M(r) = \max_{|z| = r} |f(z)|$.
   (a) Prove $M(r)$ is a nondecreasing function of $r \geq 0$.
   (b) Is it true that $\lim_{r \to \infty} M(r) = \infty$? In other words, can a nonconstant entire analytic function be bounded? Note that this assertion is weaker than that of Exercise 3.

5. (a) Prove the uniqueness assertion appearing at the end of this section.
   (b) See Exercise 8 following Section 4.3 for a proof that does not use the theory of maxima.

6. Let f be a nonconstant analytic function on a domain Ω. Prove that if $f(\Omega)$ is contained in the closed disc $|w| \leq r$, then it is in fact contained in the open disc $|w| < r$.

7. *The Schwarz Lemma.* This is a virtuoso application of the Maximum Modulus Principle. The statement is : *Let f be a nonconstant analytic function on the unit disc $|z| < 1$, satisfying $|f(z)| < 1$ and $f(0) = 0$. Then either f is a rotation through a fixed angle φ, $f(z) = e^{i\varphi} z$, or f contracts toward the origin: $|f'(0)| < 1$ and $|f(z)| < |z|$ for $0 < |z| < 1$.* You are to prove this in several steps, as follows:
   (a) The key idea is the application of the Maximum Modulus Principle to $g(z) = f(z)/z$. Note that $\lim_{z \to 0} g(z) = f'(0)$. It is a fact that defining $g(0) = f'(0)$ yields an analytic function $g(z)$ at $z = 0$ and hence in the full disc $|z| < 1$ (proof in Exercises to Section 4.10 and in Chapter 6; assume the fact for now).
   (b) Prove that $|g(z)| \leq 1$. *Hint:* Show that $M(g; r) = \max_{|z| = r} |g(z)| < 1/r$. But as r increases to 1, $1/r$ decreases to 1. Since $M(g; r)$ is nondecreasing (why?), it is ≤ 1 for all $r < 1$.
   (c) First case: If $M(g; r) = 1$ for some r, then (Exercise 6) g is constant, whence $f(z)/z = e^{i\varphi}$ for all $|z| < 1$.
   (d) Second case: $|g(z)| < 1$ implies $|f(z)| < |z|$ for $z \neq 0$ and $|f'(0)| < 1$.

8. Verify that the polynomials $f(z) = cz^n$ (with $0 < |c| \leq 1$ and $n \geq 1$) satisfy the hypotheses of the Schwarz Lemma (Exercise 7). Which of these $f(z)$ are rotations? contractions?

9. *A detail in the proof of the Maximum Modulus Principle.* Given $f(z)$ analytic on Ω, such that $f(\Omega)$ lies in the upper half-plane $u > 0$ (this may be weakened considerably). Suppose that $\ln|f(z)|$ is constant in Ω. Prove that $f(z)$ is constant in Ω. *Hint:* $\log f(z)$ is constant in Ω (why?). Review the relevant exercises in Chapter 3.

10. Let Ω be a bounded domain and f a function continuous on $\overline{\Omega}$ and analytic in Ω. Prove that, for all $\zeta \in \Omega$,
$$|f(\zeta)| \leq \max_{z \in \partial\Omega} |f(z)|,$$
and if the equality sign holds for one ζ in Ω, then f is constant on $\overline{\Omega}$.

Section 4.8 THE FUNDAMENTAL THEOREM OF ALGEBRA

This result, so often used in algebra, seems to require a certain amount of real or complex analysis for its proof. It is possible to obtain the theorem as a corollary of several other theorems of this chapter. Our first proof, given here, uses the Minimum Modulus Principle.

FUNDAMENTAL THEOREM OF ALGEBRA

Let

$$p(z) = a_n z^n + \cdots + a_1 z + a_0$$

be a nonconstant polynomial of degree $n \geq 1$ with complex coefficients. Then the equation $p(z) = 0$ has a root z in $\mathbb{C}$.

Proof: We assume $p(z)$ never vanishes (never equals zero) and obtain a contradiction. Let $m(r)$, the "minimum modulus," be defined for $r \geq 0$ by

$$m(r) = \min_{|z|=r} |p(z)|.$$

Since $p(z) \neq 0$, the Minimum Modulus Principle assures us that $m(r)$ *decreases* as $r \to \infty$ (think about this).

But we will prove in the next paragraph that $|p(z)| \to \infty$ as $r = |z| \to \infty$. This implies in particular that $m(r) \to \infty$ as $r \to \infty$ and yields the desired contradiction.

To show $|p(z)| \to \infty$ as $|z| \to \infty$, we factor

$$p(z) = z^n \left(a_n + \frac{a_{n-1}}{z} + \cdots + \frac{a_0}{z^n} \right).$$

Now, as $|z| \to \infty$, the first factor satisfies $|z^n| \to \infty$, while the second factor approaches the limit a_n (check!). Since a_n is not zero ($\infty \cdot 0$ is a delicate business), we may deduce that $\lim |p(z)| = \lim |z^n| \, |a_n| = \infty$ as $|z| \to \infty$. By our remarks in the paragraph above, the proof is complete. Done.

Comments

1. To see that the equation $p(z) = 0$ has n (not necessarily distinct) complex roots is now a problem in algebra. If $p(z_1) = 0$, then long division yields a factorization $p(z) = (z - z_1)p_1(z)$, where $p_1(z)$ is a polynomial of degree $n - 1$. This new polynomial now has a complex root, and so on.

2. The Fundamental Theorem is an existence theorem. It gives no method of finding the roots exactly. If the degree is 2, the famous quadratic formula

$$\frac{-b \pm \sqrt{b^2 - 4ac}}{2a}$$

gives both roots of $p(z) = az^2 + bz + c$. There are similar, more complicated formulas for degrees $n = 3, 4$. It was demonstrated by the young mathematicians Niels Abel (Norwegian) and Evariste Galois (French)

in the early nineteenth century that there are polynomials of degree
5, 6, 7,... whose roots cannot be constructed by starting with the
coefficients and performing only arithmetic operations and the extraction
of square, cube, and higher roots. Thus, there is no hope of a formula
like the quadratic formula in degrees 5 and higher; some equations are
not "solvable by radicals."

Thus, the moral of this section is that *roots always exist, but it may
be very difficult to write them down.*

Exercises to Section 4.8

1. Prove this corollary of the Fundamental Theorem of Algebra: If the polynomial $p(z)$ has degree n, and if c is any complex number, then there exist
   complex $z_1, \ldots, z_n$ (not necessarily distinct) satisfying $p(z_k) = c$.
2. *Towards localizing the roots.* Let $p(z) = z^n + \cdots + a_1 z + a_0$ (observe $a_n = 1$)
   so that $p(z) = (z - z_1)(z - z_2) \cdots (z - z_n)$.
   (a) Observe that $a_0 = \pm z_1 z_2 \cdots z_n$.
   (b) Prove that at least one root z_k satisfies $|z_k| \le R$, where R is the positive
   nth root of $|a_0|$. Estimates less crude than this are possible; see the applications
   following the theorem of Rouché in Paragraph 8.1.5.
3. Let $p(z)$ be a polynomial whose coefficients are *real* numbers. Prove that if z_1
   is a root of $p(z)$, then its complex conjugate $\bar{z}_1$ is also a root. *Hint:* The proof
   should be brief.

Section 4.9 LIOUVILLE'S THEOREM

This should be familiar from our study of harmonic functions.

We will say that a function f is *entire* if it is defined and analytic
at all points of $\mathbb{C}$. A function $f: \Omega \to \mathbb{C}$ is *bounded* if the image set $f(\Omega)$
is a bounded subset of the plane $\mathbb{C}$.

LIOUVILLE'S THEOREM

A bounded entire function is a constant.

Proof: If $f = u + iv$ is entire (analytic), then u and v are entire
(harmonic). If the image set $f(\mathbb{C})$ is bounded in the plane, then the sets
$u(\mathbb{C})$ and $v(\mathbb{C})$ must be bounded subsets of $\mathbb{R}$. By Liouville's Theorem for
harmonic functions, u and v are constant. Thus, f is constant. Done.

Comment

This depended only on the fact that $f = u + iv$ had real and imaginary
parts that were harmonic. We did not need the condition that f be
analytic. Which of the other theorems in this chapter are also true
for these weaker hypotheses?

Application to Growth of Entire Functions

This is a classical theme, differing somewhat from issues relating to boundary values. The natural domain of an entire function, namely $\mathbb{C}$, of course has no boundary curve for z to approach in the limit. Rather we consider the behavior of $f(z)$ as $|z|$ gets large. Questions sound like this: "If $|f(z)|$ behaves like such-and-such as $|z|$ gets large, is $f(z)$ actually equal to so-and-so?"

You might try to prove the following example of this.

ASSERTION

Let $f(z)$ be entire and $|f(z)| \leq |e^z|$ for all z in $\mathbb{C}$. Then, in fact, $f(z) = ce^z$ for some complex constant c with $|c| \leq 1$.

Thus, if a nonconstant entire function "grows" no faster than the exponential function, it *is* an exponential function. For the proof, consider the entire function $e^{-z}f(z)$.

Exercises to Section 4.9

1. Deduce that an entire (analytic) function $w = f(z)$ that maps the z-plane into the upper half w-plane (compare $v(z) > 0$) must be identically constant.

2. Prove that the image set $f(\mathbb{C})$ of a nonconstant entire function intersects every straight line in the w-plane.

3. Which result is stronger for polynomials $p(z)$: Liouville's Theorem or Exercise 1 to Section 4.8?

4. (a) Prove the assertion about $|f(z)| \leq |e^z|$ which appears at the end of the present section.
   (b) A naive student might argue that the assertion is false because he has heard that "polynomials grow more slowly than the exponential function," but surely $p(z) \neq ce^z$. Educate him by proving that if $|p(z)| \leq |e^z|$ for all z, then $p = 0$, identically. *Hint:* Look at $z = x < 0$.

5. *An alternate proof of Liouville's Theorem, independent of harmonic functions.* (See Section 4.11 for a third proof.)
   Given the bounded entire function f, prove $f(z_1) = f(z_2)$ for arbitrary z_1, z_2 in $\mathbb{C}$, as follows:
   (a) Prove $f(z_1) - f(z_2) = (1/(2\pi i)) \int_C f(z)(z_1 - z_2)(z - z_1)^{-1}(z - z_2)^{-1} \, dz$, provided $C = C(0; R)$ with $R > |z_1|, |z_2|$.
   (b) Apply the ML-inequality to prove that the integral in (a) vanishes as R gets very large, whence it is zero for all $R > |z_1|, |z_2|$.

6. *An alternate proof of the Fundamental Theorem of Algebra, using Liouville's Theorem.* Assume that the nonconstant polynomial $p(z)$ has no roots in $\mathbb{C}$. Prove that $f(z) = 1/p(z)$ is bounded and entire, and apply Liouville's Theorem.

7. *An extension of Liouville's Theorem.* Prove that if $w = f(z)$ is entire and not constant, then the image set $f(\mathbb{C})$ intersects every open set in the w-plane. *Hint:* Assume $f(\mathbb{C})$ misses the disc $D(w_0; \varepsilon)$ and consider the function $g(z) = 1/(f(z) - w_0)$.

8. *Another extension*
    (a) Suppose $f(z)$ is entire and either its real part or its imaginary part is bounded. Prove that $f(z)$ is constant.
    (b) The imaginary part of any branch of $\log z$ is bounded. Why doesn't part (a) imply that $\log z$ is constant?

Section 4.10 MORERA'S THEOREM

This theorem is the converse to the Strong Cauchy Integral Theorem, and therefore gives a sufficient condition for analyticity of $f(z)$. (Can you name other such results?) The proof of Morera's Theorem depends on the powerful fact (Theorem 7, Section 4.4) that the derivative of an analytic function is analytic. There is no mention here of harmonic functions.

MORERA'S THEOREM

Let $f(z)$ be a continuous function on some domain Ω, and suppose

$$\int_{\Gamma} f(z)\,dz = 0$$

for every simple closed (Jordan) curve Γ in Ω whose interior also lies in Ω. Then f is analytic in Ω.

Proof: The idea is to define a new function $F(z)$ in Ω which is analytic with $f(z) = F'(z)$. Since analytic functions $F(z)$ have derivatives of all orders (Section 4.4) it must be the case that $f'(z) = F''(z)$, and so $f(z)$ is analytic as claimed. Subtle, no?

A reduction of the problem: Since f is analytic in Ω if and only if it is analytic in every open subdisc of Ω (that is, analyticity is "local"), it suffices to consider the case Ω as an open disc, say $\Omega = D$.

Fix some point z_0 in D. For any ζ in D (see Figure 4.14),

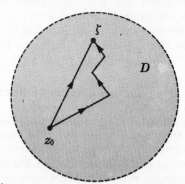

Figure 4.14

$$\int_{z_0}^{\zeta} f(z)\,dz \text{ is the same over either path}$$

define

$$F(\zeta) = \int_{z_0}^{\zeta} f(z)\, dz,$$

where the integration is carried out on the line segment in D from the fixed z_0 to the point ζ. Thus, $F(\zeta)$ is a well-defined function on D and, moreover, it is continuous at every point ζ. This latter fact is most easily proved using the hypothesis about integrals of $f(z)$; however, it will also follow from the differentiability of F, with which we now concern ourselves.

We claim that the number $\int_{z_0}^{\zeta} f(z)\, dz$ is independent of the polygonal path from z_0 to ζ. That is, we obtain the same value (namely, $F(\zeta)$) whether we integrate over the straight-line segment from z_0 to ζ or any other polygonal path from z_0 to ζ. This is true because we get zero when we integrate $f(z)\, dz$ around the closed path from z_0 to ζ over the straight-line path and then back from ζ to z_0 by another path; this last assertion follows from the hypothesis about integrals of $f(z)\, dz$ around closed curves Γ. You should check details here in the spirit of Section 4.2.

Since $F(\zeta)$ is "independent of path," the Fundamental Theorem of Calculus as stated in the Appendix to Section 4.2 guarantees that $F' = f$. By our comments at the start, the function f is analytic. Done.

Comments

1. We could have proved this immediately after Section 4.4.

2. This result (and the whole story) is remarkable in that the continuous complex differentiability (analyticity) of $f(z)$ is equivalent to $f(z)$ having "nice" integrals.

3. You might check that the only closed curves Γ that appeared in the proof were polygonal. Thus, the statement of the theorem could be strengthened: Replace "for every simple closed curve" with "for every closed polygonal path." And this can be pushed even further. Since every two-dimensional set with polygonal boundary (triangle, trapezoid, pentagon, etc.) can be cut up into triangles ("triangulated"), we may replace "for every polygonal path" in the statement of the theorem with "for every triangular path." Refer to Figure 4.15. We leave details to you.

Exercises to Section 4.10

1. The function $f(z) = 1/z^2$ satisfies $\int_{\Gamma} f(z)\, dz = 0$ if Γ is any closed curve enclosing the origin in its interior. Does Morera's Theorem imply therefore that f is analytic throughout $\mathbb{C}$, in particular at $z_0 = 0$? Explain.

2. *A preview of removable singularities.* Prove that if f is continuous in a disc D containing the point z_0 and is known to be analytic in $D - \{z_0\}$, then f is analytic at z_0 as well. Thus, the "singularity" (point of nonanalyticity) z_0 may be removed from this category. Method: Show $\int_{\Gamma} f(z)\, dz = 0$ for every loop Γ inside D, as follows:

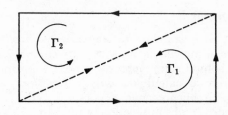

Figure   4.15

$$\int_{\text{rectangle}} = \int_{\Gamma_1} + \int_{\Gamma_2}$$

because two integrals over diagonal cancel each other (opposite directions).

(a) Argue that $\int_\Gamma f(z)\,dz$ exists, even if Γ passes through z_0.

(b) If z_0 is not on or inside Γ, then $\int_\Gamma f(z)\,dz = 0$, for a well-known reason.

(c) If z_0 is inside Γ, then the *ML*-inequality can be used to show that the integral vanishes. Refer to Exercise 5 from Section 4.3.

(d) Suppose z_0 lies on Γ. Argue that there are loops $\tilde{\Gamma}$ close to Γ but avoiding the point z_0 (whence $\int_{\tilde{\Gamma}} f(z)\,dz = 0$) such that the value of $\int_{\tilde{\Gamma}}$ comes arbitrarily close to $\int_\Gamma$, whence $\int_\Gamma$ must equal 0. Details?

We remark that the geometric details of (d) may be unwieldy and that removable singularities will be officially discussed in Chapter 6, using methods other than Morera's Theorem.

Section 4.11 THE CAUCHY INEQUALITIES FOR $f^{(n)}(z_0)$

Let f be analytic in a domain Ω as usual. Let z_0 be a point of Ω and suppose the disc $D(z_0; r)$ and its boundary, the circle $C(z_0; r)$, are contained in Ω. We write

$$M(z_0; r) = \max |f(z)| \qquad (z \in C(z_0; r)),$$

a nonnegative real number. We now state, prove, and discuss the Cauchy inequalities.

CAUCHY INEQUALITIES

Let $f, \Omega, M(z_0; r)$ *be as above. Then the nth derivative of* f *satisfies*

$$|f^{(n)}(z_0)| \le \frac{n! M(z_0; r)}{r^n}.$$

Proof: This follows easily from the Cauchy Integral Formula. We have, writing $C = C(z_0; r)$ and using the *ML*-inequality,

$$|f^{(n)}(z_0)| = \left| \frac{n!}{2\pi i} \int_C \frac{f(z)}{(z - z_0)^{n+1}}\,dz \right| \le \frac{n!}{2\pi} \frac{M(z_0; r)}{r^{n+1}} 2\pi r,$$

since the length of C is $2\pi r$. This gives the result. Done.

Comments

1. It is standard to bound or "control" the values of a function in terms of the values of its derivatives (that is, rate of change). This is the idea behind many applications of the ordinary Mean-Value Theorem of differential calculus. Analytic functions are remarkable in that the opposite phenomenon occurs as well. The complex derivatives are bounded in absolute value in terms of the values of the function itself!

2. These inequalities lead to another proof of Liouville's Theorem ("A bounded entire function is a constant") independent of the harmonic theory, as follows:

Given $|f(z)| \leq M$ for all z in $\mathbb{C}$. By the Cauchy inequality, $|f'(z_0)| \leq M(z_0; r)/r \leq M/r$ for all points z_0 and radii r. This is true as $r \to \infty$; thus, as $M/r \to 0$. Hence, $|f'(z_0)| = 0$ for all z_0. Thus, $f = $ constant. Done.

Exercises to Section 4.11

1. How many applications of the ML-inequality in this chapter can you recall? To which integrals was the inequality applied?

2. Suppose f is entire and $M(z_0; r) \leq ar^k$ for some $a > 0$, positive integer k, and all sufficiently large r. Prove that $f^{(n)}(z_0) = 0$ for all $n \geq k + 1$. *Hint:* See proof of Liouville's Theorem in this section.

3. Suppose f is an entire function such that $f^{(k+1)}(z) = 0$ for all z. Prove that f is a polynomial of degree $\leq k$.

4. *On polynomial growth.* Suppose f is entire and that, for some z_0, $M(z_0; r) \leq ar^k$ for some $a > 0$, positive integer k, and all sufficiently large r. Prove that f is a polynomial of degree $\leq k$. *Hint:* By Exercises 2 and 3, it suffices to show that for each z_1, $M(z_1; r_1) \leq a_1 r_1^k$ for some $a_1 > 0$ and all sufficiently large r_1 (where $r_1 = |z - z_1|$). To show this, use $M(z_1; r_1) \leq M(z_0; |z_1 - z_0| + r_1)$, which follows from the Maximum Modulus Principle.

   Comment: This problem points out the organic nature of analytic functions. Growth conditions on derivatives at a single point z_0 imply similar conditions at all other points and severely restrict the nature of the function. The Taylor series (Chapter 5) of f carries this theme much further.

5. *On the role of the complex integral.* How many of the results of Sections 5 through 11 actually mention the complex line integral in their statements? Contrast this with their proofs!

Section 4.12 WHAT'S AHEAD?

Though we have now caught a glimpse of several remarkable properties of analytic functions, much more remains to be seen. It will soon be necessary to construct more technical tools: power series, convergence. Before we embark on this, therefore, let us organize our search by raising two quite different (or so it seems) questions that we have not answered satisfactorily to date. These two questions motivate a great deal of what follows.

First Question: Integration of Nonanalytic Functions

We have said a great deal about $\int_\Gamma f(z)\, dz$ in the case Γ is some loop and f is analytic at all points on and inside Γ. But what if f fails to be analytic at certain points so that the Cauchy Integral Theorem (in particular) does not apply? Such integrals occur frequently in applications.

The result we are headed for here is the so-called Residue Theorem (Chapter 7). Our work on Laurent Series (Chapter 6) will tell us what certain nonanalytic functions look like.

Second Question: Mapping by Analytic Functions

Suppose $w = f(z)$ is analytic in a nice domain Ω in the z-plane. What can we say about the image set $f(\Omega)$ as a subset of the w-plane? Is it open? connected? bounded if Ω is bounded?

Another basic question: In ordinary calculus, if the derivative $f'(x_0)$ is nonzero, then $y = f(x)$ is strictly increasing or decreasing near x_0, and hence the function is one-to-one for x near x_0. Now suppose $w = f(z)$ is complex analytic and $f'(z_0) \neq 0$. Is $w = f(z)$ one-to-one in a small disc about z_0? In general, what does the complex derivative tell us about the geometric behavior of $f(z)$?

A very deep question: Given domains Ω_1, Ω_2 in the z-plane and w-plane, respectively, does there exist a one-to-one analytic mapping $w = f(z)$ of Ω_1 onto Ω_2? As we noted in the Appendix to Section 4.5, this is important in boundary value problems.

We will devote Chapter 8 to these geometric issues.

5

Analytic Functions and
Power Series

Section 5.1 SEQUENCES AND SERIES

5.1.0 Introduction

In Chapter 4 we developed the machinery of complex line integration, used this to prove the Cauchy Integral Theorem and Formula, and at last derived from these a host of marvelous consequences: infinite differentiability of analytic functions, the Maximum Modulus Principle, Liouville's Theorem, and so on. The pattern of the present chapter will be similar. We will first develop the general machinery of sequences and series. It is possible you have seen some of this already. Then we will establish Taylor's and Laurent's Theorems. These will assure us that the functions we are interested in may be written as so-called power series. Thus, for example, we will see that the familiar exponential function may be written

$$\exp z = 1 + z + \frac{z^2}{2!} + \frac{z^3}{3!} + \cdots = \sum_{k=0}^{\infty} \frac{z^k}{k!}$$

Finally, we will gain further insight into the nature of analytic functions by examining the related power series. Be encouraged that this gain in insight will justify our initial technical labors.

Let us first mention a major technical question in an informal way. Suppose we are given a *power series*, an infinite formal sum:

$$a_0 + a_1 z + a_2 z^2 + a_3 z^3 + \cdots.$$

This is an "infinitely long polynomial" in the variable z with given *coefficients* $a_0, a_1, a_2, a_3 \cdots$. What is the result of "plugging in" a particular complex number for z, say, $z = z_1$? While we know what it means to add two or three or any *finite* string of complex numbers, we must make some careful definitions before an *infinite* sum makes sense. Let us turn to this now.

5.1.1 Series of Real Numbers

Complex power series will be discussed in terms of series of complex numbers. In turn, series of complex numbers will be treated in terms of series of real numbers. We will recall here some of the basic facts about real series (essentially one definition and one theorem). All subsequent theory will depend on these.

A *real series* is an infinite formal sum

$$u_0 + u_1 + u_2 + \cdots,$$

with each *term* u_k in $\mathbb{R}$. This series is also denoted

$$\sum_{k=0}^{\infty} u_k \quad \text{or, more simply,} \quad \sum u_k.$$

We say that the series above *converges* to the *sum* u (a real number), provided

$$\lim_{n \to \infty} \sum_{k=0}^{n} u_k = u.$$

Some comments: The numbers

$$s_0 = u_0, s_1 = u_0 + u_1, \ldots, s_n = \sum_{k=0}^{n} u_k, \ldots$$

and the *partial sums* of the series. Thus, the series converges to u if and only if the sequence $s_0, s_1, \ldots, s_n, \ldots$ has limit u; that is, $\lim_{n \to \infty} s_n = u$. This is the standard language.

For example, let us look at the series

$$1 + \frac{1}{2} + \frac{1}{4} + \cdots + \left(\frac{1}{2}\right)^k + \cdots = \sum_{k=0}^{\infty} \left(\frac{1}{2}\right)^k,$$

a special case of the geometric series (of which, more later). The sequence of partial sums begins

$$s_0 = 1, \quad s_1 = 1 + \tfrac{1}{2} = \tfrac{3}{2}, \quad s_2 = 1 + \tfrac{1}{2} + \tfrac{1}{4} = \tfrac{7}{4},$$

and so on. You should convince yourself that the series converges to the sum $u = 2$. Note that the nth term $u_n = (\tfrac{1}{2})^n$ is one-half of the difference $2 - s_{n-1}$.

We have so far neglected to give an exact definition of the limit statement used to define the sum of a series. It is this:

$$\lim_{n \to \infty} \sum_{k=0}^{n} u_k = u$$

if and only if, given any $\varepsilon > 0$, no matter how small, there is an index (subscript) n_ε depending on the given ε with the property that, if $n > n_\varepsilon$, then

$$\left| \sum_{k=0}^{n} u_k - u \right| < \varepsilon.$$

In other words, the partial sums s_n eventually come arbitrarily close to the sum u.

If a series does not converge, it is said to *diverge*. For example, the series

$$1 + 1 + 1 + 1 + 1 + 1 + \cdots$$

diverges to infinity, while the series

$$1 - 1 + 1 - 1 + 1 - \cdots = \sum_{k=0}^{\infty} (-1)^k$$

diverges because its sequence of partial sums oscillates between 1 and 0, thus: $1, 0, 1, 0, 1, \ldots$, and so never tends to a unique limit.

Now we mention what is perhaps *the* crucial theoretical description of convergent series. It is an internal property that every convergent series has and no divergent series has. We say "internal" because the property makes no explicit mention of the sum of the series.

REAL CAUCHY CONVERGENCE CRITERION

The real series $\sum_{k=0}^{\infty} u_k$ is convergent $\Leftrightarrow$ given any $\varepsilon > 0$, there is an index n_ε such that, whenever $n > n_\varepsilon$ and p is any positive integer, then

$$|u_{n+1} + u_{n+2} + \cdots + u_{n+p}| < \varepsilon.$$

We discuss this briefly. First, if the series is convergent to a sum u, then the partial sums s_n and s_{n+p} must both be close to u for sufficiently large n, and hence must be close to each other. But note that the finite sum $u_{n+1} + \cdots + u_{n+p}$ is simply the difference $s_{n+p} - s_n$ and hence should be of small size.

Conversely, if the sums $u_{n+1} + \cdots + u_{n+p}$ are small in absolute value as hypothesized in the theorem, then all pairs of partial sums s_n and s_{n+p} become close for large n. It then follows from the completeness property of the real numbers (roughly, "the real axis has no gaps") that the partial sums must be close to some actual number u, which is proved to be the sum of the series. That is, the series converges. Hence, the Cauchy Convergence Criterion embodies the completeness of the real number system.

At last we may discuss series of complex numbers.

Exercises to Paragraph 5.1.1

1. *Some famous real series.* Recall the following facts from calculus and see if you can prove any at the moment. We will discuss methods for some of these in subsequent sections. We will see statement (e) often.

   (a) The harmonic series $\sum_{n=1}^{\infty} (1/n)$ diverges.

   (b) $\sum (1/n^2)$ converges (to $\pi^2/6$, in fact, though this exact value requires much more work).

   (c) More generally, if p is a real constant, then $\sum (1/n^p)$ converges if and only if $p > 1$.

   (d) The alternating harmonic series $\sum (-1)^n/n$ converges (to $\ln 2$, in fact).

   (e) If $|r| < 1$, then the geometric series $\sum_{n=0}^{\infty} r^n$ converges to $1/(1 - r)$.

2. For those who know some elementary real analysis (or topology of the real line):

   (a) Recall the definition of "Cauchy sequence."

   (b) Recall that the Completeness Axiom of the real line may be stated: "A real sequence converges if and only if it is a Cauchy sequence."

   (c) Verify that the Real Cauchy Convergence Criterion given in this section may be restated: "A real series converges if and only if its sequence of partial sums is a Cauchy sequence."

   (d) What other formulations of the Completeness Axiom do you know?

5.1.2 Series of Complex Numbers

Now we define convergence and establish a Cauchy criterion for series

$$\sum_{k=0}^{\infty} b_k$$

where the b_k are *complex* numbers. We exploit the fact that the complex term b_k decomposes into real and imaginary parts,

$$b_k = u_k + iv_k,$$

whence we write

$$\sum_{k=0}^{\infty} b_k = \left(\sum_{k=0}^{\infty} u_k \right) + i \left(\sum_{k=0}^{\infty} v_k \right).$$

Thus, we say that $\sum_{k=0}^{\infty} b_k$ *converges* to the *sum b*, written

$$\sum_{k=0}^{\infty} b_k = b$$

if and only if $b = u + iv$, where u and v satisfy

$$\sum_{k=0}^{\infty} u_k = u, \qquad \sum_{k=0}^{\infty} v_k = v.$$

Note that if the terms b_k are actually real ($v_k = 0$), then this definition reduces to that for real series.

It is a fact that $\sum b_k = b$ in the above sense if and only if

$$\lim_{n \to \infty} \sum_{k=0}^{n} b_k = b,$$

where the complex limit has a standard ε, n_ε definition. We will omit discussion of this, however, as it is not essential to our purposes.

The theoretical tool that we will use again and again to conclude that a given series converges is the following:

COMPLEX CAUCHY CONVERGENCE CRITERION

The complex series $\sum b_k$ is convergent $\Leftrightarrow$ given any $\varepsilon > 0$ there is an index n_ε such that, whenever $n > n_\varepsilon$ and p is any positive integer, then

$$|b_{n+1} + b_{n+2} + \cdots + b_{n+p}| < \varepsilon.$$

Proof: ($\Rightarrow$) Standard routine: Reduce to real case and use the triangle inequality. Thus, writing $b_k = u_k + iv_k$, we have

$$|b_{n+1} + \cdots + b_{n+p}| = |(u_{n+1} + \cdots + u_{n+p}) + i(v_{n+1} + \cdots + v_{n+p})|$$

$$\leq |u_{n+1} + \cdots + u_{n+p}| + |v_{n+1} + \cdots + v_{n+p}|,$$

since $|i| = 1$. Now we supply the details.

Suppose $\varepsilon > 0$ is given. Since $\sum b_k$ converges, the real series $\sum u_k$, $\sum v_k$ converge. By the real Cauchy criterion, given $\varepsilon' = \varepsilon/2$, there exist n^* and n^{**} (for the u_k's and v_k's, respectively) such that

$$|u_{n+1} + \cdots + u_{n+p}| < \frac{\varepsilon}{2}, \qquad |v_{m+1} + \cdots + v_{m+q}| < \frac{\varepsilon}{2},$$

provided $n > n^*$, $m > n^{**}$ for arbitrary positive p, q. Now we define n_ε to be the larger of n^* and n^{**}. Thus,

$$|u_{n+1} + \cdots + u_{n+p}| + |v_{n+1} + \cdots + v_{n+p}| < \frac{\varepsilon}{2} + \frac{\varepsilon}{2} = \varepsilon,$$

provided $n > n_\varepsilon$. The triangle inequality displayed above yields $|b_{n+1} + \cdots + b_{n+p}| < \varepsilon$, as desired.

($\Leftarrow$) To show $\sum b_k$ converges, we will show that both $\sum u_k$ and $\sum v_k$ converge, via the real Cauchy criterion. To show that $\sum u_k$ satisfies the real Cauchy criterion, let $\varepsilon > 0$ be given. By hypothesis, there is an index n_ε such that

$$|b_{n+1} + \cdots + b_{n+p}| < \varepsilon$$

for $n > n_\varepsilon$. But $u_{n+1} + \cdots + u_{n+p}$ is the real part of the complex number $b_{n+1} + \cdots + b_{n+p}$, and so (look at a triangle!)

$$|u_{n+1} + \cdots + u_{n+p}| \le |b_{n+1} + \cdots + b_{n+p}| < \varepsilon.$$

Thus, n_ε works for $\sum u_k$ and likewise for $\sum v_k$. By the remarks above, we are done.

Comments

1. This is an intrinsic criterion. It makes no mention of the sum of the series in any limit statement.

2. *Interpretation:* If, as n gets sufficiently large, the nth partial sum $b_0 + \cdots + b_n$ tends relentlessly to the sum b, then there can be no portion $b_{n+1} + \cdots + b_{n+p}$ of the difference $b - (b_0 + \cdots + b_n)$ that will be very large in absolute value.

3. We will often combine this with the Comparison Test (coming) to prove convergence of a given series.

Exercises to Paragraph 5.1.2

1. Consider $\sum_{k=1}^{\infty} b_k$, where $b_k = (i)^k/k$.
   (a) Write down the first six terms of this series.
   (b) Compute the first six partial sums $s_1, \ldots, s_6$ of this series.
   (c) Prove that this complex series converges by applying your knowledge of alternating series to its real and imaginary parts (note that this embodies our definitions of convergence).
   (d) Can you determine the actual sum of this series? (Observe, at least, that this can be much more difficult than merely proving that the sum exists!)

2. Concoct a (nonreal) complex series that diverges. *Hint:* Use a divergent real series.

3. Some metric topology of the plane (see the set of exercises to Paragraph 5.1.1 also).
   (a) Recall the definition of Cauchy sequence in the plane (in fact, in any metric space, if you know what that is).
   (b) How is the completeness of the plane formulated in terms of Cauchy sequences? *Hint:* See Exercise 2(b) to Paragraph 5.1.1.
   (c) Verify that the Cauchy criterion of the text states that "A complex series converges if and only if its sequence of (complex) partial sums is a complex Cauchy sequence."

5.1.3 Absolute Convergence

We say that the complex series $\sum b_n$ *converges absolutely* if and only if the series $\sum |b_n|$ of nonnegative real numbers converges. We will be dealing in later sections with many series that converge absolutely.

It is worth noting that it is more likely that a complex series converges than that it converges absolutely. Absolute convergence is "difficult to achieve" because an infinite set of *positive* numbers must add up to a finite sum. There is no hope of subtraction aiding convergence.

For example, you may be familiar with the *alternating harmonic series*

$$1 - \tfrac{1}{2} + \tfrac{1}{3} - \tfrac{1}{4} + \tfrac{1}{5} - \cdots.$$

It is often shown in calculus that this series converges (in fact, its sum is log 2). However, the positive *harmonic series*

$$1 + \tfrac{1}{2} + \tfrac{1}{3} + \tfrac{1}{4} + \tfrac{1}{5} + \cdots$$

does not converge. This sum is infinite. In other words, the alternating harmonic series converges, but does *not* converge absolutely.

Now we will see that the property of absolute convergence is strong enough to guarantee convergence in the ordinary sense.

THEOREM 1

If the complex series $\sum b_n$ converges absolutely, then it converges.

Proof: Note first that $|b_{n+1} + \cdots + b_{n+p}| \le |b_{n+1}| + \cdots + |b_{n+p}|$. Absolute convergence and the Cauchy criterion guarantee that the right-hand term here can be made less than any given $\varepsilon > 0$, provided only that n is chosen large enough. By the inequality noted, the same is true of the left-hand side. But, again by the Cauchy criterion, this is equivalent to convergence of the complex series. Done.

We mention that a series which converges but does not converge absolutely is said to converge *conditionally*.

Exercises to Paragraph 5.1.3

1. Prove that $\sum_{k=1}^{\infty} (i)^k/k$ converges conditionally, not absolutely (see the exercises to Paragraph 5.1.2).

2. Do the following converge absolutely, converge conditionally, or diverge?
   (a) $\sum_{n=0}^{\infty} z^n$ with $|z| < 1$,
   (b) $\sum_{n=1}^{\infty} i/n$,
   (c) $\sum_{k=1}^{\infty} (i)^k/k^2$,
   (d) $\sum_{k=1}^{\infty} (-1)^k i/k$.

3. Proof or counterexample: If the real and imaginary parts of a complex series converge absolutely, then the complex series converges absolutely.

4. True or false?

(a) The Cauchy criterion for complex series was, for us, a consequence of the criterion for real series.

(b) We used the complex Cauchy criterion to prove that absolute convergence implies convergence.

(c) The Cauchy criteria tell us the exact sum of a convergent series.

5.1.4 Further Properties of Series

We mention some facts we will be using soon. First we discuss the terms of a convergent series.

THEOREM 2

If the complex series $\sum b_k$ is convergent, then

    (i) $\lim_{k \to \infty} b_k = 0$,

    (ii) *the terms b_k are bounded; that is, there exists $M > 0$ such that $|b_k| < M$ for all k.*

Proof: (i) This follows from the Cauchy criterion by taking $p = 1$. For, given any $\varepsilon > 0$, we are thereby assured that all but a finite number of terms satisfy $|b_k| < \varepsilon$, which is the same as convergence of the sequence $\{b_k\}$ to zero.

(ii) Since all but a finite number of terms satisfy $|b_k| < \varepsilon$ for *any* positive ε, we can clearly pick a somewhat larger $M \geq \varepsilon > 0$ such that *all* terms satisfy $|b_k| < M$. Done.

We will make considerable use of the following theorem on series of nonnegative real numbers.

COMPARISON TEST

Let $\sum u_k$ and $\sum v_k$ be real series with $0 \leq u_k \leq v_k$. Then

    (i) *if $\sum v_k$ converges, then $\sum u_k$ also converges and $\sum u_k \leq \sum v_k$;*

    (ii) *if $\sum u_k$ diverges, then $\sum v_k$ also diverges.*

You should be able to prove both these statements. Use the Cauchy criterion to prove (i). Note that the divergence of the positive series $\sum u_k$ in (ii) means that this sum is infinite.

Exercises to Paragraph 5.1.4

1. Proof or counterexample: If $\lim_{k \to \infty} b_k = 0$, then $\sum b_k$ converges.

2. Proof or counterexample: If $|b_k| < M$ for all k, then $\sum b_k$ is convergent.

3. Given that $\sum r_k$ converges with real $r_k \geq 0$. Suppose $|b_k| \leq r_k$ for each k. Prove that $\sum b_k$ converges absolutely. This is a basic technique. We'll see it again.

4. Given that $\sum_{k=0}^{\infty} (1/2^k) = 2$ and $|z| < 1/2$, how much can you say about $\sum_{k=0}^{\infty} z^k$ and convergence?

5. Give a full proof of the Comparison Test.

Section 5.2 POWER SERIES

5.2.0 Introduction

Let us begin with the very important *geometric series*

$$1 + z + z^2 + z^3 + \cdots.$$

We ask: "For which values of z does this series converge to a finite complex sum?" Note that $z = 1$ yields divergence: $1 + 1 + 1 + 1 + \cdots$.

Suppose that for a particular value z the series converged to the sum $s(z)$. This would mean that the sequence of partial sums would converge to $s(z)$; that is,

$$\lim_{n \to \infty} (1 + z + z^2 + \cdots + z^n) = s(z).$$

Let us examine the special *polynomial* $1 + z + z^2 + \cdots + z^n$. We note that this polynomial may be realized as a quotient of simpler polynomials,

$$1 + z + z^2 + \cdots + z^n = \frac{1 - z^{n+1}}{1 - z}.$$

To check this, merely observe $(1 + z + \cdots + z^n)(1 - z) = 1 - z^{n+1}$.

Thus, by basic properties of limits, we have

$$s(z) = \lim_{n \to \infty} \frac{1 - z^{n+1}}{1 - z} = \frac{1}{1 - z}\left(1 - \lim_{n \to \infty} z^{n+1}\right).$$

To evaluate this last limit, note first that

$$\lim_{n \to \infty} z^n = \lim_{n \to \infty} z^{n+1}.$$

Also, $|z^n| = |z|^n$. Now suppose $|z| < 1$. Then $|z| > |z|^2 > \cdots$, and in fact,

$$\lim_{n \to \infty} |z^n| = \lim_{n \to \infty} |z|^n = 0 \qquad (|z| < 1).$$

Since $|z^n|$ is approaching zero, we see that z^n is approaching the origin, $\lim_{n \to \infty} z^n = 0$. Our conclusion is

$$s(z) = \lim_{n \to \infty} (1 + z + \cdots + z^n) = \frac{1}{1 - z} \qquad (|z| < 1).$$

As a corollary, $\sum_{k=0}^{\infty} (\frac{1}{2})^k = 2$, a result mentioned earlier.

On the other hand, if $|z| \geq 1$, then $|z^n| \geq 1$. Consequently, the nth term of the series does not approach zero, whence (Theorem 2) the series must diverge.

In summary, we have seen that

(i) the geometric series $1 + z + z^2 + \cdots$ converges for all z in the open unit disc $D(0; 1)$ centered at the origin;

(ii) for z in this disc, that is $|z| < 1$, we have a "closed form" for the sum, namely,

$$1 + z + z^2 + \cdots = \frac{1}{1 - z};$$

(iii) the series diverges for all z with $|z| \geq 1$.

We will devote the remainder of this section to obtaining similar results for arbitrary power series. We will see that they converge inside a disc of a certain radius (possibly zero or infinite) and that they represent analytic functions (say, $f(z) = 1/(1 - z)$) within that disc. Note carefully, however, that the series $1 + z + z^2 + \cdots$ converges only if $|z| < 1$, whereas the function $f(z) = 1/(1 - z)$ is defined in the entire plane except for the single point $z = 1$. Power series representation is a *local* affair!

5.2.1 The Disc of Convergence

We will discuss power series of the form

$$a_0 + a_1(z - z_0) + a_2(z - z_0)^2 + \cdots,$$

which we may write

$$\sum_{k=0}^{\infty} a_k(z - z_0)^k \qquad \text{or simply} \qquad \sum a_k(z - z_0)^k.$$

Here the point z_0 is the *center* and we say that the series is *centered at z_0 or is expanded in powers of $z - z_0$.*

The following is a very important tool. Refer to Figure 5.1.

LEMMA 3

If the series $\sum a_k(z - z_0)^k$ converges at the point z_1, then it converges absolutely for all points z such that $|z - z_0| < |z_1 - z_0|$.

Proof: Since the series $\sum a_k(z_1 - z_0)^k$ of complex numbers converges, we know by Theorem 2 that its terms are bounded,

$$|a_k(z_1 - z_0)^k| < M$$

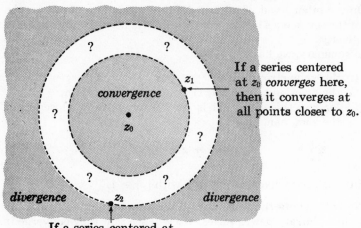

If a series centered at z_0 *converges* here, then it converges at all points closer to z_0.

If a series centered at z_0 *diverges* here, then it diverges at all points farther from z_0.

The meaning of Lemma 3

Figure 5.1

for some $M > 0$ and $k = 0, 1, \ldots$. Now we note that

$$|a_k(z - z_0)^k| = \left| a_k(z_1 - z_0)^k \left(\frac{z - z_0}{z_1 - z_0} \right)^k \right| < M \left| \frac{z - z_0}{z_1 - z_0} \right|^k = Mr^k$$

where we write

$$r = \left| \frac{z - z_0}{z_1 - z_0} \right|.$$

If $|z - z_0| < |z_1 - z_0|$, then $r < 1$ and the geometric series $\sum r^k$ converges. By comparison with $M \sum r^k$, the series $\sum |a_k(z - z_0)^k|$ of positive numbers also converges. Done.

Now we describe the region of convergence of a power series.

THEOREM 4

Given a power series $\sum a_k(z - z_0)^k$ centered at z_0, there is a value R satisfying $0 \leq R \leq \infty$, such that

    (i) *the series converges absolutely for all z in the open disc $D(z_0; R)$ if $R > 0$, or at the point $z = z_0$ only if $R = 0$;*

    (ii) *the series diverges at all points z outside the closed disc $\overline{D}(z_0; R)$, that is, for $|z - z_0| > R$.*

Note: The series may or may not diverge at a point on the circle $|z - z_0| = R$. This is often a more delicate issue.

Proof: If the series converges only for $z = z_0$, let $R = 0$ and we are done. If on the other hand, the series converges at some z_1 different from z_0, then Lemma 3 assures us that it converges in an open disc of radius (at least) $|z_1 - z_0|$. Let us form the union of all open discs centered at z_0 throughout which the series converges. This union is an open disc centered at z_0, and as such has a radius R (possibly infinite, so that the disc of convergence is the entire plane). It is now clear that the series converges at each point of this open disc $D(z_0; R)$. If, on the other hand, $|z - z_0| = R_1 > R$, then the series must diverge at z, or else, by Lemma 3, it would converge in the disc $D(z_0; R_1)$, which is strictly larger than $D(z_0; R)$. This contradicts the construction of $D(z_0; R)$. Done.

The number R whose existence is proved here is termed the *radius of convergence* of the given series. See Figure 5.2. For example, we saw at the beginning of this section that the geometric series has radius of convergence $R = 1$. Any polynomial in z is an entire function and hence converges in the disc of radius $R = \infty$, the entire plane.

Figure   5.2

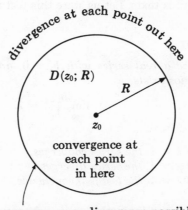

divergence at each point out here

$D(z_0; R)$

R

z_0

convergence at
each point
in here

convergence or divergence possible
at each point of circle

The disc of convergence

Note in Figure 5.2 that the radius R is measured from the point z_0 (which is the "center" of the series).

Suppose the radius of convergence of the series $\sum a_k(z - z_0)^k$ is R, with $0 < R < \infty$. Then the series converges at all points of the open disc $D(z_0; R)$ and also (possibly) at certain points on the rim of the disc, that is, on the circle $C(z_0; R)$. Usually, we are concerned with the behavior of the function defined by the series on the open disc $D(z_0; R)$ rather than at any boundary points. Thus, it is helpful to call $D(z_0; R)$ the *disc of convergence* of the series. If R is infinite, then its disc of convergence is the entire plane.

Warning

A typical series converges on its disc of convergence and at some (possibly none, possibly all) points on the rim of this disc as well. Don't make the mistake of thinking that every series converges only on an open disc (even if we speak of "the *disc* of convergence").

For The Connoisseur

The question of convergence of a series at points on the rim of the disc of convergence is a study in itself. Some elementary phenomena are treated in the exercises.

How to Compute the Radius of Convergence

It is occasionally useful to be able to compute the radius R for an explicitly given power series. We will present a method that works in most cases. It depends on the Ratio Test for convergence, which is proved in most calculus texts. Let us state this test first.

Ratio Test

Let $\sum b_k$ be a real series with $b_k > 0$, and suppose the following limit of ratios exists:

$$\lim_{k \to \infty} \frac{b_{k+1}}{b_k} = \lambda.$$

Then

    (i) *if $\lambda < 1$, the series converges.*
    (ii) *if $\lambda > 1$, the series diverges.*
    (iii) *if $\lambda = 1$, the test is inconclusive.*

We may now obtain the radius of convergence from the following result.

Theorem 5

Given $\sum a_k(z - z_0)^k$, if

$$\lim_{k \to \infty} \left| \frac{a_{k+1}}{a_k} \right| = L$$

exists, then the series has radius of convergence $R = 1/L$ if $L \neq 0$ and $R = \infty$ if $L = 0$. If the limit is infinite, then $R = 0$.

Proof: We form the ratio

$$\left| \frac{a_{k+1}(z - z_0)^{k+1}}{a_k(z - z_0)^k} \right| = \left| \frac{a_{k+1}}{a_k} \right| |z - z_0|.$$

Put

$$\lambda = \lim_{k \to \infty} \left| \frac{a_{k+1}(z - z_0)^{k+1}}{a_k(z - z_0)^k} \right| = L \cdot |z - z_0|.$$

Thus, the λ of the Ratio Test exists for this series if and only if L exists. Now $\lambda < 1$ if and only if $|z - z_0| < 1/L$ so that $R = 1/L$ is the radius of convergence. Done.

Examples

1. Given $1 - (z - 1) + (z - 1)^2 - \cdots = \sum (-1)^k(z - 1)^k$. Thus, $a_k = (-1)^k$ for all k and so $L = 1$. By the theorem, $R = 1$.

We might verify this in another way. We note that

$$\frac{1}{z} = \frac{1}{1 + (z - 1)} = \frac{1}{1 - \{-(z - 1)\}}$$

$$= 1 + \{-(z - 1)\} + \{-(z - 1)\}^2 + \cdots$$

by our work with geometric series. Removing the brackets, we see that $1/z = 1 - (z - 1) + (z - 1)^2 - (z - 1)^3 + \cdots$. Thus, the given series is actually an expansion of the function $f(z) = 1/z$ in powers of $z - 1$. But this function "blows up" at $z = 0$ only, and so we might expect that a disc of convergence centered at $z_0 = 1$ would extend only to the origin. That is, $R = 1$, as found above by the theorem.

2. Given the series

$$1 + z + \frac{z^2}{2!} + \frac{z^3}{3!} + \cdots.$$

Then

$$a_k = \frac{1}{k!}, \qquad \left| \frac{a_{k+1}}{a_k} \right| = \frac{k!}{(k + 1)!} = \frac{1}{k + 1}.$$

Thus, the limit L as k gets large is zero, and so $R = \infty$. This power series converges in the entire plane (to which familiar function? See Exercise 11, next page).

Exercises to Paragraph 5.2.1

1. Compute $\sum_{k=0}^{\infty} z^k$ in the case $z = (1 + i)/2$.
2. Given that $\sum_{k=0}^{\infty} a_k$ is a convergent series of complex numbers. We apply Lemma 3 to the power series $\sum a_k z^k$.
   (a) Argue that $\sum a_k z^k$ converges for $z = 1$.
   (b) Conclude $\sum a_k z^k$ converges if $|z| < 1$.

(c) Can you conclude $\sum |a_k z^k|$ converges if $|z| < 1$?

(d) Can you conclude $\sum |a_k z^k|$ converges if $|z| = 1$?

(e) Can you conclude $\sum a_k z^k$ diverges if $|z| = 1$?

(f) Can you conclude $\sum a_k z^k$ diverges if $|z| > 1$?

3. Given that $\sum a_k(z - i)^k$ converges at $z = 2i$ but diverges at $z = 0$.

   (a) Does the series converge at $z = (1 + i)/2$?

   (b) Does the series converge at $z = \frac{1}{2}$?

   (c) Does the series converge at $z = 1 + i$?

   (d) What is the radius R of convergence of this series?

   (e) Draw a picture of the disc of convergence, if you have not done so already.

4. Given that $\sum a_k$ converges conditionally, what is the radius of convergence of $\sum a_k z^k$?

5. Compute the radius of convergence of $\sum_{k=1}^{\infty} (-1)^k z^k/k$ without using Theorem 5. *Hint:* Exercise 4.

6. Likewise for (a) $\sum z^k/k$; (b) $\sum kz^k$.

7. Construct a series in powers of $z - 1$ which converges for $|z - 1| < 1$ and diverges for $|z - 1| > 1$.

8. Construct a series in powers of z which converges for $|z| < \frac{1}{2}$ and diverges for $|z| > \frac{1}{2}$. *Hint:* $|z| < \frac{1}{2}$ if and only if $|\zeta| < 1$, where $\zeta = 2z$.

9. *A series that converges precisely on the closed disc* $|z| \leq 1$.

   (a) Prove $\sum z^k/k^2$ converges for $|z| \leq 1$.

   (b) In fact, the convergence is absolute at all points of the closed unit disc.

   (c) Prove that the series diverges for $z = 1 + \delta$ for any real $\delta > 0$. *Hint:* Look at $(1 + \delta)^k$.

   (d) Conclude that $\sum z^k/k^2$ converges if and only if $|z| \leq 1$.

10. *A series that converges precisely on the open disc* $|z| < 1$.

    (a) Given the geometric series $\sum_{k=0}^{\infty} z^k$, recall that the nth partial sum is $s_n(z) = (1 - z^{n+1})/(1 - z)$.

    (b) Fix z with $|z| = 1$. How does z^{n+1} move in the plane as $n \to \infty$?

    (c) Prove that if $|z| \geq 1$, then there exists no complex s such that $(1 - z^{n+1})/(1 - z)$ approaches s.

    (d) Conclude that the geometric series converges precisely for $|z| < 1$ (contrast Exercise 9).

11. The series $\sum z^k/k!$ sums to e^z for all z. Proof?

12. *True or false?*

    (a) The power series $\sum a_k(z - z_0)^k$ converges on a set of points z which equals either the single point z_0 or some open disc of finite radius or the entire plane, but no other type of set.

    (b) If $\sum a_k z^k$ is convergent for $z = 1$, then it is convergent for all z with $|z| < 1$.

    (c) If $\sum a_k z^k$ is convergent for $z = 1$, then it is convergent at all points on the circle $|z| = 1$.

    (d) If $\sum a_k z^k$ is convergent for $z = 1$, then it is convergent for $z = -1$.

    (e) Some power series converge at all points of some open disc and at certain points on the rim of this disc as well, and at no other points.

    (f) A power series that converges on a certain set of points z defines a function on that set.

    (g) There are power series that converge on a set of points which is exactly equal to the closed disc $|z| \leq 1$.

(h) If the series $\sum a_k z^k$ diverges at $z = i$, then it diverges at $z = 1 + i$ also.

(i) If the series $\sum b_k(z - i)^k$ diverges at $z = 1$, then it diverges at $z = 0$ also.

5.2.2 Uniform Convergence of Power Series

We are led to this by asking for the properties of the function defined by a convergent power series. Such a series may be thought of as the limit of an "infinite sequence of polynomials." Hence, we begin by discussing sequences of functions and their limits.

Let $f_0(z), f_1(z), f_2(z), \ldots$ be a sequence of complex-valued functions, each defined for all z in the one domain Ω. We denote such a sequence $\{f_n\}$. We do not yet require that the functions be analytic or even continuous.

The limit of a sequence of functions (if it exists) should be a function. We define a function

$$f = \lim_{n \to \infty} f_n$$

by telling what $f(z_1)$ is for each z_1 in Ω. Namely, $f(z_1) = \lim_{n \to \infty} f_n(z_1)$. Note that $\{f_n(z_1)\}$ is a sequence of complex numbers, not functions, once we choose z_1. If $\lim_{n \to \infty} f_n(z_1)$ converges for each fixed z_1 in Ω, then the limit function f is defined on all Ω.

For example, let $f_n(z) = 1 + z + \cdots + z^n$. If Ω is the unit disc $D(0; 1)$, then we know that

$$f(z) = \lim_{n \to \infty} f_n(z) = \frac{1}{1 - z}.$$

Note that it is essential that we restrict ourselves to the unit disc for this limit statement to be true.

Now on to uniform convergence. The definition $f(z_1) = \lim_{n \to \infty} f_n(z_1)$ involved a separate limiting process at each $z = z_1$. We say that $\{f_n\}$ converges to f *uniformly on* a subset S of Ω if and only if, given any $\varepsilon > 0$, there is an index n_ε such that if $n > n_\varepsilon$, then $|f(z) - f_n(z)| < \varepsilon$ for every z in S. The crux here is that the index n_ε depends only on the given sequence $\{f_n\}$ and on the given $\varepsilon > 0$, *but not on the point z.* The same n_ε works for all points z in S.

In this situation we say that f is the *uniform limit* of $\{f_n\}$, or that f is *uniformly approximated* by $\{f_n\}$ (always on some specified set such as S, of course).

Culture: It is an important theme of analysis that if each function f_n has a nice property and if the f_n converge uniformly (crucial!) to a limit f, then it is reasonable to hope that f also enjoys this property. The next two theorems exemplify this theme.

THEOREM 6

The uniform limit of continuous functions is continuous (all defined on the same set S, as usual).

Proof: Let $\lim_{n\to\infty} f_n = f$ uniformly on S. To show f is continuous at z_1 in S, suppose $\varepsilon > 0$ is given. What we eventually wish to conclude is that $|f(z) - f(z_1)| < \varepsilon$ for certain z. Hence, we use the old trick of rewriting our absolute value:

$$|f(z) - f(z_1)| = |f(z) - f_n(z) + f_n(z) - f_n(z_1) + f_n(z_1) - f(z_1)|$$

$$\leq |f(z) - f_n(z)| + |f_n(z) - f_n(z_1)| + |f_n(z_1) - f(z_1)|.$$

Now we show that each term here can be made less than $\varepsilon/3$ by appropriately restricting n and z. Note that this will prove the theorem.

Given $\varepsilon/3$, uniform convergence assures us there is an index n^* such that if $n > n^*$ then both

$$|f(z) - f_n(z)| < \frac{\varepsilon}{3} \quad \text{and} \quad |f_n(z_1) - f(z_1)| < \frac{\varepsilon}{3}.$$

Having chosen $n > n^*$, we use the continuity of f_n to select a $\delta > 0$ with the property that

$$|f_n(z) - f_n(z_1)| < \frac{\varepsilon}{3}$$

whenever $|z - z_1| < \delta$. But choice of this particular δ assures us that $|f(z) - f(z_1)| < \varepsilon$. That is,

$$\lim_{z\to z_1} f(z) = f(z_1),$$

so f is continuous at z_1. Done.

We will use this result soon to prove the continuity of power series functions. In the meantime, however, let's see how uniform convergence allows us to interchange two of the limiting processes of analysis, namely, convergence of sequences and line integration. This will be applied later on via our integral representation theorems.

THEOREM 7

Let $\{f_n\}$ be a sequence of functions continuous on a curve Γ, and suppose $f = \lim_{n\to\infty} f_n$ uniformly on Γ. Then

$$\lim_{n\to\infty} \int_\Gamma f_n(z)\, dz = \int_\Gamma \lim_{n\to\infty} f_n(z)\, dz = \int f(z)\, dz.$$

Proof: Since the difference of two integrals is the integral of the difference of the integrands, it suffices to prove that

$$\lim_{n \to \infty} \left| \int_\Gamma (f_n(z) - f(z))\, dz \right| = 0.$$

By our standard inequality for the integral, we know that

$$\left| \int_\Gamma (f_n(z) - f(z))\, dz \right| < ML,$$

where L is the length of Γ and $M > |f_n(z) - f(z)|$ for all z on Γ. Now uniform convergence allows us to choose M arbitrarily small by taking the index n large enough. This can happen only if the absolute value of the integral has zero limit. Done.

Now we will see how the above discussion applies to power series. Suppose the series $\sum a_k(z - z_0)^k$ converges in a disc $D(z_0; R)$, where as usual we allow the radius R to be infinite. Thus, the power series defines a complex-valued function $f(z)$ in $D(z_0; R)$:

$$f(z) = \sum a_k(z - z_0)^k.$$

As usual we define the nth partial sum of the series by

$$s_n(z) = a_0 + a_1(z - z_0) + \cdots + a_n(z - z_0)^n.$$

Thus, the sequence $\{s_n(z)\}$ of polynomials converges to the function $f(z)$ on $D(z_0; R)$. Now we may ask whether this convergence is in fact uniform. The very important answer is this:

Theorem 8

Let $f(z) = \sum a_k(z - z_0)^k$ in $D(z_0; R)$. Then on each smaller closed disc $\bar{D}(z_0; \rho)$ with $\rho < R$, the sequence $\{s_n\}$ of partial sums converges uniformly to the function f.

Note: We say that a power series converges *uniformly on closed discs* inside the disc of convergence. No explicit mention of the partial sums, since we are mainly interested in the series.

In order to prove Theorem 8 we will first establish a general lemma that is useful in proving uniform convergence of series. It is this:

Weierstrass M-Test

Let $f_0(z) + f_1(z) + \cdots$ be a series whose terms are functions defined on a plane set S. Suppose that $M_0 + M_1 + \cdots$ is a series with positive terms and that M_n is a number such that $|f_n(z)| \leq M_n$ for all z in S

and each $n = 0, 1, \ldots$. If the series $M_0 + M_1 + \cdots$ converges, then the series $f_0(z) + f_1(z) + \cdots$ converges uniformly and absolutely throughout S.

Note: The convergence is uniform because the "dominating" series $M_0 + M_1 + \cdots$ is clearly independent of z.

Proof: Let z be any point of S. We have by hypothesis

$$|f_{n+1}(z) + \cdots + f_{n+k}(z)| \le |f_{n+1}(z)| + \cdots + |f_{n+k}(z)|$$

$$\le M_{n+1} + \cdots + M_{n+k}.$$

Given any $\varepsilon > 0$, convergence of the series of M's yields an index n_ε such that if $n > n_\varepsilon$ and $k > 0$, then the right-hand side of this equality is less than ε. Since z is arbitrary, the convergence is uniform (and obviously absolute). Done.

Obviously, the M-test could not be used to prove uniform convergence of a nonabsolutely convergent series. However, we have already seen that power series do converge absolutely inside their disc of convergence, so all is well.

Proof of Theorem 8: This is a reworking of the proof of Lemma 3. We choose a point z_1 in the large disc but not in the small one:

$$\rho < |z_1 - z_0| < R.$$

Since the series converges at $z = z_1$, its terms are bounded there by some $M > 0$:

$$|a_k(z_1 - z_0)^k| < M.$$

Now let z be in the closed disc, $|z - z_0| \le \rho$. We have

$$|a_k(z - z_0)^k| = |a_k(z_1 - z_0)^k| \left| \frac{z - z_0}{z_1 - z_0} \right|^k < M_k = Mr^k,$$

provided $r = \rho/|z_1 - z_0|$. Thus, $r < 1$. Since $M \sum r^k$ is a convergent geometric series, the M-test gives uniform convergence. Done.

COROLLARY 9

Let $f(z) = \sum a_k(z - z_0)^k$ be a function given by a convergent power series on the disc $D(z_0; R)$. Then f is continuous at each point of $D(z_0; R)$.

Proof: Use Theorems 6 and 7 and the fact that the partial sums of the series are polynomials and hence continuous. Done.

We will soon show that the power series function $f(z)$ is not merely continuous but actually analytic.

Exercises to Paragraph 5.2.2

1. Let $f_n(z) = z^n$ for $n = 0, 1, 2, \ldots$.
   (a) Fix z_1 satisfying $|z_1| < 1$. What is $\lim_{n \to \infty} f_n(z_1)$?
   (b) Describe the function $f = \lim_{n \to \infty} f_n$ on the unit disc $|z| < 1$. Can you give an explicit formula for f?
   (c) Prove that $\{f_n\}$ converges to f uniformly on any closed disc $|z| \leq \rho < 1$ strictly inside the open unit disc. Proceed as follows: Given ρ and given $\varepsilon > 0$, show there exists n_ε such that if $n > n_\varepsilon$, then $\rho^n < \varepsilon$. Conclude $|z|^n < \varepsilon$ for $|z| \leq \rho$.
   (d) Show that the convergence of $\{f_n\}$ to f is *not* uniform on the full open disc $|z| < 1$. Proceed as follows: Given any $\varepsilon > 0$ and any index n, prove there exists a point z (close to the unit circle) such that $|z^n| > \varepsilon$.

2. Let $f = \lim_n f_n$ uniformly, with all functions defined on a plane set S. True or false?
   (a) If each f_n is continuous, then f is continuous.
   (b) If all but a finite number of the f_n are continuous, then f is continuous.
   (c) If f is continuous, then each f_n is continuous.
   (d) If each f_n is a polynomial in z, then f is likewise.
   (e) If each f_n is a polynomial of degree ≤ 3, then f is likewise.

3. *Uniform convergence of real functions.* Suppose $f_n = \lim_{n \to \infty} f_n$ where $f_n(x)$ and $f(x)$ are real-valued functions defined on a subset (S) of the x-axis. Convince yourself that this has the following pictorial interpretation: Given any $\varepsilon > 0$, then the graphs of all but a finite number of the f_n lie in a "belt" of width 2ε drawn about the graph $y = f(x)$ of the limit function. This belt is the set
$$\{(x, y) \mid x \in S, f(x) - \varepsilon < y < f(x) + \varepsilon\}.$$

4. Let the complex function f be the uniform limit of f_n on a plane set S. Prove that if the f_n are each bounded on S, $|f_n(z)| < B_n$ for all z in S, then $|f(z)|$ is bounded in S. This is false if the limit is not uniform!

5. (a) The polynomials $f_n(z) = 1 + z + \cdots + z_n$ converge to which rational function $f(z)$ on the open disc $|z| < 1$?
   (b) Prove that this convergence is not uniform on the full open disc $|z| < 1$. *Hint:* Use Exercise 4.
   (c) Prove directly (do not appeal to Theorem 8) that $\lim f_n = f$ is uniform on closed discs $|z| \leq \rho < 1$. *Hint:* Write $f_n(z) = (1 - z^{n+1})/(1 - z)$ and look at $|f_n(z) - f(z)|$.
   (d) Does $f(z) = \lim f_n(z)$ for $|z| > 1$?
   (e) Compute $\int_{C(1;1)} dz/(1 - z)$. *Hint:* Chapter 4.
   (f) Compute $\lim_{n \to \infty} \int_{C(1;1)} (1 + z + \cdots + z^n)\, dz$.
   (g) In view of Theorem 7, why are the answers in (e) and (f) different?

6. Given that $\sum_{k=0}^{\infty} z^k/k!$ converges to e^z for all z, compute
$$\lim_{n \to \infty} \int_a^b \left[1 + z + \left(\frac{z^2}{2!} \right) + \cdots + \left(\frac{z^n}{n!} \right) \right] dz.$$

Note that each integral is independent of the path from a to b.

7. Uniform convergence was initially defined for sequences of functions. How did we relate it to power series? What does it mean to say that a power series converges uniformly on some plane set?

8. *Topics for meditation*
   (a) The M-test is a variant of the Comparison Test.
   (b) Roughly speaking, uniform convergence of power series on closed discs was established by comparing the series to a certain geometric series.
   (c) Suppose $\lim_{n \to \infty} f_n = f$ on a set S. What reasonable extra conditions (on $\{f_n\}$, S, etc.) will guarantee that the convergence is in fact uniform? Can you discover a useful theorem?

9. *A Cauchy criterion for uniform convergence of* $\{g_n\}$. Prove the following internal characterization: The sequence $\{g_n\}$ converges uniformly on S to its limit if and only if, given any $\varepsilon > 0$, there exists n_ε such that $n, m > n_\varepsilon$ implies $|g_n(z) - g_m(z)| < \varepsilon$ for all z in S.
   Note that we essentially used this in our proof of the M-test, taking $g_n(z) = f_0(z) + \cdots + f_n(z)$.

10. (a) Let $\{f_n\}$ be a sequence of entire analytic functions which converges uniformly on $\mathbb{C}$ to the zero function. Prove that all but a finite number of the f_n are identically constant.
    (b) Show by example that (a) is false if we replace $\mathbb{C}$ by a disc $|z| < \rho$.

5.2.3 Uniform Convergence and Analyticity

We now know that a power series is the uniform limit of a sequence of polynomials, namely, its partial sums. These polynomials are each analytic. The next result will enable us to conclude the analyticity of the function defined by the power series.

THEOREM 10

Let $\{f_n\}$ be a sequence of analytic functions all defined on a domain Ω. Let $f(z) = \lim_{n \to \infty} f_n(z)$ uniformly on Ω. Then f is also analytic on Ω.

Proof: We will use Morera's Theorem to conclude the analyticity of f. By Theorem 7,

$$\lim_{n \to \infty} \int_\Gamma f_n(z)\, dz = \int_\Gamma f(z)\, dz$$

where Γ is any closed curve contained in any disc inside Ω. Since $f_n(z)$ is analytic, its contour integral around Γ vanishes (why?). Thus the integral of $f(z)$ around Γ vanishes. Since Γ was typical, Morera's Theorem gives the result. Done.

COROLLARY 11

A power series defines an analytic function at each point inside its open disc of convergence.

Proof: Every point of the disc of convergence (open!) is interior to a closed subdisc. We leave the rest to you. Done.

A Delicate Point

The power series might converge at some point on the rim of its disc of convergence. It does not always follow, however, that the function represented by the convergent series is analytic at this point on the rim. Points on the rim must be handled separately. See the Exercises for various phenomena on the rim.

Power Series and Derivatives

Our original definition of complex analyticity involved existence and continuity of the derivative. Hence, we are led to ask about the derivative of a power series. Given the series

$$a_0 + a_1(z - z_0) + a_2(z - z_0)^2 + \ldots,$$

we may differentiate it term-by-term in a formal way, obtaining a power series

$$a_1 + 2a_2(z - z_0) + 3a_3(z - z_0)^2 + \ldots.$$

Now we ask, "If the first series converges for $z = z_1$, does the second converge there also?" And here is a somewhat more subtle question: "If the first series represents an analytic function $f(z)$, does the second series converge to $f'(z)$?" To answer these questions, let us first consider the following general theorem.

THEOREM 12

Let the sequence $\{f_n\}$ of functions analytic on the domain Ω converge uniformly on each closed subdisc of Ω to the analytic function f. Then the sequence $\{f'_n\}$ of derivatives converges uniformly on each closed subdisc of Ω to the derivative f'.

Proof: Let $\overline{D}$ be any closed subdisc of Ω. We must show that for any $\varepsilon > 0$, there is an index n_ε such that if $n > n_\varepsilon$, then $|f'(\zeta) - f'_n(\zeta)| < \varepsilon$ for all ζ in $\overline{D}$.

Since we have some information about $f(z) - f_n(z)$, we use the Integral Formula,

$$f'(\zeta) - f'_n(\zeta) = \frac{1}{2\pi i} \int_\Gamma \frac{f(z) - f_n(z)}{(z - \zeta)^2}\, dz,$$

which immediately yields the standard estimate

$$|f'(\zeta) - f'_n(\zeta)| \le \frac{1}{2\pi} \max_{z \in \Gamma} \frac{|f(z) - f_n(z)|}{|z - \zeta|^2} \cdot \text{length } \Gamma.$$

Here, Γ is a circle in Ω, which has the same center but a slightly larger radius than $\overline{D}$. In fact, let $d > 0$ be the smallest distance from Γ to $\overline{D}$; that is,

$$d = \text{radius } \Gamma - \text{radius } \overline{D}.$$

Now let $\varepsilon^* = \varepsilon 2\pi \, d^2/\text{length } \Gamma$. Since $\{f_n\}$ converges uniformly on the closed subdisc bounded by Γ to its limit f, we know there is an index n^* such that $n > n^*$ implies $|f(z) - f_n(z)| < \varepsilon^*$ for all z on Γ. Thus the standard estimate above assures us that if $n > n^*$, then

$$|f'(\zeta) - f'_n(\zeta)| < \frac{1}{2\pi} \frac{\varepsilon^*}{d^2} \cdot \text{length } \Gamma = \varepsilon$$

for every ζ in $\bar{D}$. Thus, if we define n_ε to be n^*, we have satisfied the requirement for uniform convergence on $\bar{D}$. Done.

Now we may answer our question about term-by-term differentiation of power series.

COROLLARY 13

Let $f(z) = \sum a_k(z - z_0)^k$ have disc of convergence $D(z_0; R)$. Then the differentiated power series $\sum k a_k(z - z_0)^{k-1}$ converges to the derivative $f'(z)$ and has the same disc of convergence $D(z_0; R)$.

Proof: Because $f(z)$ is an analytic function, we know first that $f'(z)$ is defined on $D(z_0; R)$. Now we consider the partial sums

$$a_1 + 2a_2(z - z_0) + \cdots + na_n(z - z_0)^{n-1}.$$

On the one hand, these are the derivatives of the partial sums of the power series that defined f. By Theorem 12, these derivatives converge to $f'(z)$ on $\Omega = D(z_0; R)$. On the other hand, these derivatives clearly approach the series $\sum k a_k(z - z_0)^{k-1}$, which thereby equals $f'(z)$ and converges on $D(z_0; R)$.

To see that the derived series does not converge in any open disc larger than $D(z_0; R)$, let z_1 be some point such that $|z_1 - z_0| > R$. Now let k be an integer larger than $|z_1 - z_0|$. We have

$$|a_k(z_1 - z_0)^k| < |k a_k(z_1 - z_0)^{k-1}|.$$

Since the series obtained from terms on the left diverges, so does the series with terms given by the right-hand side. Done.

Moral of Section 5.2 Convergent power series yield analytic functions.

In the next section we will learn that analytic functions yield convergent power series!

Exercises to Paragraph 5.2.3

1. *On Theorem 10.* True or false?
   (a) If f is the uniform limit of any sequence of polynomials (not necessarily the partial sums of a power series) defined on a domain Ω, then f is analytic in Ω.

(b) If $f = \lim f_n$, with each f_n analytic on a domain Ω, and if the convergence is uniform on every closed subdisc of Ω, then f is analytic on Ω.

(c) If the series $f = \sum g_n$ is uniformly convergent in a domain Ω, and if each g_n is analytic in Ω, then f is analytic in Ω.

2. *A different proof of Theorem 10.* Given that f is the uniform limit of analytic functions f_n on Ω, we will show that f' exists and is continuous.

(a) Prove that $\{f'_n\}$ converges uniformly on closed subdiscs Ω to some limit (call it g). *Hint:* Write f'_n in terms of f_n by means of an integral formula and show that the sequence of integrals satisfies a uniform Cauchy criterion (all this on a closed subdisc of Ω).

(b) Observe that g is continuous on Ω.

(c) Prove that g is the derivative of f on Ω. *Hint:* Fundamental Theorem.

3. *A third proof of Theorem 10*

(a) In the difference quotient $(f(z) - f(z_0))/(z - z_0)$, replace $f(z)$ by $\lim_{n \to \infty} f_n(z)$, etc., so that $f'(z_0)$ is expressed as a double limit.

(b) Argue that uniform convergence permits the interchange of $\lim_{z \to z_0}$ and $\lim_{n \to \infty}$ in (a). This is the point of uniform convergence.

(c) Deduce that $f'(z_0)$ exists and equals $\lim_{n \to \infty} f'_n(z_0)$.

(d) Conclude f' is continuous, whence f is analytic.

4. *A fourth proof of Theorem 10*

(a) Show that $f(z) = (2\pi i)^{-1} \int_\Gamma f(\zeta)(\zeta - z)^{-1} \, d\zeta$ by writing $f_n(z)$ in terms of the Cauchy Integral Formula and then letting n approach ∞. Verify details.

(b) Differentiate under the integral sign to prove analyticity.

5. Invent another proof of Theorem 10.

6. Theorem 10 implies that the uniform limit of complex polynomials on a plane domain Ω must be analytic (and hence a very nice function). Contrast this with the following two facts true of real functions.

    (i) *Weierstrass Approximation Theorem.* Every real function $y = f(x)$ continuous on an interval $a \le x \le b$ is the uniform limit of real polynomials on that interval.

    (ii) There exist many continuous real functions that are not differentiable at any point.

Granted (i) and (ii), answer the following (true or false?):

(a) The uniform limit of complex polynomials on a plane domain Ω is continuously complex differentiable.

(b) The uniform limit of real polynomials on a real interval is continuously real differentiable.

(c) The Weierstrass Approximation Theorem fails spectacularly for continuous complex functions.

(d) The Weierstrass Approximation Theorem for real continuous functions follows from Theorem 10 by restricting things to the x-axis.

(e) A real power series $\sum a_k x^k$ convergent for $|x| < R$ is the uniform limit of its partial sums on subintervals $|x| \le r < R$.

(f) Every continuous real $f(x)$ has a convergent real power series expansion.

7. *Real analytic functions*

(a) Let $\sum a_k(x - x_0)^k$, with a_k, x, x_0 real, converge at $x = x_1 \ne x_0$. Argue that the complexified series $\sum a_k(z - x_0)^k$ converges (at least) for all z satisfying $|z - x_0| < |x_1 - x_0|$.

(b) Deduce that the real series converges on an interval symmetric with respect to the center x_0.

(c) Deduce that the function $\sum a_k(x - x_0)^k$ is the uniform limit of real polynomials and is infinitely (real) differentiable on an open interval of convergence.

(d) The real function $f(x) = 1/(1 + x^2)$ is well behaved for all real x. Why does its power series $1 - x^2 + x^4 - \cdots$ converge on no open interval larger than $|x| < 1$?

8. Given that $f(1) = 0$ and $f'(z) = \sum_{k=0}^{\infty} (-1)^k(z - 1)^k$.

(a) Expand $f(z)$ in powers of $z - 1$.

(b) The function $f(z)$ is famous. What is it?

9. *Producing analytic functions*

(a) Let $\sum a_k$ be any convergent series of complex numbers. Argue that $\sum a_k z^k$ converges in the disc $|z| < 1$ (at least) and hence yields an analytic function there.

(b) Think of some ways of producing convergent series $\sum a_k$ of scalars.

(c) How would you obtain convergent series $\sum a_k(z - z_0)^k$ with z_0 arbitrary?

(d) Name some other ways of generating analytic functions.

10. *Why closed subdiscs?* Here is an example of a series that converges uniformly on $|z| < 1$, but whose derivative does not converge uniformly on $|z| < 1$.

(a) Show that

$$f(z) = \sum_{k=1}^{\infty} \frac{z^k}{k^2}$$

converges uniformly on $|z| < 1$ (in fact, on $|z| \leq 1$). *Hint: M*-test.

(b) Show that

$$f'(z) = \sum_{k=1}^{\infty} \frac{z^{k-1}}{k}$$

does not converge uniformly on $|z| < 1$. *Hint:* It suffices to show that the *real* series $f'(x)$ does not converge uniformly on the *real* open interval $-1 < x < 1$. Note that the series diverges at $z = x = 1$. Use this to show that, given any index N, no matter how large, the "tail end" $\sum_N^{\infty} (x^{k-1}/k)$ can be made larger than any given number ε by choosing x sufficiently close to 1. Thus, convergence is not uniform in x.

(c) Prove that the series in (b) does converge uniformly in z, provided $|z| \leq r < 1$. Note that in this case we cannot choose x arbitrarily close to 1, as we did in (b).

Section 5.3 ANALYTIC FUNCTIONS YIELD POWER SERIES

5.3.0 Introduction

In this section we will begin with an analytic function (contrast Section 5.2, where we began with a power series). We have three objectives:

1. Proof that $f(z)$ analytic in a neighborhood of z_0 implies a representation $f(z) = a_0 + a_1(z - z_0) + a_2(z - z_0)^2 + \cdots$.

2. Interpretation of the coefficients $a_0, a_1, \ldots$ in terms of f.

3. Methods for obtaining the power series (that is, the a_k's) for an explicit function f.

5.3.1 The Coefficients

It is perhaps simplest to deal with objective (2) first. Hence, we will begin by supposing that f does have a power series representation.

THEOREM 14

Suppose the analytic function $f(z)$ has a power series representation

$$f(z) = a_0 + a_1(z - z_0) + a_2(z - z_0)^2 + \cdots$$

valid in some open disc centered at z_0. Then the coefficients are given by

$$a_k = \frac{f^{(k)}(z_0)}{k!} \qquad (k = 0, 1, 2, \ldots).$$

Note: It follows that f is represented by only one power series centered at z_0, namely, its *Taylor series*:

$$f(z_0) + f'(z_0)(z - z_0) + \cdots + \frac{f^{(k)}(z_0)}{k!} (z - z_0)^k + \cdots.$$

Here, of course, $f^{(0)} = f, f^{(1)} = f'$, etc.

Proof: Let $z = z_0$ in the equation $f(z) = a_0 + a_1(z - z_0) + \cdots$ to justify $a_0 = f(z_0)$. The terms involving $a_1, a_2, \cdots$ all vanish.

To isolate the unknown a_1, differentiate both sides of this equation:

$$f'(z) = a_1 + 2a_2(z - z_0) + \cdots.$$

Again, letting $z = z_0$ yields $a_1 = f'(z_0)$.

Continuing in the fashion, we obtain in general

$$f^{(k)}(z) = (1 \times 2 \times \cdots \times k)a_k + (2 \times 3 \times \cdots \times (k + 1))a_{k+1}(z - z_0) + \cdots.$$

Plugging in z_0 for z yields $(k!)a_k = f^{(k)}(z_0)$. Done.

Example

Let us suppose that $f(z) = e^z$ does have a power series expansion centered at $z_0 = 0$ (it does). Thus,

$$e^z = a_0 + a_1z + a_2z^2 + \cdots.$$

How do we obtain the coefficients a_k?

By the theorem,

$$a_k = \frac{f^{(k)}(0)}{k!}.$$

Since $f^{(k)}(z) = e^z$ for all k, and since $e^0 = 1$, we conclude $a_k = 1/k!$, and thus we obtain the famous expansion (see Paragraph 5.1.0)

$$e^z = 1 + z + \frac{z^2}{2!} + \frac{z^3}{3!} + \cdots.$$

Exercises to Paragraph 5.3.1

1. Compute the general form of the derivative $f^{(k)}(z_0)$ for the given function $f(z)$ and center z_0, and thence the general coefficient a_k of the Taylor expansion $f(z) = \sum a_k(z - z_0)^k$.
   (a) $f(z) = e^{2z}, z_0 = 0$;
   (b) $f(z) = \sin z, z_0 = 0$;
   (c) $f(z) = z, z_0 = 3$;
   (d) $f(z) = e^z, z_0 = 1$;
   (e) $f(z) = z^2, z_0 = i$;
   (f) $f(z) = \sin z, z_0 = \pi/2$.

2. In Section 5.2 we argued algebraically that $f(z) = 1/(1 - z)$ is represented by the Maclaurin series $1 + z + z^2 + \cdots$ for $|z| < 1$. Obtain this series now by using the formula for a_k of Theorem 14.

3. Suppose $f(z) = 5$ identically in an open disc about z_0. Is it possible that $f(z) = 5 + a_1(z - z_0) + a_2(z - z_0)^2 + \cdots$ with some $a_k \neq 0, k \geq 1$?

4. Justify the term-by-term differentiation of power series in the proof of Theorem 14. *Hint:* See Section 5.2.

5. (a) *Some algebra of power series.* Verify that we add, subtract, multiply, and divide power series (same center z_0) as if they were polynomials. Thus,

$$\sum a_k(z - z_0)^k \pm \sum b_k(z - z_0)^k = \sum (a_k \pm b_k)(z - z_0)^k,$$

$$\sum a_k(z - z_0)^k \sum b_k(z - z_0)^k = \sum_k \left\{ \sum_{j=0}^{k} a_j b_{k-j} \right\} (z - z_0)^k,$$

$$\frac{\sum a_k(z - z_0)^k}{\sum b_k(z - z_0)^k} = \left(\frac{a_0}{b_0} \right) + \left\{ \left(\frac{a_1}{b_0} \right) - \left(\frac{a_0 b_1}{b_0^2} \right) \right\} (z - z_0) + \cdots;$$

with $b_0 \neq 0$ in the last equation.

   It may be necessary to invoke absolute convergence here to justify certain rearrangements.

   (b) Suppose now that $f(z)$, $g(z)$ are analytic at z_0 with Taylor expansions $\sum a_k(z - z_0)^k$, $\sum b_k(z - z_0)^k$, respectively. Verify that the functions $f(z) \pm g(z), f(z)g(z), f(z)/g(z)$ have Taylor expansions equal to the sum, difference, product, and quotient, respectively, of the given expansions.

6. Suppose we had *defined* the exponential function by its power series, $\exp z = \sum_{k=0}^{\infty} z^k/k!$ Using series methods (see Section 5.2), how would you prove
   (a) $\exp z$ is entire analytic,
   (b) $\exp z$ is its own derivative,
   (c) $\exp z$ has period $2\pi i$?
   Which definition do you prefer?

5.3.2 The Taylor Expansion

Now we will see that the hypothesis of Theorem 14 actually does hold, that analytic functions may be represented by power series.

THEOREM 15

Let f be analytic in a domain Ω, and let z_0 be a point of Ω. Let $D(z_0; \rho)$ be the largest open disc centered at z_0 and contained entirely inside Ω. Then, for z in $D(z_0; \rho)$, we have the representation

$$f(z) = \sum_{k=0}^{\infty} \frac{f^{(k)}(z_0)}{k!} (z - z_0)^k.$$

Note: The series here is called the *Taylor series for f at z_0* (or *Maclaurin series* in the case $z_0 = 0$). It clearly depends on z_0. Also, it is possible that the series converges on some' disc larger than $D(z_0; \rho)$, since in particular Ω was not specified to be the "maximal" domain of definition for f.

Proof: Given z_0, hence $D(z_0; \rho)$ and a point z in this disc, we write by the Cauchy Integral Formula:

$$f(z) = \frac{1}{2\pi i} \int_{\Gamma} \frac{f(\zeta)}{\zeta - z} \, d\zeta.$$

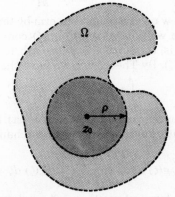

Figure 5.3

Here, Γ is a circle $C(z_0; \rho_1)$ centered at z_0 with $|z - z_0| < \rho_1 < \rho$. Thus, z is inside Γ and Γ is inside Ω.

Now we manipulate the integrand here. We have

$$\frac{1}{\zeta - z} = \frac{1}{\zeta - z_0 - (z - z_0)} = \frac{1}{\zeta - z_0} \cdot \frac{1}{1 - ((z - z_0)/(\zeta - z_0))}.$$

Now note that $|(z - z_0)/(\zeta - z_0)| < 1$. Thus we have a geometric series

$$\frac{1}{1 - ((z - z_0)/(\zeta - z_0))} = \sum_{k=0}^{\infty} \left(\frac{z - z_0}{\zeta - z_0}\right)^k,$$

whence our integrand becomes a series:

$$\frac{f(\zeta)}{\zeta - z} = \frac{f(\zeta)}{\zeta - z_0} \sum \left(\frac{z - z_0}{\zeta - z_0}\right)^k = \sum f(\zeta) \frac{(z - z_0)^k}{(\zeta - z_0)^{k+1}}.$$

We wish to integrate this series term-by-term. To justify term-by-term integration, note first that the geometric series here, being a power series, converges uniformly in $(z - z_0)/(\zeta - z_0)$ on those closed discs for which $|(z - z_0)/(\zeta - z_0)| < 1$. Since z, z_0 are fixed, this geometric series converges uniformly in ζ, and thus our series

$$\sum f(\zeta) \frac{(z - z_0)^k}{(\zeta - z_0)^{k+1}}$$

converges uniformly in ζ to the integrand. It follows that we may integrate this series term-by-term (see discussion below) around Γ so that

$$f(z) = \sum \left\{ \frac{1}{2\pi i} \int_\Gamma \frac{f(\zeta)}{(\zeta - z_0)^{k+1}} \, d\zeta \right\} (z - z_0)^k.$$

But we recognize the expression in the brackets as the Cauchy integral representation of $f^{(k)}(z_0)/k!$. Done.

We first comment that our proof here duplicated the work of Theorem 14, showing

$$a_k = \frac{f^{(k)}(z_0)}{k!}.$$

Let us review our argument on term-by-term integration (used in the proof). Suppose we have $g(\zeta) = \sum g_k(\zeta)$ converging uniformly on some set of ζ's. That is, the sequence whose nth term is $\sum_{k=0}^{n} g_k(\zeta)$ converges uniformly to $g(\zeta)$. By Theorem 7, we have that

$$\int_\Gamma g(\zeta) \, d\zeta = \lim_{n \to \infty} \int_\Gamma \sum_{k=0}^{n} g_k(\zeta) \, d\zeta,$$

where convergence is uniform on Γ (at least). Since the sum on the right-hand side is finite, however, we may interchange it with the line integral to obtain

$$\int_\Gamma g(\zeta) \, d\zeta = \lim_{n \to \infty} \sum_{k=0}^{n} \int_\Gamma g_n(\zeta) \, d\zeta = \sum_{k=0}^{\infty} \int g_n(\zeta) \, d\zeta,$$

that is, term-by-term integration. Let us state this result formally.

Lemma on Term-by-Term Integration

Let $\sum g_k(\zeta)$ converge uniformly in ζ on a curve Γ. Then

$$\int_\Gamma \sum g_k(\zeta)\,d\zeta = \sum \int_\Gamma g_k(\zeta)\,d\zeta.$$

The existence of a Taylor expansion for all functions analytic in a neighborhood of z_0 has the following simple consequence, which you may find startling.

Local Identity Theorem

Let $f(z)$ and $g(z)$ be analytic in a domain Ω containing the point z_0, and suppose that both functions and all their derivatives have equal values at the point z_0; that is,

$$f^{(k)}(z_0) = g^{(k)}(z_0) \qquad (k = 0, 1, \dots)$$

Then $f(z) = g(z)$ for all points z in the largest open disc centered at z_0 and contained in Ω.

Proof: It is easy to see that $f(z)$ and $g(z)$ are represented by the same Taylor series centered at z_0. Hence, they are equal as functions in the disc of convergence of this series. Done.

Question

If, as just above, $f^{(k)}(z_0) = g^{(k)}(z_0)$ for $k = 0, 1, \dots$, can we conclude the stronger result that $f = g$ throughout their common domain of definition Ω and not merely on some disc, as in the result above? See Theorem 18 (p.233).

Exercises to Paragraph 5.3.2

1. *On Theorem 15.* True or false?
   (a) If $f(z)$ is *entire*, that is, analytic in all of $\mathbb{C}$, and if z_0 is any point of $\mathbb{C}$, then $f(z)$ is represented by a Taylor series centered at z_0 and convergent for all z in $\mathbb{C}$.
   (b) Theorem 15 guarantees that $\log z$ may be expanded in positive powers of z.
   (c) $f(z)$ is analytic at z_0 if and only if it has a Taylor expansion in powers of $z - z_0$ convergent in some open disc centered at z_0.
   (d) A function analytic at z_0 may have two different Taylor series expansions centered at z_0.
   (e) If f is analytic at z_0 and $f^{(k)}(z_0) = 0$ for $k = 0, 1, 2, \dots$, then $f(z) = 0$ for all z in some open disc centered at z_0.
2. Given that $f(z)$ is analytic at z_0 (and hence in some domain containing z_0), we expand it in Taylor series, $f(z) = \sum a_k(z - z_0)^k$. True or false?
   (a) The set of all points at which the series converges is an open disc of finite radius or the entire plane.

(b) If the series converges in the open disc $D(z_0; R)$, with $R < \infty$, but in no open disc of larger radius, then f fails to be analytic at some point on the rim $C(z_0; R)$.

(c) $f(z)$ must fail to be analytic at some points on the rim of the (maximal) disc of convergence of $\sum a_k(z - z_0)^k$, even if the series converges at these points.

(d) The principal branch of $\log z$ is analytic at $z_0 = 1$, but is not analytic at the origin.

(e) The Taylor expansion $\sum a_k(z - 1)^k$ of the principal branch of $\log z$ has $D(1; 1)$ as open disc of convergence.

(f) If f is analytic at the origin and $f^{(k)}(0) = 1$ for $k = 0, 1, 2, \ldots$, then f is entire, and in fact $f(z) = e^z$.

3. From your knowledge of $f(z)$ below, predict without computation the radius of convergence of its Taylor expansion about the given center z_0.

(a) $f(z) = 1/(1 - z)$, $z_0 = 0$;

(b) $f(z) = 1/(1 - z)$, $z_0 = 3i$;

(c) $f(z) = \sin z$, $z_0 = \pi/6$;

(d) $f(z) = 1/\sin z$, $z_0 = \pi/2$;

(e) $f(z) = \sin(1/z)$, $z_0 = i$;

(f) $f(z) = \cot z$, $z_0 = \pi/2$.

Hint: What is the nearest point to z_0 at which f fails to be analytic?

4. *Parseval's identity and corollaries.* Suppose $f(z) = \sum_n a_n(z - z_0)^n$ for $|z - z_0| < R$. We write $z - z_0 = re^{i\theta}$ (use polar coordinates centered at z_0). Prove the identity

$$\frac{1}{2\pi} \int_0^{2\pi} |f(z_0 + re^{i\theta})|^2 \, d\theta = \sum_n |a_n|^2 r^{2n}$$

for $r < R$, as follows:

(a) Show

$$f(z)\overline{f(z)} = \sum_{m,n} a_m \bar{a}_n r^{m+n} e^{i(m-n)\theta}.$$

(b) Verify $\int_0^{2\pi} e^{ik\theta} \, d\theta = 0$ if k is a nonzero integer.

(c) Integrate both sides of (a) over $0 \leq \theta \leq 2\pi$ to obtain the identity. Can you justify the multiplication of series in (a) and the term-by-term integration in (c)?

(d) *Corollary: Liouville's Theorem again.* Prove that, as $r \to \infty$, the right-hand side of the equality becomes arbitrarily large unless $a_n = 0$, $n \geq 1$. Deduce that $|f(z)|$ cannot be bounded in $\mathbb{C}$.

(e) *Corollary: The Maximum Modulus Principle again.* Prove that if f is not constant in a disc about z_0, then $|f(z_0 + r_1 e^{i\theta_1})|^2 > |f(z_0)|^2 = |a_0|^2$ for each r_1, $0 < r_1 < R$, and some θ_1 depending on r_1. Conclude that $|f(z)|$ has no interior maxima.

5. *Bessel's inequality.* Let f be as in Exercise 4, and define

$$M(r; z_0) = \max_{|z - z_0| = r} |f(z)|.$$

Prove that

$$M(r; z_0)^2 \geq \sum_n |a_n|^2 r^{2n}.$$

6. *Polynomial growth.* Prove (see Section 4.11) that if $f(z)$ is entire (analytic for all z) and if $|f(z)| \leq |az^k|$, provided only $|z|$ is sufficiently large, then $f(z)$ is a polynomial of degree $\leq k$.

7. *Nonanalyticity on the circle of convergence.* Given that $f(z)$ is analytic in the disc $|z| < r$ (at least).

(a) Recall that f is represented by a series in powers of z which converges to $f(z)$ at all z satisfying $|z| < r$. Simply quote a theorem from the text for this.

(b) Suppose the series in (a) that represents f converges in $|z| < r$, but in no larger open disc $|z| < r_1$ with $r_1 > r$. Prove that the function f fails to be analytic at some point on the rim $|z| = r$. *Hint:* If f is analytic at a point on the rim, it is analytic in a small open disc about that point. If we assume this happens at all points of the rim $|z| = r$, then f must be analytic in some larger open disc (details?). Now apply (a) to obtain a contradiction.

8. *An enlightening example.* Demonstrate that analyticity of f and convergence of its Taylor series are not equivalent on the circle of convergence as follows. Take the slit plane $\Omega = \{z = re^{i\theta} \mid r > 0, -\pi < \theta < \pi\}$, and define f on $\Omega \cup \{0\}$ by $f(z) = z \log z$ for z in $\Omega, f(0) = 0$. Here, $\log z = \log re^{i\theta} = \log r + i\theta$ with $-\pi < \theta < \pi$ (that is, principal branch).

(a) Prove that f is continuous at the origin and hence at all points of the set $\Omega \cup \{0\}$ (not a domain!). Use the fact that $z \log z \to 0$ as $z \to 0$ in $\Omega \cup \{0\}$.

(b) Observe that f is not analytic at the origin. *Hint:* f can't be extended to points on the negative x-axis so as to be continuous in an open neighborhood of the origin.

(c) Check that the Taylor expansion of $f(z)$ about $z_0 = 1$ is given by

$$f(z) = (z - 1) + \sum_{n=2}^{\infty} \frac{(-1)^n (z-1)^n}{n(n-1)}.$$

Hint: Use $z = 1 + (z - 1)$, $\log z = \sum_{n=1}^{\infty} (-1)^{n+1}(z-1)^n/n$.

(d) We know that this Taylor expansion converges to $f(z)$ for $|z - 1| < 1$ at least. Show also that this series converges at $z = 0$ and in fact that its sum there is $0 = f(0)$. *Hint:* Show

$$\sum_{n=2}^{\infty} \frac{1}{n(n-1)} = \left(1 - \frac{1}{2}\right) + \left(\frac{1}{2} - \frac{1}{3}\right) + \cdots = 1$$

by telescoping (write out the first few partial sums!).

(e) Conclude that even if a Taylor expansion of $f(z)$ converges at a point ζ on the circle of convergence to the correct value $f(\zeta)$, the function f may fail to be analytic at this point.

9. *A summary.* Much of what we have done is contained in the following theorem. How many of the equivalences can you prove?

THEOREM

> Let D be an open disc. The following assertions are all equivalent and merely state that $f(z) = u(z) + iv(z)$ is analytic in D.
>
>   (i) $f(z)$ has a continuous complex derivative in D;
>   (ii) $u, v \in \mathscr{C}^1(D)$ and $u_x = v_y, u_y = -v_x$;
>   (iii) $\int_\Gamma f(z)\,dz = 0$ for every Jordan curve Γ in D;
>   (iv) $f(z)$ is represented in a neighborhood of every point of D by a convergent Taylor series centered at the point.

5.3.3 How to Expand Functions in Taylor Series

The situation is this: We know $f(z)$ in some way (by a formula, say). It is analytic at $z = z_0$. To study f near z_0, we want the expansion

$$f(z) = a_0 + a_1(z - z_0) + a_2(z - z_0)^2 + \cdots;$$

that is, we want the coefficients a_k. Here are four methods.

First Method: Take Derivatives

Theorem 14 tells us

$$a_k = \frac{f^{(k)}(z_0)}{k!}.$$

For some functions $f(z)$, the general kth derivative $f^{(k)}$ is readily computed.

Example: Let $f(z) = e^z$, $z_0 = 0$. Then, as we saw in Paragraph 5.3.1, each $a_k = 1/k!$, so that

$$e^z = 1 + z + \frac{z^2}{2!} + \cdots = \sum_{k=0}^{\infty} \frac{z^k}{k!}.$$

Example: Let $f(z) = \sin z$ and let $z_0 = 0$. We know $f'(z) = \cos z$, $f''(z) = -\sin z$, $f'''(z) = -\cos z$, $f^{(4)}(z) = \sin z$, so that

$$f(0) = 0, \quad f'(0) = 1, \quad f''(0) = 0, \quad f'''(0) = -1, \quad f^{(4)}(0) = 0, \ldots.$$

It follows that coefficients with even index vanish, $a_{2n} = 0$, while

$$a_{2n+1} = \frac{(-1)^n}{(2n + 1)!}.$$

Thus we get the Taylor series

$$\sin z = z - \frac{z^3}{3!} + \frac{z^5}{5!} - \cdots = \sum_{n=0}^{\infty} \frac{(-1)^n}{(2n + 1)!} z^{2n+1}.$$

Second Method: Substitution in Known Series

This is widely applicable, and best described by examples.

Example: Given $f(z) = e^{-z^2}$, $z_0 = 0$. Expand $f(z)$ in Taylor series centered at z_0.

Well, we know

$$e^{\zeta} = 1 + \zeta + \frac{\zeta^2}{2!} + \cdots = \sum_{k=0}^{\infty} \frac{\zeta^k}{k!}.$$

for any complex ζ. Now replace ζ with $-z^2$. We obtain immediately the desired series, namely,

$$f(z) = e^{-z^2} = 1 + (-z^2) + \frac{(-z^2)^2}{2!} + \cdots$$

$$= 1 - z^2 + \frac{z^4}{2!} - \cdots = \sum_{n=0}^{\infty} (-1)^n \frac{z^{2n}}{n!}.$$

Example: Expand $f(z) = 1/z$ in Taylor series centered at $z_0 = 1$. We alter the by-now-familiar geometric series

$$\frac{1}{1-\tau} = 1 + \tau + \tau^2 + \cdots \qquad (|\tau| < 1).$$

To do this, we put $1/z$ in the form $1/(1-\tau)$ so that τ looks like $z - z_0$, that is, $z - 1$. This is easy:

$$\frac{1}{z} = \frac{1}{1 + (z-1)}.$$

The idea here is $z = (z-1) + 1$. Thus, we have

$$\frac{1}{z} = \frac{1}{1 - (-1)(z-1)} = 1 - (z-1) + (z-1)^2 - (z-1)^3 + \cdots$$

$$= \sum_{k=0}^{\infty} (-1)^k (z-1)^k.$$

This converges when $|\tau| = |z-1| < 1$. We worked this out before following Theorem 5.

Another Example: Expand $f(z) = 5z/(3 + z^2)$ in powers of z. To do this, we deal first with $1/(3 + z^2)$, as follows:

$$\frac{1}{3 + z^2} = \frac{1}{3(1 + \frac{1}{3}z^2)} = \frac{1}{3}(1 - \frac{1}{3}z^2 + (\frac{1}{3}z^2)^2 - \cdots)$$

$$= \frac{1}{3} \sum_{n=0}^{\infty} (-\frac{1}{3})^n z^{2n}.$$

Having this, ordinary multiplication gives us

$$f(z) = \frac{5z}{3 + z^2} = \frac{5}{3} \sum_{n=0}^{\infty} \left(\frac{-1}{3}\right)^n z^{2n+1}.$$

Note: Once again we used a variant of the geometric series (for $1/(1 + \tau)$ this time). This is much easier than computing derivatives!

Third Method: Integrating or Differentiating Known Series

We illustrate this method with examples.

Example: Expand $f(z) = \log z$ in powers of $z - 1$.

Well, we know $f'(z) = 1/z, f(1) = 0$. Also, we found above the expansion of $1/z$ in powers of $z - 1$, namely,

$$\frac{1}{z} = 1 - (z - 1) + (z - 1)^2 - \cdots$$

in the open disc $|z - 1| < 1$. Integrating this series term-by-term along any path in the open disc form $z_0 = 1$ to z, we obtain

$$\log z = \int_1^z \frac{dz}{z} = (z - 1) - \tfrac{1}{2}(z - 1)^2 + \tfrac{1}{3}(z - 1)^3 - \cdots$$

$$= \sum_{k=1}^{\infty} \frac{(-1)^{k+1}}{k}(z - 1)^k.$$

Example: Let $f(z) = \cos z$. Expand it in powers of z.

We recall that $\cos z$ is the derivative of $\sin z$ and, moreover, we computed by the first method that

$$\sin z = z - \frac{z^3}{3!} + \frac{z^5}{5!} - \cdots = \sum_{n=0}^{\infty} \frac{(-1)^n}{(2n + 1)!}z^{2n+1}.$$

Differentiating term-by-term, we see that

$$\cos z = 1 - \frac{z^2}{2!} + \frac{z^4}{4!} - \cdots = \sum_{n=0}^{\infty} \frac{(-1)^n}{(2n)!}z^{2n}.$$

Note: The series for $\log z$ and $\cos z$ obtained in the last two examples (by integration and differentiation, respectively) converge in the same disc as the series used to obtain them. This verifies Corollary 13 of this chapter.

Fourth Method: Solve Differential Equations

This is a subject in itself. Let us content ourselves with a very simple example.

Example: Find an analytic function $f(z)$ that satisfies the differential equation $f'' - f = 0$ and the further conditions $f(0) = 1, f'(0) = 0$. Such an equation might have occurred in the study of some physical phenomenon, for example. To solve it, we write a series with "undetermined coefficients":

$$f(z) = a_0 + a_1 z + a_2 z^2 + \cdots + a_k z^k + \cdots.$$

We know

$$a_k = \frac{f^{(k)}(0)}{k!}.$$

What is $f^{(k)}(0)$?

Well, we are told that $f(0) = 1$, $f'(0) = 0$. To get $f''(0)$, observe that the differential equation requires that $f''(z) = f(z)$. Thus, $f''(0) = f(0) = 1$. Likewise $f'''(z) = f'(z)$—by differentiating both sides of $f''(z) = f(z)$ say—so that $f'''(0) = f'(0) = 0$. Continuing in this fashion, we see that all even derivatives equal 1, all odd derivatives vanish, whence

$$a_{2k} = \frac{1}{k!}, \qquad a_{2k+1} = 0.$$

Thus, a solution (unique!) to the problem is given by

$$f(z) = 1 + \frac{z^2}{2!} + \frac{z^4}{4!} + \cdots + \frac{z^{2k}}{(2k)!} + \cdots.$$

It is not hard to verify that this series converges in the entire z-plane. The function $f(z)$ is commonly known as the *hyperbolic cosine*, $f(z) = \cosh z$.

Exercises to Paragraph 5.3.3

1. Expand $f(z)$ in Taylor series about z_0 in several ways. You may utilize the expansions given in the text.
   (a) $f(z) = e^{\pi z}$, $z_0 = 0$;
   (b) $f(z) = \sin \pi z$, $z_0 = 0$;
   (c) $f(z) = 1/(2 + z)$, $z_0 = 0$;
   (d) $f(z) = 1/(2 + z)$, $z_0 = -1$;
   (e) $f(z) = 1/(1 + 2x)$, $z_0 = 0$;
   (f) $f(z) = \pi i + (1/z)$, $z_0 = i$.
2. Determine by inspection of $f(z)$ the radius of convergence in each case in Exercise 1.
3. *The hyperbolic sine.* Let $f(z)$ be the solution of the differential equation $f''(z) - f(z) = 0$ satisfying $f(0) = 0$, $f'(0) = 1$. We write $f(z) = \sinh z$.
   (a) Expand $\sinh z$ in powers of z by using the differential equation.
   (b) Verify that $\sinh z = (e^z - e^{-z})/2i$.

5.3.4 An Application: The Zeros of an Analytic Function

The use of Taylor series allows us to approach this general problem: Describe the set of *zeros* of an analytic function, that is, the set of points z with the property that $f(z) = 0$. If $f(z)$ is a polynomial, these are just the *roots* of the polynomial.

One reason this problem is important is this: If we are studying the mapping properties of an analytic function $w = g(z)$, then the zeros of the analytic function $g'(z)$ play a central role. (Think of calculus!) These are the so-called *critical points* of $g(z)$.

For example, $f(z) = z^2 + 1$ has two zeros, $z = \pm i$, while the function e^z has no zeros. The function $\sin z$ has an infinite set of zeros; can you find all of them? See Exercise 2 to Paragraph 3.4.3.

Suppose $z = z_0$ is a zero of the analytic function $f(z)$, and that $f(z)$ is not identically zero near z_0. Then, by the Local Identity Theorem, we may write the Taylor series centered at z_0 as

$$f(z) = a_n(z - z_0)^n + a_{n+1}(z - z_0)^{n+1} + \cdots,$$

with $a_n \neq 0$ for some $n \geq 1$. In other words, we have for some n,

$$f(z_0) = f'(z_0) = \cdots = f^{(n-1)}(z_0) = 0,$$

but $f^{(n)}(z_0) \neq 0$.

Now we wish to treat $f(z)$ as if it were a polynomial. Factoring out $(z - z_0)^n$, we have

$$f(z) = (z - z_0)^n(a_n + a_{n+1}(z - z_0) + a_{n+2}(z - z_0)^2 + \cdots)$$

$$= (z - z_0)^n g(z),$$

where $g(z)$ is defined by the power series $a_n + a_{n+1}(z - z_0) + \cdots$. Note that this new power series converges at each z for which the original series for f converged, because its partial sums are those for $f(z)$ divided by $(z - z_0)^n$. Hence, $g(z)$ is analytic in an open disc around the point z_0.

In fact, if $f(z)$ is analytic in some domain Ω containing z_0 (larger, perhaps, than the disc of convergence of the following series, $a_n(z - z_0)^n + a_{n+1}(z - z_0)^{n+1} + \cdots$), then $g(z)$ is analytic in Ω also. This is because we have identically

$$g(z) = \frac{f(z)}{(z - z_0)^n}$$

and both numerator and denominator are analytic in Ω (and, moreover, the quotient $g(z)$ is well behaved at z_0).

We now scrutinize $g(z)$. First observe that

$$g(z_0) = a_n \neq 0.$$

In essence, we have factored out the "zero-ness" of f at z_0. But even more is true. Since $g(z)$ is analytic, it is certainly continuous, and since $g(z_0) \neq 0$, there is an open disc of some radius centered at z_0 such that $g(z) \neq 0$ for any z in that disc. (This follows readily from the definition of continuity. Think about it. Or see Exercise 9 to Paragraph 3.2.1.) Let us call this new disc D.

Now we ask: "For which z in D is $f(z) = 0$?"

Since $0 = f(z) = (z - z_0)^n g(z)$ implies either $(z - z_0)^n = 0$, or else $g(z) = 0$ (which doesn't happen), it follows that $f(z) = 0$ for z in D if and only if $z = z_0$. (Note that $(z - z_0)^n = 0$ if and only if $z = z_0$.)

Thus we have proved

THEOREM 16

Let $f(z)$ be analytic in some domain Ω containing the point z_0, and suppose $f(z_0) = 0$. Then there is an open disc D centered at z_0 such *that either $f(z) = 0$ for all z in D or else $f(z) \neq 0$ for all $z \neq z_0$ in D. Thus, the zeros occur either in solid discs or at isolated points.*

This result may be reformulated as follows.

COROLLARY 17

Let f be analytic in a domain Ω, and let $z_1, z_2, \ldots$ be an infinite sequence of distinct points of Ω with the properties

(i) *$z^* = \lim_{n \to \infty} z_n$ exists and is a point of Ω;*
(ii) *$f(z_n) = 0$ for $n = 1, 2, \ldots$.*

Then $f(z) = 0$ for all z in some open disc containing z^*.

Proof: By continuity, $f(z^*) = 0$. By hypothesis, there are points z_n arbitrarily close to z^* such that $f(z_n) = 0$. Theorem 16 allows only one conclusion: that f is identically zero in some open disc about z^*. Done.

Thus, for example, if f vanishes along some small piece of curve in Ω, then it must vanish in a two-dimensional open subset containing the curve. This is true because a curve will surely contain a sequence $z_1, z_2, \ldots$ of the type described in the corollary.

Remark: Corollary 17 is not the best possible result, as our next theorem will show.

Exercises to Paragraph 5.3.4

1. Meditate upon the following important facts, supplying as much of the argument as you can.
   (a) A function analytic at z_0 is uniquely determined in a neighborhood of z_0 by its Taylor series centered at z_0.
   (b) If $f(z)$ is analytic at z_0 and not identically constant near z_0, then there exists an integer $n \geq 1$ and function $g(z)$ analytic at z_0 such that $f(z) = f(z_0) + (z - z_0)^n g(z)$ and $g(z_0) \neq 0$. Compare "polynomial" with n-fold root at z_0.
   (c) If $f(z)$ is as in (b), then there exists an open disc about z_0 such that $f(z) \neq f(z_0)$ for z in this disc except when $z = z_0$.
      Fact: Most of our applications of power series will really be applications of 1(b) above. See Exercise 11 to Paragraph 3.2.3.
2. Let $f(z)$ be analytic in some domain Ω and not identically zero on any disc inside Ω. The phrase "The zeros of f in Ω are isolated" implies which of the following?
   (a) f has only finitely many zeros in Ω.
   (b) The zeros of f occur at points not in Ω.
   (c) If $f(z_n) = 0$ for a sequence $\{z_n\}$ of distinct points of Ω, then this sequence does not have a limit point in $\mathbb{C}$.
   (d) Same as (c) with final $\mathbb{C}$ replaced by Ω.

(e) If Ω is a Jordan domain (open disc, say) with f defined and continuous on the closure $\overline{\Omega}$ and free of zeros on $\partial\Omega$, then f has only finitely many zeros inside Ω.

3. *An interesting example.* Discuss the following:

(a) The function $f(z) = \sin(1/z)$ has zeros at the points $z_n = 1/n\pi$, $n = \pm 1$, $\pm 2, \ldots$.

(b) The sequence $z_n = 1/n\pi$, $n = 1, 2, \ldots$, tends to the origin $z = 0$.

(c) Why doesn't the existence of the limit point $z = 0$ contradict the isolatedness of the zeros of $\sin(1/z)$? For $z = 0$ is *not* isolated from the set $\{z_n\}$.

(d) Can it be shown that $\lim_{z\to 0} \sin(1/z) = 0$? *Hint:* Consider the graph of $y = \sin(1/x)$ near $x = 0$.

4. Let $\sum a_n x^n$ be a power series with all a_n real. Suppose the series converges for real x precisely in the interval $-R < x < R$ for some $R > 0$.

(a) Prove that $\sum a_n z^n$ converges in $|z| < R$ but in no larger open disc. *Hint:* Section 5.2.

(b) Can the series in (a) converge at any points on the circle $|z| = R$?

(c) Prove that the real analytic function $f(x)$ represented on $-R < x < R$ by $\sum a_n x^n$ has a *unique* extension to a complex analytic function $f(z)$ on the open disc $|z| < R$. Thus, there is only one reasonable complex e^z, $\sin z$, $\cos z$, etc., namely, that defined by us.

5.3.5 A Topological Bonus

The key to Corollary 17 was an analytic fact, namely the existence of power series. We combine this with a simple topological observation about connectedness to prove the following stronger and more satisfying theorem. See Figure 5.4.

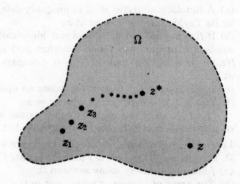

Figure   5.4

If $f(z_1) = f(z_2) = \cdots = 0$, then $f = 0$ throughout Ω

THEOREM 18

Let f be analytic in a domain Ω, and let $z_1, z_2, \ldots$ be an infinite sequence of distinct points of Ω with the properties that

(i) *$z^* = \lim_{n \to \infty} z_n$ exists and is a point of Ω;*
(ii) *$f(z_n) = 0, n = 1, 2, \ldots$.*

Then $f = 0$ identically throughout the entire domain Ω.

Proof: First we decompose Ω into three disjoint subsets:

$\Omega_0 = \{z \in \Omega \mid f \text{ vanishes on some } D(z; r)\}$;
$\Omega_1 = \{z \in \Omega \mid f(z) = 0 \text{ but } f \neq 0 \text{ in some } D(z; r) - \{z\}\}$;
$\Omega_2 = \{z \in \Omega \mid f(z) \neq 0\}$.

By Corollary 17, the point z^* is in Ω_0, the set of nonisolated zeros of f. Also, Ω_0 is clearly open.

By continuity, Ω_2 is open. Moreover, it is crucial to note that $\Omega_1 \cup \Omega_2$ is open, since all its points are interior points (check!).

Thus, the domain Ω is the disjoint union of two open sets, Ω_0 and $\Omega_1 \cup \Omega_2$. Since Ω is connected, $\Omega_1 \cup \Omega_2$ must be empty. This means $f = 0$ on all of Ω. Done.

Remark: This answers the question raised in Paragraph 5.3.2 following the Local Identity Theorem. We see now that if two functions f and g defined in the same domain Ω have identical power series expansions centered at one point z_0 in Ω, then the functions are equal throughout Ω, $f = g$. This shows the "rigid" nature of analytic functions. If you alter or deform the function near one point, then you must alter the function everywhere if it is to remain analytic.

Exercises to Paragraph 5.3.5

1. *On Theorem 18.* Let f, g be analytic on a domain Ω. Which of the following conditions imply $f = g$ identically on Ω? Give proofs or counterexamples.
   (a) $f(z_0) = g(z_0)$ for some point z_0 of Ω.
   (b) f and g have identical power series centered at the point z_0 of Ω.
   (c) $f(z_n) = g(z_n)$ for a sequence $\{z_n\}$ of distinct points of Ω.
   (d) $f^{(k)}(z_0) = g^{(k)}(z_0)$, $k = 0, 1, 2, \ldots$, at some point z_0 of Ω.
   (e) $f(z) = g(z)$ for all z on a curve Γ inside Ω.
   (f) $f(z_n) = g(z_n)$ for a sequence $\{z_n\}$ of distinct points of Ω which has a limit point in Ω.
2. *On the number of zeros.* Let Γ be a simple closed (Jordan) curve in the plane, and suppose f is analytic on and inside the curve Γ. In Chapter 7 we will inquire about the number of zeros of f inside Γ. Suppose f has no zeros on Γ itself. Prove that the number of zeros of f inside Γ is finite. *Hint:* This requires a bit of basic topology. If there were infinitely many zeros inside Γ, then the set of

zeros would have a "point of accumulation" z^*, which would lie on or inside Γ (Bolzano–Weierstrass Theorem). But $f(z^*) = 0$ (why?) so that z^* is not on the boundary curve Γ. Now apply Theorem 18 to prove finiteness.

3. *True or false?*

(a) If two entire functions are equal on the unit disc, then they are equal throughout $\mathbb{C}$.

(b) Two successive branches of log z agree at all points of the negative x-axis.

(c) If two analytic functions have the same set of zeros in Ω, then they are equal throughout Ω.

(d) If $f(z)$ is defined and continuous on the closed upper half-plane $y \geq 0$, analytic for $y > 0$, and equal to 0 on the full x-axis, then f vanishes identically throughout the closed upper half-plane. (Guess!)

(e) A function analytic in Ω is fully determined by its values at any sequence of points in Ω which converges to a point of Ω.

4. *Analytic continuation.* Let f_1, f_2 be analytic on Ω_1, Ω_2 respectively and suppose that the overlap $\Omega_1 \cap \Omega_2$ is not empty and that $f_1 = f_2$ on $\Omega_1 \cap \Omega_2$. Then f_2 is called an *analytic continuation* of f_1 to Ω_2, and vice versa.

(a) Prove that the analytic continuation of f_1 to Ω_2 is unique (if it exists).

(b) Let $f_1(z) = \sum_{k=0}^{\infty} z^k$ in the disc Ω_1 given by $|z| < 1$. Determine the unique analytic continuation of f_1 to the domain $\Omega_2 = \mathbb{C} - \{1\}$ (which in this case contains Ω_1).

(c) Let Ω_1 be the upper half-plane of points $z = re^{i\theta}$ with the agreement $r > 0, 0 < \theta < \pi$, and let $f_1(z) = \log r + i\theta$ in Ω_1. Let Ω_2 be the right half-plane of points $z = re^{i\theta}$ with $r > 0$ and $-\pi/2 < \theta < \pi/2$. Determine the analytic continuation of f_1 to Ω_2. Note that $\Omega_1 \cap \Omega_2$ is the open first quadrant.

(d) Let $\Omega_3 = \{z = re^{i\theta} \mid r > 0, -\pi < \theta < \pi\}$, a slit plane, and $f_3(z) = \log r + i\theta$. Let $\Omega_4 = \{z = re^{i\theta} \mid r > 0, 0 < \theta < 2\pi\}$. Verify that f_3 has no analytic continuation to Ω_4. Note that the overlap $\Omega_3 \cap \Omega_4$ is very large and also that $\Omega_3 \cup \Omega_4 = \mathbb{C} - \{0\}$. Contrast (c) with (d).

5. *The Monodromy Theorem for log z.* Prove this special case. Let D be a disc and f a branch of log z defined on D (which therefore does not contain the origin). Suppose Ω_1, Ω_2 are domains such that $\Omega_1 \cap \Omega_2 \supset D$ and $\Omega_1 \cup \Omega_2$ is simply connected. If f_1, f_2 are analytic continuations of f to Ω_1, Ω_2, respectively, then in fact $f_1 = f_2$ on all of $\Omega_1 \cap \Omega_2$. Hence, f_1, f_2 are analytic continuations of each other and so define a unique analytic continuation of f to the domain $\Omega_1 \cup \Omega_2$. In brief, *analytic continuations to simply connected $\Omega \supset D$ are unique (if they exist)*.

6. *The Schwarz Reflection Principle.* This, too, is in the spirit of analytic continuation. Let $f(z)$ be continuous in the closed upper half-plane $y \geq 0$, analytic in $y > 0$, and real-valued on the x-axis $y = 0$. We "reflect" f across the x-axis as follows, obtaining a function analytic in the entire z-plane.

(a) Prove that $F(z)$, defined for all $z = x + iy$ by

$$F(z) = \begin{cases} f(z), & y \geq 0 \\ \overline{f(\bar{z})}, & y < 0 \end{cases},$$

is continuous everywhere. Check the x-axis especially.

(b) Prove that $F'(z)$ exists and is continuous for z in the lower half-plane $y < 0$, and in fact $F'(z) = \overline{f'(\bar{z})}$.

(c) Prove that F is analytic at each point on the x-axis as well. This is nontrivial.

(d) Argue that F is the unique analytic extension of f to the entire z-plane.

(e) Suppose the given function f vanished identically on the x-axis. Prove that f vanished everywhere. (See Exercise 3(d) above.)

(f) Consider $f(z) = \sin z$ in the closed upper half-plane $y \geq 0$. What is $f(\bar{z})$ for z in the lower half-plane?

7. Find all entire analytic functions which agree with e^x on the x-axis.

8. *Proof or counterexample.* If two real harmonic functions $u_0(x, y)$, $u_1(x, y)$ agree on a curve contained in Ω, their common domain of definition, then they are equal at all points of Ω.

9. Let $F(x)$ be a real-valued function defined for $-\infty < x < \infty$ and equal to a power series (real coefficients) convergent for all x, $F(x) = \sum a_n x^n$. That is, F is real analytic.

(a) Prove there exists a complex analytic function defined for all z which agrees with F when $z = x$. *Hint:* Find a concrete expression for the new complex function.

(b) Argue that the function found in (a) is unique.

(c) If we denote the entire function found in (a) by $f(z)$, prove that $f(\bar{z}) = \overline{f(z)}$.

6

Singular Points and
Laurent Series

Section 6.1 THE THREE TYPES OF ISOLATED
SINGULARITY

6.1.0 Introduction

We are sometimes faced with the following situation: We are given a plane domain Ω containing the point z_0 and are given also a function f, which is known to be defined and analytic at all points of Ω except the point z_0. That is, f is analytic on the domain $\Omega' = \Omega - \{z_0\}$. We ask, "What can happen at z_0?" For instance, is it always possible to define the value $f(z_0)$ so that f becomes analytic at the missing point z_0 as well, and hence analytic in all of Ω? The answer to this is clearly "No," as the example $f(z) = 1/z$, $z_0 = 0$ shows. Hence, we are led to make some definitions, analyzing the different phenomena that may occur.

Some nonisolated singularities appear in the exercises to the next paragraph.

6.1.1 Isolated Singularities of an Analytic Function

As above, f is analytic on $\Omega' = \Omega - \{z_0\}$, where z_0 is an interior point of the domain Ω. In this case we say that z_0 is an *isolated singularity* (or *singular point*) for the function f. Now we define three types of isolated singularity and then show that each type actually occurs in nature. We have throughout this discussion f analytic on $\Omega' = \Omega - \{z_0\}$.

We say that z_0 is a *removable singularity* for f, provided there is some finite complex number w_0 such that

$$\lim_{z \to z_0} f(z) = w_0 \qquad (z \in \Omega').$$

On the other hand, z_0 is a *pole* of f, provided

$$\lim_{z \to z_0} |f(z)| = \infty \qquad (z \in \Omega').$$

As usual, this means that given any positive lower bound m (no matter how large), there is a small punctured disc centered at z_0 such that $|f(z)| > m$ for all z in the punctured disc. The absolute value $|f(z)|$ is arbitrarily large, provided z is sufficiently close to z_0.

Finally, we say that the isolated singularity z_0 is an *essential singularity* of f if it is neither of the two types just defined. One thing we must do is show that essential singularities occur. Later we will give a somewhat more complete picture (the Casorati-Weierstrass Theorem) of the behavior of an analytic function in the neighborhood of an essential singularity.

Examples of Isolated Singularities

Removable Singularities: It is easy to give many examples of this type. Just let f be any analytic function on any domain Ω. Let z_0 be any point of Ω. Then f is analytic on $\Omega' = \Omega - \{z_0\}$ and $\lim_{z \to z_0} f(z) = f(z_0) = w_0$, so that z_0 is a removable singularity.

Such examples miss the point, however. But consider the following: Let $f(z) = (\sin z)/z$. This is clearly defined and analytic on the domain $\Omega' = \mathbb{C} - \{0\}$, the punctured plane. The point $z_0 = 0$ is an isolated singularity for $(\sin z)/z$. Moreover, it is not immediately clear what happens to $f(z)$ as z approaches $z_0 = 0$. Does $\lim_{z \to 0} (\sin z)/z$ exist as a finite number? Is it infinite? Neither of these?

To deal with this puzzle, we appeal to Chapter 5 to write

$$f(z) = \frac{\sin z}{z} = \frac{1}{z}\left(z - \frac{z^3}{3!} + \frac{z^5}{5!} - \cdots\right) = 1 - \frac{z^2}{3!} + \frac{z^4}{5!} - \cdots.$$

Thus, we readily compute that $\lim_{z \to 0} f(z) = 1$. Hence, $z_0 = 0$ is a removable singularity for $f(z) = (\sin z)/z$. If we *define* $f(0) = 1$, the function f is now defined and continuous for *all* z, the origin included. Moreover, f is even analytic at $z_0 = 0$ because it is represented by the Taylor series $1 - (z^2/3!) + (z^4/5!) - \cdots$ there. This example is worthy of meditation.

Poles: The function $f(z) = 1/z$ has a pole at the origin $z_0 = 0$, for it is clear that $1/|z|$ becomes arbitrarily large as z tends to the origin. Likewise, $g(z) = 1/z^2$ and $h(z) = 1/(z - 1)$ have poles at $z_0 = 0$ and $z_0 = 1$, respectively.

Essential Singularities: Let $f(z) = e^{1/z}$. This is defined and analytic in the punctured plane $\mathbb{C} - \{0\}$. We claim that the isolated singularity $z_0 = 0$ is neither removable nor a pole and therefore essential. To see this, first let z approach 0 along the positive x-axis (from the right). We get

$$\lim_{z \to 0+} e^{1/z} = \lim_{x \to 0+} e^{1/x} = \lim_{t \to +\infty} e^t = +\infty.$$

Thus, $z_0 = 0$ is certainly not a removable singularity for $f(z)$. Now we show that neither is it a pole. To see this, simply let z approach the origin along the *negative* real axis (from the left). We obtain a finite limit:

$$\lim_{z \to 0^-} e^{1/z} = \lim_{x \to 0^-} e^{1/x} = \lim_{t \to -\infty} e^t = 0.$$

You should check this last limit, using the definition of the exponential function. It follows that $z_0 = 0$ is not a pole for $f(z)$, either. This example shows that essential singularities *do* exist and must be dealt with in order to understand isolated singularities. In fact, the Casorati-Weierstrass Theorem will show that $e^{1/z}$ behaves even more wildly than we have indicated here.

Exercises to Paragraph 6.1.1

1. Locate all isolated singularities of the following functions and classify each singularity according to type (see examples in the text).
   (a) $1/(z^2 + z)$.
   (b) e^{-z}.
   (c) e^{1/z^2}.
   (d) $(\sin z)/z^2$.
   (e) $1/\sin z$.
   (f) $(z + 1)/(z^2 + 3z + 2)$.

2. *A nonisolated singularity*
   (a) Verify that $\sin(1/z)$ is analytic except at $z = 0$ and vanishes precisely at the points $z_n = 1/n\pi$, $n = \pm 1, \pm 2, \ldots$. We have seen this already.
   (b) Verify that $f(z) = 1/\sin(1/z)$ is defined and analytic except at $z = 0$ and the points z_n given in (a).
   (c) Verify that the points z_n are poles of $f(z)$. Use either the Taylor series for $\sin(1/z)$ in powers of $z - z_n$ or (much quicker) Exercise 3 below. Are the points z_n isolated from each other?
   (d) Conclude that the origin $z = 0$ is a singularity (point of nonanalyticity) of $f(z)$ which is *not* isolated from the poles z_n.

   For another type of function with a nonisolated singularity, see Exercise 6.

3. (a) Recall the Taylor series argument that $f(z)$, if nonconstant and analytic at z_0, may be written as $f(z_0) + (z - z_0)^n g(z)$ with n a nonnegative integer, $g(z)$ analytic at z_0 and $g(z_0) \neq 0$ (whence n is uniquely determined).
   (b) Prove that if $f(z)$ is nonconstant and analytic at z_0 with $f(z_0) = 0$, then the function $g(z) = 1/f(z)$ has a pole at z_0. Why, first of all, is z_0 an isolated singularity of $g(z)$?

4. Give examples of each of the following:
   (a) A nonrational function with a pole at $z_0 = 0$.
   (b) A function with an essential singularity at $z_0 = 1$.

(c) Different functions $f(z), g(z)$, each with poles at $z_0 = 0$, such that the function $f(z) - g(z)$ has no singularities (technically speaking, a removable singularity).

(d) A polynomial $p(z)$ such that $f(z) = p(z)/z^3 e^z$ satisfies $\lim_{z \to 0} f(z) = 5$.

(e) A polynomial $p(z)$ such that $f(z) = 1/p(z)$ has poles at $z = \pm 1, \pm i$ and at no other points in $\mathbb{C}$.

(f) A function $f(z)$ with one pole (at $z_0 = 0$) and one essential singularity (at $z_1 = 1$).

5. Verify that every isolated singularity of a rational function $f(z) = A(z)/B(z)$ in lowest terms is a pole. (Note that "lowest terms" helps avoid removable singularities.)

6. *Logarithmic singularities.* Let $z_0 \in \Omega$, and suppose $f(z) = \log(z - z_0) + g(z)$, where $g(z)$ is analytic throughout Ω.

(a) Observe that f is not analytic at z_0.

(b) Argue that f is not even continuous in a punctured disc about z_0, owing to the necessity of choosing a branch of the logarithm. Thus, the "branch point" z_0 is a nonisolated singularity of f.

6.1.2 The Removability of Removable Singularities

The Laurent series is a useful tool in studying isolated singularities. Even before we introduce these series, however, we can justify the name "removable singularity." We will now see that what happened in the removable singularity example for $f(z) = (\sin z)/z$ actually happens in general; that is, a removable singularity may be regarded as a point of analyticity. The function seemed to be nonanalytic at z_0 only because of a lack of information, not because of any pathology.

THEOREM 1

Let $f(z)$ be analytic on the domain $\Omega' = \Omega - \{z_0\}$ where z_0 is an interior point of Ω, and suppose that z_0 is a removable singularity for f; that is, $\lim_{z \to z_0} f(z) = w_0$ exists. Then, if we define $f(z_0) = w_0$, the function f is analytic at z_0 also, and hence is analytic in all of Ω.

Note: It is clear that defining $f(z_0) = w_0$ produces a function *continuous* at z_0. However, analyticity at z_0 is less clear and must be proved.

Proof: We choose a circle $C(z_0; R)$ contained in Ω' and then define, for $|z - z_0| < R$, a "new" function:

$$F(z) = \frac{1}{2\pi i} \int_{C(z_0; R)} \frac{f(\zeta)}{\zeta - z} \, d\zeta.$$

The theorem will follow from two assertions: First, $F(z)$ is analytic for $|z - z_0| < R$ (in particular, at z_0); and second, $F(z) = f(z)$ provided $0 < |z - z_0| < R$ so that F is the extension of f to z_0.

First, it is immediate that $F(z)$ is analytic as asserted, for it is an integral of Cauchy type (just differentiate under the integral sign!). See Chapter 4.

Second, let $0 < |z_1 - z_0| < R$. To show $F(z_1) = f(z_1)$, write

$$F(z_1) - f(z_1) = \frac{1}{2\pi i} \int_{C(z_0;R)} \frac{f(\zeta)}{\zeta - z_1} \, d\zeta - \frac{1}{2\pi i} \int_{C(z_1;r_1)} \frac{f(\zeta)}{\zeta - z_1} \, d\zeta.$$

This follows from the definition of F and the Integral Formula for f. Note that the small circle $C(z_1; r_1)$ does not contain the "questionable" point z_0. See Figure 6.1.

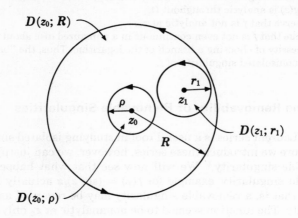

Figure   6.1

Now we refer this to z_0, claiming in fact that

$$F(z_1) - f(z_1) = \frac{1}{2\pi i} \int_{C(z_0;\rho)} \frac{f(\zeta)}{\zeta - z_1} \, d\zeta.$$

Here, $C(z_0; \rho)$ does not intersect the other small circle $C(z_1; r_1)$. This formula follows from the fact that $f(\zeta)/(\zeta - z_1)$ is an analytic function in the Jordan domain

$$D(z_0; R) - \bar{D}(z_1; r_1) - \bar{D}(z_0; \rho)$$

(that is, points in the big open disc but not in either small closed disc) whose boundary consists of three oriented circles,

$$C(z_0; R) - C(z_1; r_1) - C(z_0; \rho).$$

Here, the minus signs denote negative (clockwise) orientation. The Cauchy Integral Theorem for Jordan domains (Chapter 4) yields

$$\int_{C(z_0;R)} - \int_{C(z_1;r_1)} - \int_{C(z_0;\rho)} = 0,$$

which proves that $F(z_1) - f(z_1)$ equals the integral around $C(z_0; \rho)$, as claimed.

We are done if we can show that this last integral vanishes. To show that this is indeed the case, note that $|f(\zeta)|$ is bounded for $|\zeta - z_0| = \rho$ sufficiently small because $\lim f(\zeta) = w_0$ is finite. (Important observation! See Riemann's Criterion below.) Also, $|\zeta - z_1| \geq |z_1 - z_0| - \rho$ because ζ on $C(z_0; \rho)$ is bounded away from z_1. Thus, the integrand satisfies

$$\left| \frac{f(\zeta)}{\zeta - z_1} \right| < M$$

for some upper bound $M > 0$ and all sufficiently small radii $\rho > 0$. The *ML*-inequality applied to the integral around $C(z_0; \rho)$ now yields

$$|F(z_1) - f(z_1)| < \frac{1}{2\pi} M \cdot 2\pi\rho = M\rho$$

for arbitrarily small $\rho > 0$. This can happen only if $F(z_1) = f(z_1)$. Done.

Comment on the Proof

This was an orchestration of themes from the complex integral calculus of Chapter 4: defining a function ($F(z)$) as an integral of Cauchy type, representing another function ($f(z)$) as an integral, sliding the path of integration across a region of analyticity, bounding an integral by the *ML*-inequality. Once again, there is no mention of integration in the *statement* of the theorem!

A Useful Extension of Theorem 1

In the preceding proof of Theorem 1, we did not actually require that $\lim_{z \to z_0} f(z) = w_0$ exist. Note that the same conclusion holds, given only the seemingly weaker hypothesis that $f(z)$ is bounded in some punctured disc $0 < |z - z_0| < R$. Thus, we may conclude the following stronger theorem.

RIEMANN'S CRITERION FOR REMOVABLE SINGULARITIES

Let $f(z)$ be analytic in a punctured domain $\Omega' = \Omega - \{z_0\}$. Then the isolated singularity z_0 is removable if and only if $|f(z)|$ is bounded in some punctured disc about z_0. In this case $\lim_{z \to z_0} f(z) = w_0$ exists, and defining $f(z_0) = w_0$ produces a function analytic in all of Ω.

Remark

Thus, $|f(z)|$ must be unbounded in every punctured neighborhood of an isolated *essential* singularity. However, it is not true that

$$\lim_{z \to z_0} |f(z)| = +\infty,$$

for this happens only at poles. We'll apply this in Section 6.4.

Exercises to Paragraph 6.1.2

1. (a) Verify that $\cos(\pi z/2) = (z-1)g(z)$ with $g(z)$ analytic and $g(1) = -\pi/2$.
   (b) Verify that $\lim_{z\to 1}\{\cos(\pi z/2)/(z-1)\} = -\pi/2$, so that $f(z) = \cos(\pi z/2)/(z-1)$ has a removable singularity at $z_0 = 1$.
   (c) How should we define $f(1)$ so as to remove the singularity?
   (d) Verify that $f(z)$ is now analytic at $z_0 = 1$. *Hint:* Imitate what we did with $(\sin z)/z$.

2. Suppose $f(z) = (z-z_0)^h f_1(z)$, $g(z) = (z-z_0)^k g_1(z)$ with f_1, g_1 analytic and nonzero at z_0 and h, k positive integers. Note $f(z_0) = g(z_0) = 0$.
   (a) Find a necessary and sufficient condition on the pair h, k that the quotient $f(z)/g(z)$ have a removable singularity at z_0.
   (b) For which pairs h, k is $\lim_{z\to z_0} f(z)/g(z)$ finite and nonzero?
   (c) What happens if your condition in (a) is not satisfied by h, k?

3. Prove the following fact, a slight variation of one used in the proof of Theorem 1: If $f(z)$ is analytic in a punctured disc $0 < |z-z_0| < R$ and, moreover, $|f(z)|$ is bounded there, then $\int_{C(z_0;r)} f(z)\,dz = 0$ for $r < R$.
   Note that this generalizes (and utilizes) the Cauchy Integral Theorem.

4. (a) Prove: If $f(z)$ has an isolated removable singularity at z_0, then
$$\int_{C(z_0;r)} f(z)\,dz = 0$$
   for sufficiently small r. This is not new!
   (b) Compute $\int_C ((\sin z)/z)\,dz$, where C is the positively oriented unit circle.
   (c) Give an example of a function $f(z)$ with an isolated nonremovable singularity at the origin such that $\int_C f(z)\,dz \neq 0$. (We have seen one example often.)
   (d) Give an example as in (c), but satisfying $\int_C f(z)\,dz = 0$.
   (e) Why doesn't your example in (d) contradict Morera's Theorem in the case Ω is a disc about the origin? For your $f(z)$ in (d) is *not* analytic at the origin.

5. *On the proof of Riemann's criterion.* Verify this key observation: That the integral
$$\frac{1}{2\pi i}\int_{C(z_0;\rho)} \frac{f(\zeta)}{\zeta - z_1}\,d\zeta$$
   vanishes, provided $f(\zeta)$ is bounded near the isolated singularity z_0. In the statement of Theorem 1 we assumed that f actually had a limit at z_0, that is, that z_0 was a removable singularity for f. The observation here yields a simpler characterization of removable singularity.

6. *Further descriptions of analyticity.* In less scrupulous moments we agree that f is analytic at an isolated removable singularity z_0 (because the singular nature can be removed by defining $f(z_0) = \lim_{z\to z_0} f(z)$, etc.). Verify that if Ω is a domain containing the points $z_0, z_1, \ldots, z_n$, the following statements are equivalent:
   (a) f is analytic in Ω.
   (b) f is continuous in Ω and analytic in the punctured domain
$$\Omega - \{z_0, z_1, \ldots, z_n\}.$$

(c) f is analytic in $\Omega - \{z_0, z_1, \ldots, z_n\}$ and bounded in a small punctured disc centered at each of the singularities $z_0, z_1, \ldots, z_n$.

Note that statement (c) does not say that f is even defined at $z_0, z_1, \ldots, z_n$. Yet this happens to come true.

Section 6.2 LAURENT SERIES

6.2.0 Introduction

We will now study the behavior of $f(z)$ at an isolated singularity z_0 by expanding $f(z)$ in a series centered at z_0 (sound familiar?). This series will not in general be a Taylor series $a_0 + a_1(z - z_0) + a_2(z - z_0)^2 + \cdots$ because Taylor series yield analytic functions, whereas $f(z)$ is not analytic at a pole or essential singularity.

The series we will obtain will involve negative (as well as positive) powers of $z - z_0$. A series consisting of negative powers looks like

$$b_0 + b_1(z - z_0)^{-1} + b_2(z - z_0)^{-2} + \cdots + b_k(z - z_0)^{-k} + \cdots;$$

that is,

$$b_0 + \frac{b_1}{z - z_0} + \frac{b_2}{(z - z_0)^2} + \cdots + \frac{b_k}{(z - z_0)^k} + \cdots.$$

Observe that such a series shows little promise of converging to a finite sum when $z = z_0$ because the denominators $(z - z_0)^k$ vanish. We ask: "For which values of z does such a series converge?"

To answer this question, let us write

$$\zeta = (z - z_0)^{-1} = \frac{1}{z - z_0}.$$

The series above become ordinary power series in ζ:

$$b_0 + b_1\zeta + b_2\zeta^2 + \cdots + b_k\zeta^k + \cdots.$$

But the theory of power series which we saw in Chapter 5 assures us that the series just above converges for all complex ζ satisfying $|\zeta| < R$ for some radius R. Hence, the original series in $(z - z_0)^{-1}$ converges for those z satisfying $1/(|z - z_0|) < R$; that is,

$$|z - z_0| > R_1 = \frac{1}{R}.$$

In other words, the original series converges for all z *outside* some closed disc of radius R_1 centered at z_0. If $R = 0$, then the original series diverges for all z (except $z = \infty$).

A very simple and very important example of this phenomenon is the series $1/(z - z_0)$ consisting of a single term. This series converges for all

z such that $|z - z_0| > R_1 = 0$, that is, $z \neq z_0$, as is clear from the arithmetic of the situation. Actually, we will be dealing with a more general series (positive and negative powers!) such as

$$\cdots + \frac{1}{z^2} + \frac{1}{z} + 1 + z + z^2 + \cdots.$$

We write such a series in either of the forms

$$\sum_{n=-\infty}^{+\infty} c_n(z - z_0)^n \quad \text{or} \quad \sum_{n=0}^{\infty} a_n(z - z_0)^n + \sum_{n=1}^{\infty} \frac{b_n}{(z - z_0)^n}.$$

This series will be said to converge at a point z if and only if *both* parts,

$$\sum_{n=0}^{\infty} a_n(z - z_0)^n \quad \text{and} \quad \sum_{n=1}^{\infty} \frac{b_n}{(z - z_0)^n},$$

converge at z. Thus, the domain of convergence will be annular.

6.2.1 The Laurent Expansion

We will be concerned with functions defined and analytic in an *annulus* (ring) centered at z_0, that is, the set of points z satisfying the inequalities

$$R_1 < |z - z_0| < R_2$$

for nonnegative radii R_1, R_2 (Figure 6.2). Note that if $R_1 = 0$, the annulus reduces to a punctured disc: The point z_0 is deleted. Thus, annuli appear naturally in a treatment of isolated singular points of functions.

We will now see that a function analytic in an annulus is represented by a so-called *Laurent series* in positive and negative powers of $z - z_0$.

Figure   6.2

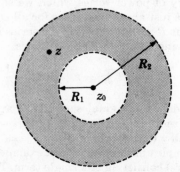

THEOREM 2

Let $f(z)$ be analytic in the annulus $R_1 < |z - z_0| < R_2$. Then $f(z)$ is represented by a Laurent series

$$f(z) = \sum_{n=-\infty}^{+\infty} c_n (z - z_0)^n$$

which converges to $f(z)$ throughout the annulus. Moreover, the coefficients are given by

$$c_n = \frac{1}{2\pi i} \int_C \frac{f(\zeta)}{(\zeta - z_0)^{n+1}} \, d\zeta \qquad (n = 0, \pm 1, \pm 2, \ldots)$$

where C is any circle centered at z_0 and contained in the annulus.

Some Preliminary Comments

1. Suppose f is actually analytic in the full disc of radius R_2 centered at z_0. For negative n, the coefficients are obtained by integrating functions of the form $f(\zeta)(\zeta - z_0)^k$ with k nonnegative ($k = -1 - n$). These integrals around C all vanish (why?) as we would hope: The Laurent series reduces to the Taylor series in the analytic case!

2. The value of c_n is independent of our choice of C because the integrand $f(\zeta)(\zeta - z_0)^{-n-1}$ is analytic between any two circles contained in the annulus and centered at z_0.

3. One usually does not compute Laurent series by integrating for the c_n. We will see better ways. On the other hand, if we know the Laurent series, then we can read off some integrals, namely, those for the c_n. This is a useful idea. For instance, $2\pi i c_{-1}$ gives us an integral of $f(\zeta)$, something we have long wanted. But we are getting ahead of ourselves....

Proof of Theorem 2: This will be quite similar to the proof of Taylor's Theorem in Chapter 5.

Given z in the annulus $R_1 < |z - z_0| < R_2$, the Cauchy Integral Formula for Jordan domains (see Section 4.3, including the exercises) gives

$$f(z) = \frac{1}{2\pi i} \int_{C_2} \frac{f(\zeta_2)}{\zeta_2 - z} \, d\zeta_2 - \frac{1}{2\pi i} \int_{C_1} \frac{f(\zeta_1)}{\zeta_1 - z} \, d\zeta_1.$$

Here, $C_k = C(z_0; r_k)$ with $R_1 < r_1 < |z - z_0| < r_2 < R_2$, so that z lies between the concentric circles C_1 (inner) and C_2. See Figure 6.3. Note that we use ζ_1, ζ_2 for the typical points on C_1, C_2, and that both circles are given the standard counterclockwise orientation.

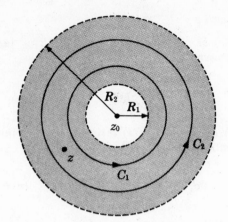

Figure   6.3

Now we deal with the integral over C_1, manipulating the denominator $\zeta_1 - z$ in order to introduce a series in $z - z_0$. We have (compare Taylor's Theorem)

$$\zeta_1 - z = (\zeta_1 - z_0) - (z - z_0)$$

$$= -(z - z_0)\left\{1 - \frac{\zeta_1 - z_0}{z - z_0}\right\}.$$

Since $|\zeta_1 - z_0| < |z - z_0|$, we are led to introduce a geometric series:

$$\frac{1}{\zeta_1 - z} = \frac{-1}{z - z_0}\sum_{k=0}^{\infty}\left(\frac{\zeta_1 - z_0}{z - z_0}\right)^k.$$

Now we operate formally with the integral over C_1; thus,

$$-\frac{1}{2\pi i}\int_{C_1}\frac{f(\zeta_1)}{\zeta_1 - z}\,d\zeta_1 = \frac{1}{2\pi i}\int_{C_1}f(\zeta_1)\sum_{k=0}^{\infty}\frac{(\zeta_1 - z_0)^k}{(z - z_0)^{k+1}}\,d\zeta_1$$

$$= \sum_{k=0}^{\infty}\left\{\frac{1}{2\pi i}\int_{C_1}f(\zeta_1)(\zeta_1 - z_0)^k\,d\zeta_1\right\}\frac{1}{(z - z_0)^{k+1}}$$

$$= \sum_{n=-1}^{-\infty}c_n(z - z_0)^n,$$

where $n = -(k + 1)$ and

$$c_n = \frac{1}{2\pi i}\int_{C_1}\frac{f(\zeta_1)}{(\zeta_1 - z_0)^{n+1}}\,d\zeta_1 = \frac{1}{2\pi i}\int_C\frac{f(\zeta)}{(\zeta - z_0)^{u+1}}\,d\zeta.$$

The last equal sign holds because the value c_n of the integral does not change if we replace C_1 by any other similarly oriented circle C in the annulus of analyticity (compare the statement of the theorem). We leave it to you to justify the interchange of $\sum$ and $\int$ carried out just above; it's quite similar to what we did in proving Taylor's Theorem in Chapter 5.

Thus, the integral over C_1 gives the negative powers $c_{-1}(z - z_0)^{-1}$, $c_{-2}(z - z_0)^{-2}$, etc. We leave it to you to prove that the integral over C_2 gives the series in nonnegative powers of $z - z_0$ with the "correct" coefficients. There is nothing essentially new here. Done.

Question

Can you recall any other situations in Chapter 5 and this chapter where the geometric series emerged at a crucial moment? Note that we began our discussion of series with this particular type.

Exercises to Paragraph 6.2.1

1. Consider a series

$$\sum_{n=0}^{\infty} a_n(z - z_0)^n + \sum_{n=1}^{\infty} b_n(z - z_0)^{-n}.$$

(a) Starting with the last sentence of Paragraph 6.2.0, prove that the set of points z for which the series converges will consist of an open annulus $R_1 < |z - z_0| < R_2$ (where possibly $R_1 = 0$, or $R_1 = R_2$, or $R_2 = \infty$) together with (perhaps) certain points on either rim of the annulus. Use your knowledge of Taylor series.

(b) Is it possible that the series converges for no z whatever? This is indicated by $R_1 = R_2$.

(c) Prove that the series converges at z_0 (the center) if and only if all b_n vanish (compare an ordinary power series).

(d) Does the series determine an analytic function at each z inside an open annulus $R_1 < |z - z_0| < R_2$ of convergence?

(e) Argue that the function in (d) fails to be analytic at points on each rim of the annulus $R_1 < |z - z_0| < R_2$. *Hint:* How did we do this for Taylor series?

2. *On Laurent's Theorem 2.* True or false?

(a) If $f(z)$ is represented by the series $z^{-1} + a_0 + a_1z +$ (higher powers of z) convergent in some punctured disc about the origin, then f has a nonremovable isolated singularity at the origin.

(b) The function $f(z)$ in (a) has a pole at the origin.

(c) Given $f(z)$, z_0, and the annular radii R_1, R_2 of Theorem 2, the Laurent series obtained in the theorem is unique.

(d) The function $f(z)$ may have different Laurent series centered at z_0, depending on the annulus of convergence selected.

(e) If $f(z)$ has an isolated singularity at z_0, then it may be expanded in a Laurent series centered at z_0 and it may be convergent in some punctured disc $0 < |z - z_0| < R_2$.

(f) If a Laurent series for $f(z)$ convergent in some ring $r_1 < |z - z_0| < r_2$ is actually a Taylor series (no negative powers of $z - z_0$), then this series actually converges in the full disc $|z - z_0| < r_2$ (at least).

(g) If the last conclusion holds, then $f(z)$ has at worst removable singularities in the full disc $|z - z_0| < r_2$ and may be considered analytic throughout this disc.

3. The function $f(z) = 1/(1 - z)$ has a pole at $z_0 = 1$. This is its only singularity.
   (a) What is the Taylor series of $f(z)$ in powers of z?
   (b) What is the Laurent series of $f(z)$ convergent in the punctured plane $0 < |z - 1| < \infty$?
   (c) What is the Laurent series of $f(z)$ convergent in the annulus $1 < |z| < \infty$? Note that the center of this series is a point of analyticity; yet, the series *must* contain negative powers of z. *Hint:* Write $1/(1 - z)$ in terms of $1/z$.

4. Suppose we are given $f(z) = \sum\limits_{n=-\infty}^{\infty} c_n(z - z_0)^n$, a Laurent series convergent

   in $0 < |z - z_0| < R$. Argue that $\int_{C(z_0;\, r)} f(z)\, dz = 2\pi i c_{-1}$, provided $0 < r < R$.
   This is the key to the Residue Theorem, which will enable us to integrate functions that fail to be analytic at isolated points. See Chapter 7.

5. Does Laurent's Theorem 2 imply that $\log z$ may be expanded in positive and negative powers of z convergent in some annulus $R_1 < |z| < R_2$? Discuss.

6. Let $f(z)$ be analytic, $0 < |z - z_0| < R$, with a pole at z_0. Let $C = C(z_0; r)$, $0 < r < R$. Does the Cauchy Integral Formula yield an infinite value in this case? What *is* the value of

$$\frac{1}{2\pi i} \int_C \frac{f(z)}{z - z_0}\, dz?$$

(You should be able to deduce immediately from elementary theory of the integral that this value is finite.)

6.2.2 How to Compute Laurent Series

As we mentioned in Paragraph 6.2.1 ("Comments"), the Laurent coefficients c_n $(n = 0, \pm 1, \pm 2, \ldots)$ are usually obtained by methods other than the integral formulas of Theorem 2. In particular, common sense and knowledge of related Taylor series work in many cases. Here are some examples.

Example 1

Expand $f(z) = 1/z$ in Laurent series centered at $z_0 = 0$.

This is solved by noting that $1/z$ *is* the Laurent expansion of $f(z)$. The only nonzero coefficient is $c_{-1} = 1$. Done.

Note that this series converges for $|z| > 0$.

Example 2

Expand the same function $f(z) = 1/z$ in Laurent series centered at $z_0 = 1$ and convergent near this point.

This time we note that $f(z)$ does not have a singularity at $z_0 = 1$. Hence, its Laurent series there will actually be a Taylor series (no *negative* powers of $z - 1$). In fact, we have seen already that

$$\frac{1}{z} = \frac{1}{1 + (z - 1)} = 1 - (z - 1) + (z - 1)^2 - \cdots \qquad (|z - 1| < 1).$$

Example 3

Expand $g(z) = e^{1/z}$ in Laurent series centered at $z_0 = 0$.

We recall that $e^{1/z}$ has an essential singularity at the origin. To get the Laurent expansion, we first write

$$e^{\tau} = 1 + \tau + \frac{\tau^2}{2!} + \frac{\tau^3}{3!} + \cdots \qquad (|\tau| < \infty)$$

Now let $\tau = 1/z$. This yields

$$g(z) = e^{1/z} = 1 + \frac{1}{z} + \frac{1}{2! \, z^2} + \frac{1}{3! \, z^3} + \cdots \qquad (|z| > 0)$$

That is,

$$e^{1/z} = \sum_{n=0}^{\infty} \frac{z^{-n}}{n!}.$$

Example 4

Expand $f(z) = 1/(z - 1)$ in powers of $1/z$ (that is, $z_0 = 0$) valid for $1 < |z|$. See Figure 6.4.

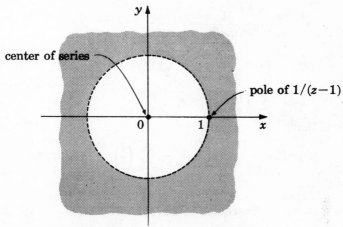

center of series

pole of $1/(z-1)$

$$1/(z-1) = \sum_{n=1}^{\infty} z^{-n} \text{ convergent for } |z| > 1$$

Figure 6.4

Note first that $f(z)$ is *not* analytic for all $|z| \leq 1$, since it has a pole at $z = 1$. We therefore expect an honest Laurent series in an annulus outside the closed unit disc.

To get some powers of $1/z$ into the discussion, we observe that

$$\frac{1}{z - 1} = \frac{1}{z} \cdot \frac{1}{1 - (1/z)} = \frac{1}{z}\left(1 + \frac{1}{z} + \frac{1}{z^2} + \cdots\right) \qquad (|1/z| < 1).$$

That is,

$$f(z) = \frac{1}{z-1} = \frac{1}{z} + \frac{1}{z^2} + \frac{1}{z^3} + \cdots \qquad (|z| > 1)$$

We comment that this series certainly does *not* represent $f(z)$ *inside* the unit disc. See what happens, for example, when we try to substitute $z = 0$ on both sides of this expansion!

Example 5

Expand $g(z) = 1/(z^2 + 1)$ in Laurent series convergent in a punctured disc around the pole $z_0 = i$.

We note first that $g(z) = 1/(z - i)(z + i)$. We wish to expand this in positive and negative powers of $z - i$. It makes sense to expand the factor $1/(z + i)$ in powers of $z - i$, and then multiply this expansion by $1/(z - i)$ to get the expansion for $g(z)$.

As usual, we alter the geometric series for $1/(1 - \tau)$ with a shrewdly chosen τ involving $z - i$. We observe that

$$
\begin{aligned}
\frac{1}{z+i} &= \frac{1}{2i + (z-i)} = \frac{1}{2i} \cdot \frac{1}{1 + (1/2i)(z-i)} \\
&= \frac{-i}{2} \cdot \frac{1}{1 - (i/2)(z-i)} \\
&= \frac{-i}{2} \left(1 + \frac{i}{2}(z-i) - \frac{1}{2^2}(z-i)^2 - \frac{i}{2^3}(z-i)^3 + \cdots \right) \\
&= \frac{-i}{2} \sum_{n=0}^{\infty} \left(\frac{i}{2} \right)^n (z-i)^n.
\end{aligned}
$$

It follows that

$$g(z) = \frac{1}{z-i} \cdot \frac{1}{z+i} = - \sum_{n=0}^{\infty} \left(\frac{i}{2} \right)^{n+1} (z-i)^{n-1}.$$

Exercises to Paragraph 6.2.2

1. Decide the nature of the singularity (if any) at $z_0 = 0$ for the following $f(z)$. If the function is analytic or the singularity isolated, expand the function in appropriate powers of z convergent in a punctured disc $0 < |z| < R$.
   (a) $\sin z$,
   (b) $\sin(1/z)$,
   (c) $(\sin z)/z$,
   (d) $(\sin z)/z^2$,
   (e) $1/\sin(1/z)$,
   (f) $z\sin(1/z)$,
   (g) $\log z$.
2. Expand $g(z) = 1/(z^2 + 1)$ as follows (see Example 5):
   (a) In powers of z convergent for $|z| < 1$.

(b) In powers of z convergent for $|z| > 1$.

(c) In powers of $z + i$ convergent for $0 < |z + i| < 2$.

3. We consider the not uncommon situation $f(z) = 1/g(z)$ where the behavior of $g(z)$ is known near z_0 (and not identically zero). Argue the following:

(a) If g is analytic at z_0 with $g(z_0) \neq 0$, then f is analytic at z_0 with $f(z_0) \neq 0$.

(b) If $g(z_0) = 0$, then f has a pole at z_0. See also Exercise 3 to Paragraph 6.1.1.

(c) If g has a pole at z_0, then f has a zero (removable singularity) at z_0.

(d) If g has an essential singularity at z_0, then f has either an isolated essential singularity or a nonisolated singularity at z_0. Give examples of each phenomenon.

(e) Where relevant, the Laurent series for $f(z)$ may be obtained from the Laurent series for $g(z)$ by long division.

(f) Find the first few terms of the Laurent expansion for $f(z) = 1/\sin z$ about the origin. Use (e).

(g) Likewise for $f(z) = 1/e^{1/z}$. *Hint:* There are a couple of methods available here.

Section 6.3 POLES

In this section we will correspond the three types of isolated singular points of $f(z)$ with three types of Laurent series.

Let us begin by recalling that an isolated singularity z_0 of an analytic function f is termed a pole if and only if

$$\lim_{z \to z_0} |f(z)| = +\infty.$$

The following simple observation is worth singling out.

Lemma 3

Let z_0 be a pole for the analytic function $f(z)$. Then the function $g(z) = 1/f(z)$ is analytic at z_0 and has an isolated zero there. Conversely, if z_0 is an isolated zero of an analytic function $g(z)$, then the function $f(z) = 1/g(z)$ has a pole there.

Proof: Given that $\lim_{z \to z_0} |f(z)| = \infty$, it is clear that $\lim_{z \to z_0} g(z) = 0$ for $g(z) = 1/f(z)$. This means that $g(z)$ has a removable singularity at z_0 and so, by Theorem 1, becomes analytic at z_0 by defining $g(z_0) = 0$. This is an *isolated* zero because z_0 was an isolated pole for $f(z)$. The proof of the converse is quite similar. Done.

Let's push this observation a bit further. Since $g(z) = 1/f(z)$ has an isolated zero at z_0, we may factor it uniquely as follows (see Section 5.3):

$$g(z) = (z - z_0)^m g_1(z),$$

where g_1 is analytic in a neighborhood of z_0, $g_1(z_0) \neq 0$, and m is a positive integer. This tells us that

$$f(z) = \frac{f_1(z)}{(z - z_0)^m},$$

where $f_1(z) = 1/g_1(z)$ is analytic and nonzero at z_0. You should convince yourself that f "blows up" as $z \to z_0$ precisely because of the denominator $(z - z_0)^m$. The positive integer m is termed the *order* of the pole at z_0. Every pole has a finite order $m \geq 1$ (proof?). Poles of order $m = 1$ are often called *simple* poles. Thus $1/z$ has a simple pole at $z_0 = 0$.

Now on to Laurent series. If z_0 is an isolated singularity for an analytic function $f(z)$, then we may expand $f(z)$ in a Laurent series convergent in some punctured disc centered at z_0, that is, convergent for all z with

$$0 < |z - z_0| < R_2.$$

We will speak of this expansion as *the Laurent expansion of $f(z)$ about z_0* (or *at z_0*).

Thus, let

$$f(z) = \sum_{n=-\infty}^{+\infty} c_n(z - z_0)^n$$

by the Laurent expansion of $f(z)$ about z_0. The *polar part* or *principal part* of this expansion is the portion consisting of negative powers of $z - z_0$; that is

$$\sum_{n=-1}^{n=-\infty} c_n(z - z_0)^n \quad \text{or} \quad \sum_{m=1}^{\infty} c_{-m}(z - z_0)^{-m}.$$

Clearly, one of the following three cases must occur:

1. All $c_{-m} = 0$, $m = 1, 2, \ldots$, so that the Laurent series is actually a Taylor series; the polar part is identically zero.

2. The polar part of the Laurent series is not identically zero as in (1), but only a finite number of the coefficients c_{-m} are different from zero.

3. Infinitely many of the c_{-m} are different from zero.

Of course a term $c_{-m}/(z - z_0)^m$ is identically zero if and only if $c_{-m} = 0$.

The point of the next theorem is that these three cases correspond to the three types of isolated singularity. Thus, if we know the Laurent series, we need not take limits to decide the nature of the singularity.

THEOREM 4

Let z_0 be an isolated singularity for the analytic function $f(z)$, and suppose $\sum_{m=1}^{\infty} (c_{-m}/(z - z_0)^m)$ is the polar part of the Laurent series for $f(z)$ about z_0. Then

    (i) *z_0 is a removable singularity if and only if all $c_{-m} = 0$, as in case (1) above.*

(ii) z_0 *is a pole of order* $h > 0$ *if and only if the polar part is a finite sum*

$$\frac{c_{-h}}{(z - z_0)^h} + \cdots + \frac{c_{-1}}{z - z_0}$$

with $c_{-h} \neq 0$, *as in case* (2) *above.*

(iii) z_0 *is an essential singularity if and only if the polar part has infinitely many nonzero terms, as in case* (3) *above.*

Proof: (i) From Theorem 1 it is clear that z_0 is a removable singularity if and only if its Laurent series reduces to a Taylor series.

(ii) Now suppose z_0 is a pole of order h. By Lemma 3, $g(z) = 1/f(z)$ has a zero at z_0. Hence, we get a Taylor expansion with zero constant term:

$$g(z) = b_h(z - z_0)^h + b_{h+1}(z - z_0)^{h+1} + \cdots,$$

with $b_h \neq 0$, $h \geq 1$. That is, $b_0 = b_1 = \cdots = b_{h-1} = 0$. Now we factor, just as with polynomials:

$$g(z) = (z - z_0)^h(b_h + b_{h+1}(z - z_0) + \cdots)$$
$$= (z - z_0)^h g_1(z),$$

where $g_1(z)$ is analytic and nonzero at z_0. In fact, $g_1(z_0) = b_h \neq 0$. Since $f(z) = 1/g(z)$, we may write

$$f(z) = (z - z_0)^{-h} f_1(z),$$

where $f_1(z) = 1/g_1(z)$ is analytic and nonzero at z_0. In fact, $f_1(z_0) = 1/b_h$. Thus, $f_1(z)$ has a Taylor expansion

$$f_1(z) = a_0 + a_1(z - z_0) + a_2(z - z_0)^2 + \cdots$$

with $a_0 = 1/b_h \neq 0$. Multiplying this series by the factor $(z - z_0)^{-h}$ yields the Laurent series for $f(z)$, namely,

$$f(z) = \frac{a_0}{(z - z_0)^h} + \frac{a_1}{(z - z_0)^{h-1}} + \cdots + a_h + \cdots.$$

The polar part here has only a finite number of nonzero terms. This proves case (2).

Conversely if case (2) holds, then $f(z)$ looks like

$$f(z) = c_{-h}(z - z_0)^{-h} + c_{-h+1}(z - z_0)^{-h+1} + \cdots + c_0 + \cdots$$
$$= (z - z_0)^{-h}(c_{-h} + c_{-h+1}(z - z_0) + \cdots),$$

with $c_{-h} \neq 0$. Clearly, now

$$\lim_{z \to z_0} |f(z)| = \lim_{z \to z_0} |z - z_0|^{-h} |c_{-h} + c_{-h+1}(z - z_0) + \cdots|$$

is infinite and so z_0 is a pole for $f(z)$.

(iii) By the process of elimination, z_0 is an essential singularity precisely when case (3) holds. This completes the proof of Theorem 4.

Some Examples Revisited

Let us verify the theorem with some now-familiar illustrations. We have seen that $(\sin z)/z$ has a removable singularity at $z_0 = 0$, that $1/z$ has a pole at $z_0 = 0$ and that $e^{1/z}$ has an essential singularity at $z_0 = 0$. We have also obtained these expansions:

$$\frac{\sin z}{z} = 1 - \frac{z^2}{3!} + \frac{z^4}{5!} - \frac{z^6}{7!} + \cdots \qquad (a \ Taylor \ series)$$

$$\frac{1}{z} = \text{itself} \qquad (a \ finite \ number \ of \ negative \ powers \ of \ z)$$

$$e^{1/z} = 1 + \frac{1}{z} + \frac{1}{2! \, z^2} + \cdots \qquad (infinitely \ many \ negative \ powers)$$

These results coincide with the three cases of the theorem.

Further Discussion

By now it should be clear that $f(z)$ has an isolated zero at z_0, or $g(z)$ has a pole at z_0, if and only if we may factor these functions thus:

$$f(z) = (z - z_0)^k f_1(z),$$

$$g(z) = (z - z_0)^{-h} g_1(z) = \frac{g_1(z)}{(z - z_0)^h},$$

with k, h positive integers and the factors f_1, g_1 functions analytic and nonzero in a neighborhood of z_0. The integers k, h are the orders of the zero and the pole, respectively, at z_0. This order gives a measure of "how fast" the function value approaches zero (or infinity in the case of a pole) as z approaches z_0. Thus, the functions $1/z$ and $1/z^2$ have poles of orders 1 and 2 at $z_0 = 0$, respectively. Just so, the second function "blows up" more rapidly than the first as z approaches the origin, since

$$\left|\frac{1}{z}\right| < \left|\frac{1}{z^2}\right| \qquad (|z| < 1).$$

It is worthy of note that *these orders (of zeros or poles) are always integers*, provided the function is analytic in some (possibly punctured) disc about z_0. This follows from expandability into powers (positive and negative) of $z - z_0$. Note in contrast that the "function" $f(z) = z^{1/2}$, which has a zero "of order 1/2" at $z_0 = 0$, is *not* single-valued analytic in a neighborhood of $z_0 = 0$.

Let us conclude this section on poles by remarking on the behavior of a function at an (isolated) pole. Actually, this behavior is not so very bad (in contrast, as we will see, with the case of essential singularities).

If we were willing to accept ∞ as an allowable value, then at a pole z_0 we might define $f(z_0) = \infty$, and so

$$\lim_{z \to z_0} f(z) = f(z_0) = \infty\,;$$

that is, the function f would be continuous at z_0 and in fact might even be considered to have a removable singularity there. In brief, poles are "infinite-valued" removable singularities.

We mention the Riemann sphere in this context. As a set, the Riemann sphere $\sum = \mathbb{C} \cup \{\infty\}$ = the complex numbers with infinity adjoined. It is not difficult to put the points of $\sum$ in one-one correspondence with the points of an ordinary sphere (hollow globe), with ∞ corresponding to the North Pole (whence the term "pole"?). Thus, we are led to the study of functions $f: \Omega \to \sum$ rather than $f: \Omega \to \mathbb{C}$.

Exercises to Section 6.3

1. True or false?

   (a) If $f(z)$ has a zero of order k at z_0, then $1/f(z)$ has a pole of order k there.

   (b) $\lim_{z \to 0} |e^{1/z}| = +\infty$.

   (c) If $f(z)$ has a pole of order k at z_0, then there exists a polynomial $p(z)$ such that $f(z) - [p(z)/(z - z_0)^k]$ is analytic at z_0.

   (d) If $f(z)$ is analytic at the origin, then there exists an integer n such that $g(z) = f(z)/z^m$ has a pole at the origin, provided $m > n$.

   (e) If $f(z)$ has an essential singularity at z_0, then $\lim_{z \to z_0} f(z)$ does not exist.

   (f) If $p(z)$ is a nonconstant polynomial, then $e^{1/p(z)}$ has essential singularities at the roots of $p(z)$.

   (g) The function given by $\sum_{n=0}^{\infty} (-1)^n/\{(2n)!\, z^{2n}\}$ has a pole at the origin.

   (h) If f, g have poles of order h, k at the point z_0, then $f(z)g(z)$ has a pole of order $h + k$ at z_0.

2. *Meromorphic functions.* The function $f(z)$ is called *meromorphic in* Ω provided its only (non-removable) singularities there are (isolated) poles. If $\Omega = \mathbb{C}$, f is usually called *meromorphic*. We concentrate on the case $\Omega = \mathbb{C}$.

   (a) Which of the following are meromorphic? Polynomials, rational functions, entire functions, $e^{1/z}$, $\sin(1/z)$, $(\sin z)/z$, $1/\sin z$, e^z/z. $1/\sin(1/z)$.

   (b) If f is meromorphic, is its reciprocal $1/f$ meromorphic?

   (c) Prove that if the meromorphic function $f(z)$ has only finitely many poles in $\mathbb{C}$, then $f(z)$ is a quotient [entire function]/[polynomial].

   (d) Observe that the meromorphic functions may be thought of as analytic mappings $f: \mathbb{C} \to \Sigma$ (= the Riemann sphere $\mathbb{C} \cup \{\infty\}$ mentioned in the text).

   Culture: A famous problem was "Construct all meromorphic functions." This has been solved by the work of several mathematicians, among them Weierstrass, Mittag-Leffler, Hadamard. Results: (i) Every meromorphic function is the quotient of two entire functions (a huge generalization of (c) above); (ii) Hadamard gave a characterization of entire functions as products of polynomials, exponentials, and so-called infinite products.

3. *Rational functions and partial fractions.* We write $f(z) = p(z)/q(z)$, where the polynomials p, q have no common linear factors (roots).

   (a) Prove that $\lim_{z \to \infty} f(z)$ is 0, $c \neq 0$, or ∞ according as the degree of $p(z)$ is $<$, $=$, or $>$ the degree of $q(z)$.

(b) Prove that a rational function has the form $f(z) = f_1(z) + (A(z)/B(z))$ with f_1, A, B polynomials such that $A(z)$, $B(z)$ have no common linear factors and either $A \equiv 0$ or the degree of A is strictly less than the degree of B. *Hint:* Long division by $q = B$ with remainder A.

(c) Prove that the function $A(z)/B(z)$ in (b) has finitely many poles.

(d) If $\varphi_1(z), \ldots, \varphi_m(z)$ are the various polar parts (negative powers in the Laurent series) of $A(z)/B(z)$ at its m poles $z_1, \ldots, z_m$, then $(A(z)/B(z)) - \sum_{j=1}^{m} \varphi_j(z)$ is a polynomial that approaches zero as z approaches ∞. *Hint:* Recall from (b) that $A(z)$ has smaller degree than $B(z)$ and that $\varphi_j(z)$ consists of finitely many negative powers of $z - z_j$.

(e) Apply a famous theorem to deduce that $A(z)/B(z) = \sum_{j=1}^{m} \varphi_j(z)$ (= a sum of partial fractions).

(f) Conclude [rational function] = [polynomial] + [finite sum of fractions $c_{jk}/(z - z_j)^k$]. Note that complex numbers are required for this partial fraction decomposition.

(g) As a corollary, prove that a complex rational function has an antiderivative that is the sum of a polynomial and finitely many terms of the form $c \log(z - z_j)$.

4. *Behavior at $z = \infty$.* The behavior of the meromorphic (in $\mathbb{C}$) function $f(z)$ at $z = \infty$ is defined to be $\lim_{z \to \infty} f(z)$ or (same thing) $\lim_{\zeta \to 0} F(\zeta)$, where $F(\zeta) = f(1/\zeta)$, provided this limit makes sense. Thus, we say that $z = \infty$ is a *removable singularity* (hence point of analyticity), *pole*, *essential singularity*, or *nonisolated singularity* for $f(z)$ according as $\zeta = 0$ is such for $F(\zeta)$.

(a) Describe the behavior at $z = \infty$ of each of the following meromorphic functions: $1/z$, e^z, $\sin z$, $(z^3 + 1)/z^2$, $1/\sin z$, $(3z^2 + 4)/(2z^2 + z + 1)$.

(b) If $f(z) = p(z)/q(z)$ is a rational function (in lowest terms), then f equals 0, $c \neq 0$, ∞ at $z = \infty$, according as the degree of $p(z)$ is less than, equal to, or greater than the degree of $q(z)$, respectively. See Exercise 3(a).

(c) Prove that the meromorphic $f(z)$ has only finitely many poles in each disc $|z| \leq r$. *Hint:* Otherwise the poles of f would cluster about some nonisolated nonremovable singularity in $|z| \leq r$. We have seen this (Chapter 5) for zeros.

(d) Hence, if f has infinitely many poles $z_1, z_2, \ldots, z_m, \ldots$ in $\mathbb{C}$, then $\lim_{m \to \infty} z_m = \infty$.

(e) Prove that $z = \infty$ is a nonisolated singularity for f if and only if f has infinitely many poles in $\mathbb{C}$. *Hint:* $\zeta = 1/z$.

(f) Deduce that $z = \infty$ is an isolated (possibly removable) singularity for $f(z)$ if and only if $f(z) = $ [entire function]/[polynomial].

5. *The natural functions on the Riemann sphere.* If f is meromorphic, then we think of it as $f : \mathbb{C} \to \Sigma$. If, moreover, f is rational, then we have $f : \Sigma \to \Sigma$, since $z = \infty$ is either a removable singularity (finite value) or pole for f. We ask "Which meromorphic functions $f(z)$ other than the rational functions have a removable singularity or a pole at $z = \infty$ and thus determine functions $f : \Sigma \to \Sigma$ from the sphere to the sphere?"

(a) Prove that if $f : \Sigma \to \Sigma$ is meromorphic, then also its reciprocal $1/f : \Sigma \to \Sigma$ is meromorphic, provided we admit the function that is identically equal to ∞.

(b) Use (a) and Exercise 4(f) to prove that if $f : \Sigma \to \Sigma$ is meromorphic and not identically equal to ∞, then f is a rational function of z.

Moral: The natural meromorphic (or "analytic," if we think of poles as ∞-valued removable singularities) mappings $f : \Sigma \to \Sigma$ are the rational functions, together with the constant function ∞. Or, recalling the notion of Riemann surface from Chapter 3, we conclude that the Riemann surface (natural domain of definition) of a rational function is the sphere Σ.

6. If the rational function $f(z) = p(z)/q(z)$ in lowest terms, then the *degree* of f is defined to be the larger of the degrees of $p(z)$ and $q(z)$. Prove that if f has degree d and if $w \in \Sigma$, then there exist exactly d solutions z in Σ to $f(z) = w$, provided we count properly. For example, $f(z) = 1/z^2$ has a double root (compare $f(z) = 0$) at $z = \infty$, as predicted by our assertion.

   This indicates how considering $f: \Sigma \to \Sigma$ rather than $\Omega \to \mathbb{C}$ or $\mathbb{C} \to \Sigma$ leads to certain more graceful statements about the rational function f.

Section 6.4 ESSENTIAL SINGULARITIES

   You will recall that z_0 (as always, an interior point of a plane domain Ω) is defined to be an essential singularity of f if it is neither an isolated removable singularity nor a pole. Theorem 4 characterizes essential singularities for us in terms of the Laurent series of f at z_0; this series must have infinitely many nonzero terms when expanded about an essential singularity. Now we ask about the *behavior* of f at an essential singularity z_0: "What does f do for z near z_0?" It is worth noting that we employed a "behavioral" *definition* for removable singularities and poles, namely, that f had a certain limit at z_0. In the present case we must prove a *theorem*, the Casorati–Weierstrass Theorem.

   We begin by recalling the more general criterion for removable singularities, due to Riemann. See Paragraph 6.1.1.

FACT

The isolated singularity z_0 of an analytic function f is a removable singularity $\Leftrightarrow |f(z)|$ is bounded in some punctured disc centered at z_0.

Note: This statement doesn't involve the existence of $\lim_{z \to z_0} f(z)$.

CASORATI–WEIERSTRASS THEOREM

In every open neighborhood, no matter how small, of an isolated essential singularity z_0, the function f assumes values $f(z)$ arbitrarily close to every complex number w. That is, if D' is any punctured disc centered at z_0 and $D(w; \rho)$ is any disc about any point w, then there exists some $z \in D'$ such that $f(z) \in D(w; \rho)$.

   Proof: The point w and discs D', $D(w; \rho)$ are given arbitrarily. Assume $|f(z) - w| \geq \rho$ for all $z \in D'$. Define $g(z) = 1/(f(z) - w)$. Then g has an isolated singularity at z_0 and, moreover, $|g(z)| \leq \rho^{-1}$ for all $z \in D'$. By Riemann's Criterion, z_0 is a removable singularity for $g(z)$. Hence, it is possible to remove the singularity by defining $g(z_0) = w_0$ for some finite w_0. Now if $w_0 \neq 0$, then $1/g(z) = f(z) - w$ (and hence $f(z)$) has a *removable* singularity at z_0, while if $w_0 = 0$, then $1/g(z) = f(z) - w$ (and hence $f(z)$, as you should check) has a *pole* at z_0. This contradicts the hypothesis that z_0 is an essential singularity of f. Thus, $|f(z) - w| < \rho$ for some z in D', as claimed. Done.

Thus, the function f takes every small punctured disc D' centered at the essential singularity z_0 and maps it "all over" the w-plane so that, in the language of topology, the set $f(D')$ of images is "dense" in the w-plane.

Illustration

We know that $f(z) = e^{1/z}$ has an essential singularity at $z_0 = 0$. We will show that if D' is any punctured disc centered at the origin and w is any nonzero complex number, then there is a z in D' such that $e^{1/z} = w$. This is even stronger than the conclusion of Casorati–Weierstrass: It shows equality rather than proximity.

What does $f(z) = e^{1/z}$ do to the disc D'? Note first that the mapping that sends z to $1/z$ will send the small disc D' onto the *exterior* of some other closed disc. That is, for $\zeta = 1/z$, if $0 < |z| < r$ then $\infty > |\zeta| > 1/r$. But we know that the exponential function e^ζ is periodic, mapping every horizontal strip of height 2π onto the set $\mathbb{C} - \{0\}$. Surely the exterior of each closed disc contains not only one but an infinity of such horizontal strips. It follows that $e^\zeta = e^{1/z} = w$ is satisfied for infinitely many z in D', provided only $w \neq 0$ (since e^ζ is never zero).

Postscript

Actually, this example is typical. For a deep theorem of Picard states that in every punctured disc about an isolated essential singularity, an analytic function *assumes* every complex value w with at most one exception!

Exercises to Section 6.4

1. Construct a function analytic in all of $\mathbb{C}$ except for essential singularities at the two points $z_0 = 0$, $z_1 = 1$.

2. The function $f(z) = 1/(z - 1)$ is represented by the Laurent series $z^{-1} + z^{-2} + z^{-3} + \cdots$, provided $|z| > 1$. A naive student, observing that this series has infinitely many negative powers of z, concludes that the point $z_0 = 0$ is an essential singularity for f. Point out the flaw in his argument.

3. The function $f(z) = (e^{iz} - e^{-iz})/2iz$ has an isolated singularity at the origin. Prove that $f(z)$ is bounded near the origin. Deduce that the origin is a removable singularity for f. *Hint:* Use series. Or, what is $f(z)$?

4. How many applications of the ML-inequality can you detect in this chapter?

5. Proof or counterexample: If $f(z)$ and $g(z)$ have a pole and essential singularity, respectively, at the point z_0, then the product $f(z)g(z)$ has an essential singularity at z_0.

6. Prove, using the Casorati–Weierstrass Theorem, that if f has an essential singularity at z_0 and if w is any complex value whatever, then there exists a sequence $\zeta_1, \zeta_2, \ldots$ such that $\lim_{n \to \infty} \zeta_n = z_0$ and $\lim_{n \to \infty} f(\zeta_n) = w$.

7. *Behavior at infinity revisited*

(a) Suppose f is analytic in $\mathbb{C}$ except for isolated singularities (poles or essential). How should we define the behavior of f at $z = \infty$? Recall that we dealt with this in the exercises to Section 6.3, and that f may assume a finite value or have a singularity (not necessarily isolated) at infinity.

(b) Argue that a nonconstant polynomial has a pole at infinity, and that e^z has an essential singularity there.

(c) What is the nature of the singularity of $e^{1/z}$ at $z = \infty$?

(d) Prove, using Casorati–Weierstrass, that a nonconstant, periodic, entire function has an essential singularity at infinity. (Compare Exercise 11(c) also.)

(e) Classify the singularities of $\sin z$, $\cos z$, e^{-z}, $\cosh z$ at $z = \infty$ (using (d) when applicable).

8. Prove that if z_0 is a pole of $g(z)$, then it is an essential singularity of $f(z) = e^{g(z)}$. Is this true at $z_0 = \infty$ as well?

9. Construct a function with infinitely many distinct isolated essential singularities in $\mathbb{C}$. *Hint:* Use Exercise 8.

10. *Meromorphic functions at infinity.* Let $f(z)$ be meromorphic (isolated *poles* only) in $\mathbb{C}$. See the exercises to Section 6.3. As usual, we exclude many-valued "functions" such as $\sqrt{z}$, $\log z$.

(a) Prove (once again!) that f has infinitely many poles in $\mathbb{C}$ if and only if it has a nonisolated singularity at infinity.

(b) Suppose f has an isolated singularity at infinity. Prove that this singularity is essential if and only if f is not a rational function.

11. *Entire functions at infinity.* Let $f(z)$ be entire.

(a) If f is not identically zero, then the reciprocal $1/f$ is meromorphic. Proof?

(b) What famous theorem may be phrased: If f (entire) is not constant, then infinity is an (isolated) pole or essential singularity, but not a removable singularity, for f.

(c) Refine (b) by proving that an entire function $f(z)$ that is not a polynomial has an essential (isolated) singularity at $z = \infty$. A very short proof follows from our power series considerations of Section 6.3. This result generalizes that of Exercise 7(b), which dealt with periodic functions.

12. *Entire transcendental functions at infinity.* These are the entire functions that are not polynomials, such as e^z.

(a) Prove that if f is entire transcendental, then in every domain $|z| > R$ (neighborhood of ∞) the values $f(z)$ come arbitrarily close to any preassigned finite complex value w_0. *Hint:* Casorati–Weierstrass for $f(1/\zeta)$.

(b) Contrast the case in which f is a nonconstant polynomial: The condition $|z| > R$ must be dropped, but $f(z) = w_0$ is actually solvable for (finitely many) z and any given w_0.

But a better theorem follows in Exercise 13.

13. *A version of Picard's First Theorem.* Prove that a nonconstant entire function f actually assumes every finite complex value, with possibly one exception. Of course $f(z) = e^z$ fails to assume the value zero.

Use the following free (and highly nontrivial) information. It is possible to define a (nonconstant) branch $\mu = \mu(w)$ of the "inverse modular function" satisfying (i) if g is entire and $g(z) \neq 0, 1$, then $\mu(g(z))$ is entire; and (ii) the value $\mu(g(z))$ lies in the upper half-plane. *Hint:* If $f(z) \neq w_0, w_1$, then $g(z) = (f(z) - w_0)/(w_1 - w_0)$ omits the values $0, 1$.

7

The Residue Theorem and the
Argument Principle

Section 7.1 THE RESIDUE THEOREM

7.1.0 Introduction

You will recall that the Cauchy Integral Theorem and Formula produced a large number of corollary results about analytic functions: the Maximum Modulus Principle, Liouville's Theorem, the Fundamental Theorem of Algebra, and so forth. In contrast, Chapters 5 and 6 on series have been of a more technical nature. But now that we have performed the labor of expanding complex functions in power series, it is time to reap the profits. This harvesting will be done in the remaining chapters.

The first results we obtain will be the Residue Theorem and the Argument Principle (Sections 7.1 and 7.2). Both of these follow readily from facts about Laurent series. Their uses are legion. The Residue Theorem is invaluable as a method of computing both real and complex integrals (Appendix) and the Argument Principle will be prominent when we discuss local conformal mapping in Chapter 8.

7.1.1 Residues and the Residue Theorem

Recall that the Cauchy Integral Theorem says that

$$\int_{\Gamma} f(z) \, dz = 0,$$

provided f is analytic on and inside the simple closed curve Γ. The Residue Theorem, which we will state in a moment, also enables us to integrate "without integrating." It is a kind of generalization of the Cauchy Integral Theorem to functions f with singularities inside Γ.

First a new concept. Let the point z_0 be an isolated singularity of f. This means that f is analytic in a punctured neighborhood of z_0. We know that f has a Laurent expansion

$$f(z) = \sum_{n=-\infty}^{+\infty} c_n(z - z_0)^n,$$

where certain of the coefficients c_n may, of course, be zero. We define the *residue of f at the point z_0* to be the coefficient (a number) c_{-1}. We denote it thus:

$$c_{-1} = \text{res}(f; z_0).$$

It is absolutely important to note that we obtain $\text{res}(f; z_0)$ from the Laurent expansion convergent in a punctured disc centered at z_0, that is, for all z satisfying a condition of the form

$$0 < |z - z_0| < R_2.$$

The usefulness of the residue as a property of the function f is indicated by the observation that

$$\text{res}(f; z_0) = c_{-1} = \frac{1}{2\pi i} \int_\Gamma f(z)\, dz,$$

where Γ is any simple closed curve containing z_0 as the *only* singular point of f inside it. See Figure 7.1. This follows, of course, from the general formula for the coefficients c_n in Laurent's Theorem 2 of Chapter 6. Thus, if we know the Laurent series of f about z_0 (or, at least, the coefficient c_{-1} of this series), and if z_0 is the only singularity of f enclosed by Γ, then we may immediately evaluate

$$\int_\Gamma f(z)\, dz = 2\pi i c_{-1} = 2\pi i\, \text{res}(f; z_0).$$

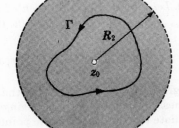

Figure 7.1

f analytic in $0 < |z - z_0| < R_2$

Let us extend this idea in a standard way. Suppose we wish to evaluate $\int_\Gamma f(z)\,dz$, where the points $z_1, \ldots, z_m$ are precisely the singularities of f inside Γ. Then we note that

$$\int_\Gamma f(z)\,dz = \int_{\Gamma_1} f(z)\,dz + \cdots + \int_{\Gamma_m} f(z)\,dz,$$

where $\Gamma_1, \ldots, \Gamma_m$ are simple closed curves and Γ_k has only one singularity of f in its interior, namely, z_k. See Figure 7.2 in the case $m = 3$. This last equality is true, you will recall, because the integral of an analytic function around the full boundary of a Jordan domain is zero.

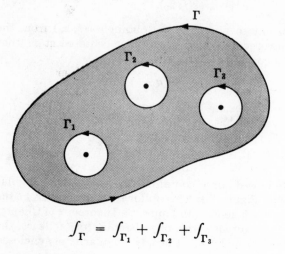

Shaded area is a 4-connected Jordan domain with boundary $\Gamma - \Gamma_1 - \Gamma_2 - \Gamma_3$

$$\int_\Gamma = \int_{\Gamma_1} + \int_{\Gamma_2} + \int_{\Gamma_3}$$

Figure 7.2

Hence we have arrived at the following result:

RESIDUE THEOREM

Let $f(z)$ be analytic in a domain Ω except for isolated singularities $z_1, \ldots, z_m$ in Ω. Let Γ be a positively oriented simple closed curve in Ω which has $z_1, \ldots, z_m$ in its interior. Then the line integral of f around Γ is given by

$$\int_\Gamma f(z)\,dz = 2\pi i \sum_{k=1}^{m} \operatorname{res}(f; z_k).$$

In brief, the integral is $2\pi i$ times the sum of the residues.

It now becomes important to locate the singular points of a function and then to calculate the residues at these points. This latter is done, of course, if we know the full Laurent series at the point, but often it is simpler to compute only the residue c_{-1} and not the full series. More on this soon.

An Illustration of the Residue Theorem

We calculate the familiar integral $\int_\Gamma (dz/z)$, where Γ is any simple closed curve about the origin. Thus, $f(z) = 1/z$, which has a pole at $z_0 = 0$. Since $f(z)$ is already given as a Laurent series, we see immediately that $\text{res}(f; 0) = 1$. The Residue Theorem now tells us that

$$\int_\Gamma \frac{dz}{z} = 2\pi i.$$

You may recall that we computed this earlier in the case Γ was the unit circle $|z| = 1$ by making the substitution $z = e^{i\theta}$ and integrating with respect to θ. We will meet this integral again.

Another Example: Two Singularities

We compute

$$\int_\Gamma \left(\frac{1}{z} + \frac{1}{1-z} \right) dz,$$

where Γ is any positively oriented loop enclosing the singularities $z_0 = 0$, $z_1 = 1$. See Figure 7.3.

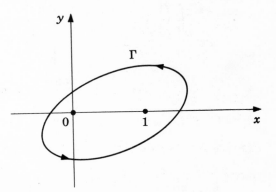

Figure 7.3

First we obtain the residues at z_0 and z_1. To obtain $\text{res}(f; z_0)$ where f is the integrand, we expand f in powers of z; thus,

$$f(z) = \frac{1}{z} + \frac{1}{1-z} = \frac{1}{z} + 1 + z + z^2 + \cdots.$$

It follows that $\text{res}(f; 0) = 1$.

To compute $\text{res}(f; 1)$, we note that

$$f(z) = \frac{-1}{z-1} + \frac{1}{1 + (z-1)} = \frac{-1}{z-1} + 1 - (z-1) + (z-1)^2 - \cdots,$$

whence $\text{res}(f; 1) = -1$.

The desired integral is, by the Residue Theorem,

$$\int_\Gamma \left(\frac{1}{z} + \frac{1}{1-z}\right) dz = 2\pi i(1 + (-1)) = 0.$$

Thus, even if f is *not* analytic at all points of the interior of Γ, its integral around Γ may vanish in some cases.

The Residue Theorem vs. the Cauchy Formula

It should in fairness be noted that the preceding examples can be dealt with in the language of Cauchy's Theorem and Formula without mention of residues. The exercises to Section 4.3, in which integrals of functions with singularities were computed using Cauchy's Formula, underscored this fact. The point is, of course, that Cauchy's Formula (for $f(z_0)$) is a special case of the Residue Theorem (for the function $f(z)/(z - z_0)$), and this special case is adequate for many practical computations.

Note in the examples below that the integrands look like integrals of Cauchy type, with denominators involving $z - z_0$. This is the key.

Example 1 (same as above)

We have

$$\int_\Gamma \left(\frac{1}{z} + \frac{1}{1-z}\right) dz = \int_\Gamma \frac{1}{z}\, dz + \int_\Gamma \frac{1}{1-z}\, dz$$

$$= \int_{\Gamma_0} \frac{1}{z}\, dz - \int_{\Gamma_1} \frac{1}{z-1}\, dz$$

$$= 2\pi i - 2\pi i$$

$$= 0$$

where Γ_0, Γ_1 are small circles around $z_0 = 0$, $z_1 = 1$, respectively.

Example 2

$$\int_{|z|=2} \frac{e^z}{z^2(z-1)}\, dz = \int_{\Gamma_0} \frac{e^z}{z^2(z-1)}\, dz + \int_{\Gamma_1} \frac{e^z}{z^2(z-1)}\, dz,$$

where Γ_0, Γ_1 are circles as in Example 1. But

$$\int_{\Gamma_1} \frac{1}{z-1}\left(\frac{e^z}{z^2}\right) dz = 2\pi i \left.\frac{e^z}{z^2}\right|_{z_1=1} = 2\pi e i.$$

By the Cauchy Formula for a derivative, we observe that

$$\int_{\Gamma_0} \frac{1}{z^2}\left(\frac{e^z}{z-1}\right) dz = 2\pi i \left.\frac{d}{dz}\left(\frac{e^z}{z-1}\right)\right|_{z_0=0}$$

$$= 2\pi i \left.\frac{e^z(z-2)}{(z-1)^2}\right|_{z_0=0}$$

$$= -4\pi i.$$

Thus, we conclude that

$$\int_{|z|=2} \frac{e^z}{z^2(z-1)}\, dz = 2\pi ei - 4\pi i = 2\pi i(e-2).$$

Example 3

Compute

$$\int_{|z|=1} \cot z\, dz = \int_{|z|=1} \frac{\cos z}{\sin z}\, dz.$$

Well, the only singularity of the integrand is at $z_0 = 0$, because $\sin z = zg(z)$ with $g(0) = 1$ (from the Taylor series). This gives us an integral of Cauchy type, namely,

$$\int_{|z|=1} \frac{\cos z}{\sin z}\, dz = \int_{|z|=1} \frac{1}{z}\left(\frac{\cos z}{g(z)}\right) dz$$

$$= 2\pi i \left(\frac{\cos z}{g(z)}\right)\Bigg|_{z_0=0}$$

$$= 2\pi i.$$

Summary

For theoretical purposes, as well as for certain knotty integrals (see the Appendix to this Chapter), the language of residues is indispensable and universally accepted. However, certain complex integrals, generally involving $(z - z_0)^k$ in the denominator of the integrand may be handled, using the Cauchy Integral Formula, without mention of residues.

Exercises to Paragraph 7.1.1

1. Compute the following integrals around the positively oriented unit circle $C(0; 1)$, applying theorems wherever possible.
    (a) $\int_C \csc z\, dz.$
    (b) $\int_C z \csc z\, dz.$
    (c) $\int_C (\tan z/z)\, dz.$
    (d) $\int_C (1/z(z^2 + 4))\, dz.$
    (e) $\int_C (1/z^2 - (1/4))\, dz.$
    (f) $\int_C e^{-z^2}\, dz.$
    You may have to read Paragraph 7.1.2 in order to compute certain residues here.

2. Suppose $f(z) = (z-1)^{-3} + 5(z-1)^{-1} - 4 + 2(z-1) + \cdots$ and that f has no singularities in the plane except $z_0 = 1$. Compute $\int_\Gamma f(z)\, dz$, where Γ is the positively oriented circle $|z-1| = 1$.

3. *A naive approach to the Residue Theorem.* Given that $f(z) = c_{-k}z^{-k} + \cdots + c_{-1}z^{-1} + c_0 + c_1z + \cdots$, compute $\int_\Gamma f(z)\, dz$ (where Γ is a sufficiently small,

positively oriented circle about the origin) by integrating the series term-by-term. Observe that the integral of each term vanishes, with one exception. Was this our method of proving the Residue Theorem?

4. *True or false?* Here Γ is a simple closed curve, $f(z)$ is analytic on Γ.
   (a) If $\int_\Gamma f(z)\, dz \neq 0$, then f is not analytic at all points on and inside Γ.
   (b) If $\int_\Gamma f(z)\, dz = 0$, then f is analytic at all points inside Γ.
   (c) If f is analytic inside Γ except at z_0, and $(z - z_0)f(z)$ is analytic everywhere inside Γ, then $\int_\Gamma f(z)\, dz \neq 0$.
   (d) It is necessary to perform integration to compute every Laurent series.
   (e) The Cauchy Integral Formula is a special case of the Residue Theorem.

5. Which of the integrals worked out in the text above can you compute using only anti-differentiation? This should help you appreciate the theory we have developed.

7.1.2 How are Residues Computed?

We illustrate some methods by examples.

First Method: Use the Laurent Expansion

Suppose that $f(z) = e^{1/z}$. Then we know that

$$f(z) = 1 + \frac{1}{z} + \frac{1}{2!\, z^2} + \cdots,$$

so that $\operatorname{res}(f; 0) = $ the coefficient of $1/z = 1$. Of course $z_0 = 0$ is the only isolated singularity of $e^{1/z}$ in $\mathbb{C}$.

Here is another example. Let $g(z) = \sin(1/z^2)$. This has an isolated singularity at $z_0 = 0$ only. The Laurent series centered at the origin is obtained from the familiar Taylor series

$$\sin \tau = \tau - \frac{\tau^3}{3!} + \frac{\tau^5}{5!} - \cdots$$

by replacing τ with $1/z^2$; thus,

$$g(z) = \sin \frac{1}{z^2} = \frac{1}{z^2} - \frac{1}{3!\, z^6} + \frac{1}{5!\, z^{10}} - \cdots.$$

We see immediately that $\operatorname{res}(g; 0) = c_{-1} = 0$. Thus, you can now verify the integrations

$$\int_\Gamma e^{1/z}\, dz = 2\pi i, \qquad \int_\Gamma \sin\left(\frac{1}{z^2}\right) dz = 0,$$

where Γ is any simple closed curve about the origin. These were quite beyond our means before we reached the Residue Theorem.

Second Method: Simple Poles

A pole z_0 of f is said to be *simple* if its order is 1, that is, if f may be expressed as

$$f(z) = \frac{c_{-1}}{z - z_0} + c_0 + c_1(z - z_0) + \cdots.$$

The basic example of this is, of course, $z_0 = 0$ for $f(z) = 1/z$. We obtain the residue at a simple pole by noticing that

$$\operatorname{res}(f; z_0) = c_{-1} = \lim_{z \to z_0} (z - z_0) f(z).$$

For example, let us examine the rational function

$$f(z) = \frac{(z - i)^3(z + 1)}{z(z - 2)^4}.$$

This has a simple pole at $z_0 = 0$ and a pole of order 4 at the point $z = 2$. Rather than expand $f(z)$ in Laurent series, we use the paragraph above to deduce

$$\operatorname{res}(f; 0) = \left. \frac{(z - i)^3(z + 1)}{(z - 2)^4} \right|_{z=0} = \frac{i}{16}.$$

Note that the factor z was first canceled from the denominator.

The Residue Theorem then gives

$$\int_{\Gamma} \frac{(z - i)^3(z + 1)}{z(z - 2)^4} \, dz = 2\pi i \left(\frac{i}{16} \right) = -\frac{\pi}{8},$$

where Γ is the unit circle, say (and hence Γ does not enclose the pole $z = 2$ as well as the origin).

An Extension to Double Poles

Suppose f has a double pole (pole of order 2) at z_0; that is,

$$f(z) = \frac{c_{-2}}{(z - z_0)^2} + \frac{c_{-1}}{z - z_0} + c_0 + c_1(z - z_0) + \cdots,$$

with $c_{-2} \neq 0$. Hence,

$$(z - z_0)^2 f(z) = c_{-2} + c_{-1}(z - z_0) + c_0(z - z_0)^2 + \cdots$$

and so we get c_{-1} by differentiating:

$$\operatorname{res}(f; z_0) = c_{-1} = \lim_{z \to z_0} ((z - z_0)^2 f(z))'.$$

This is our formula.

For example, let our function be

$$f(z) = \frac{z + 1}{z(z - i)^2}.$$

This has a simple pole at the origin and a double pole at $z_0 = i$. Now $(z - i)^2 f(z) = (z + 1)/z = 1 + (1/z)$. The first derivative of this function is $-1/z^2$, whence

$$\operatorname{res}(f; i) = \lim_{z \to i} \left(\frac{-1}{z^2} \right) = 1.$$

We leave it to you to compute the residue of f at the origin and hence the integral

$$\int_\Gamma \frac{z + 1}{z(z - i)^2} \, dz$$

for various closed curves Γ.

Exercises to Paragraph 7.1.2

1. Compute all residues inside the unit circle of each of the integrands in Exercise 1 to Paragraph 7.1.1. Compute the integrals now if you could not do it earlier.
2. Suppose that g is analytic at z_0 and define $f(z) = g(z)/(z - z_0)$.
   (a) What is the nature of the singularity of f at z_0?
   (b) What is $\operatorname{res}(f; z_0)$ in terms of g?
   (c) Compute $\int_C f(z) \, dz$, where C is a positively oriented circle about z_0 containing no other singularities of f.
   (d) What famous theorem is this?
3. (a) Locate all singularities of $f(z) = (z^3 + 2z^2 + z)^{-1}$ and compute the residues at these singularities.
   (b) Compute $\int_{|z|=2} (z^3 + 2z^2 + z)^{-1} \, dz$ (the path of integration is the positively oriented circle $C(0; 2)$).
4. Same as Exercise 3 for $f(z) = (3 + iz^4)/z^5$.
5. Suppose $f(z)$ has a simple zero at z_0 so that $1/f(z)$ has a simple pole there. Prove $\operatorname{res}(1/f; z_0) = 1/f'(z_0)$.
6. (a) Prove that the zeros of $\sin z$ occur at the points $z = k\pi$, $k = 0, \pm 1, \pm 2, \ldots$ and, moreover, all are simple. We have seen this already.
   (b) Prove $\operatorname{res}(1/\sin z; k\pi) = (-1)^k$, using Exercise 5.
   (c) Compute the integrals $\int_{C(k\pi; 1)} \csc z \, dz$, $k = 0, \pm 1, \ldots$.
7. Compute $\int_\Gamma (z + 1)z^{-1}(z - i)^{-2} \, dz$ in these cases (see text):
   (a) $\Gamma = C(0; 1/2)$.
   (b) $\Gamma = C(i; 1/2)$.
   (c) $\Gamma = C(0; 2)$.

Section 7.2 THE ARGUMENT PRINCIPLE

7.2.0 Introduction

This section contains some surprises. Here is the situation we will be considering: Ω is a plane domain as usual, and Γ is a piecewise-smooth positively oriented simple closed curve inside Ω having the property that every point inside Γ is a point of Ω. This is certainly the case if Ω is "simply connected," that is, has no holes or punctures whatever. The function f is analytic in Ω except perhaps for a finite number of poles. However, f has no zeros and no poles on the curve Γ.

We will consider these three numbers:

(i) the integral

$$\frac{1}{2\pi i} \int_\Gamma \frac{f'(z)}{f(z)}\, dz;$$

(ii) the difference between the number of zeros and the number of poles of f inside the curve Γ;

(iii) the number of times the image curve $f(\Gamma)$ wraps around the origin in the w-plane.

See Figure 7.4.

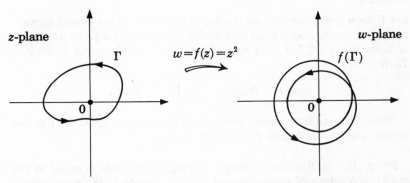

f has a double zero $f(\Gamma)$ wraps twice about the origin

Figure   7.4

It is clear that the last two numbers are integers. We will show that all three numbers are equal! This is the Argument Principle. To deal with (iii), we must introduce the notion of "winding number" of a curve about a point. But let us begin by relating (i) and (ii).

7.2.1   Counting Zeros and Poles; $N_0(\Gamma)$ AND $N_\infty(\Gamma)$

As mentioned in Paragraph 7.2.0, f is analytic at all points on and inside the closed curve Γ except for a finite number of poles inside Γ. We require also that f have no zeros on Γ itself. We introduce the notation

$N_0(\Gamma)$ = number of zeros (properly counted) of f inside Γ,

$N_\infty(\Gamma)$ = number of poles (properly counted) of f inside Γ.

Here "properly counted" means that a simple zero is counted once, a double zero twice, a zero of order k is counted k times, and likewise for poles. *We always use this convention when speaking of the number of zeros, poles, roots, and so on.*

For example, if

$$f(z) = \frac{(z - \frac{1}{2})^3}{z^4(z - 2)^6}$$

and Γ is the unit circle about the origin, then $N_0(\Gamma) = 3$, $N_\infty(\Gamma) = 4$.

Note that because f is not identically zero (since $f(z) \neq 0$ on Γ), its zeros are isolated. This implies that $N_0(\Gamma)$ is finite (why?). You may argue similarly for $N_\infty(\Gamma)$.

The integrand $f'(z)/f(z)$ is the *logarithmic derivative* of f. It is the derivative of $\log f(z)$, where the latter is defined.

The first half of the Argument Principle follows.

LEMMA 1

Let Γ be a piecewise-smooth positively oriented simple closed curve and suppose f is analytic on and inside Γ, except for a finite number of poles inside Γ, but with no poles or zeros on the curve Γ itself. Then

$$N_0(\Gamma) - N_\infty(\Gamma) = \frac{1}{2\pi i} \int_\Gamma \frac{f'(z)}{f(z)} \, dz.$$

Note: If f is analytic everywhere, then the integral gives the number of zeros exactly, since then $N_\infty(\Gamma) = 0$.

Proof: By the Residue Theorem, the right-hand side is equal to the sum of the residues of the integrand $f'(z)/f(z)$ inside Γ. We must obtain these residues. To do this, note first that the integrand has singularities (possibly removable) precisely where f has zeros or poles.

Let us compute a residue of $f'(z)/f(z)$ at a zero z_0 of f. We may write, thanks to our usual factorization,

$$f(z) = (z - z_0)^h f_1(z),$$
$$f'(z) = h(z - z_0)^{h-1} f_1(z) + (z - z_0)^h f'_1(z),$$

where h is a positive integer and $f_1(z_0) \neq 0$. Thus, the integrand is

$$\frac{f'(z)}{f(z)} = \frac{h}{z - z_0} + \frac{f'_1(z)}{f_1(z)}.$$

Since $f_1(z)$ is analytic and nonzero in some open set about z_0, we conclude that the residue of $f'(z)/f(z)$ at the zero z_0 of the original function f is just the order h of this zero. If we add up the residues obtained at each zero of f inside Γ, we clearly get the total number of zeros (properly counted) of f inside Γ. This gives $N_0(\Gamma)$.

On the other hand, suppose f has a pole of order k at some point z_1 inside Γ. Then we have, as before,

$$f(z) = (z - z_1)^{-k} f_1(z),$$

$$f'(z) = -k(z - z_1)^{-k-1} f_1(z) + (z - z_1)^{-k} f'_1(z),$$

with $f_1(z_1)$ finite and different from zero. Thus we have

$$\frac{f'(z)}{f(z)} = \frac{-k}{z - z_1} + \frac{f'_1(z)}{f_1(z)},$$

whence the residue of $f'(z)/f(z)$ at z_1 is $-k$. Thus, the sum of the residues at poles of f is $-N_\infty(\Gamma)$.

Since all singularities of $f'(z)/f(z)$ have been accounted for, the Residue Theorem assures us that the line integral equals $N_0(\Gamma) - N_\infty(\Gamma)$ as claimed. Done.

An Example

Let $f(z) = z^2$, and let Γ be the unit circle. We see that $f'(z)/f(z) = 2z/z^2 = 2/z$, so that

$$\frac{1}{2\pi i} \int_\Gamma \frac{f'(z)}{f(z)} = \frac{2}{2\pi i} \int_\Gamma \frac{dz}{z} = 2.$$

This is just what Lemma 1 predicts, since $f(z)$ clearly has a double zero inside Γ and no poles anywhere.

It is worth noting how often we have used the explicit computation

$$\frac{1}{2\pi i} \int_\Gamma \frac{dz}{z} = 1,$$

which we worked out with our bare hands in Chapter 4 long ago.

Exercises to Paragraph 7.2.1

1. Suppose $f(z)$ is analytic in the plane except for poles at $z = 1 + i$, $i/2$, 0 of orders 3, 2, 1, respectively. Suppose also that f has zeros at $z = 5i$ and $z = 1/2$ of orders 7 and 1, respectively. Evaluate

$$\frac{1}{2\pi i} \int_\Gamma \frac{f'(z)}{f(z)} \, dz$$

in the following cases:
   (a) Γ is the unit circle with its standard positive parametrization.
   (b) $\Gamma = C(0; 2)$ with a standard simple positive parametrization.

2. We verify Lemma 1 in the special case $f(z) = z^k$, where k is any integer. Let C be the positively oriented unit circle.
   (a) By observation, $N_0(C) - N_\infty(C) = k$.
   (b) Calculate $(2\pi i)^{-1} \int_C (f'(z)/f(z)) \, dz$ directly and obtain the same value k.
   (c) Check that the computation in (b), in a more general version, of course, is at the heart of the proof of Lemma 1.

3. Compute $\int_{|z|=1} \cot z \, dz$, first by using residues and then by observing that $\cot z = f'(z)/f(z)$ in the case that $f(z) = \sin z$, and then applying Lemma 1. Which method is simpler?

4. Likewise for $\int_{|z|=1} \tan z \, dz$.

5. Suppose (standard situation) that f is analytic on and inside the Jordan curve Γ except for a finite number of poles inside Γ, and that f has no zeros or poles on Γ itself. Let $\{z_j\}$ be the zeros (of orders $h_j > 0$) inside Γ, and let $\{\zeta_m\}$ be the poles (of orders $k_m > 0$) inside Γ. Prove, by imitating the proof of Lemma 1, the formula

$$\frac{1}{2\pi i} \int_\Gamma \frac{zf'(z)}{f(z)} \, dz = \sum_j h_j z_j - \sum_m k_m \zeta_m.$$

6. Generalize Exercise 5. If g is analytic on and inside Γ, and f, z_j, ζ_m are as in Exercise 5, then

$$\frac{1}{2\pi i} \int_\Gamma \frac{g(z)f'(z)}{f(z)} \, dz = \sum_j h_j g(z_j) - \sum_m k_m g(\zeta_m).$$

7. To what does the preceding formula reduce when $f(z) = z - z_0$?

7.2.2 The Winding Number

Given a closed curve Γ and point z_0 not on the curve, we want to define a number $n(\Gamma; z_0)$ which tells us how many times the curve Γ winds around the point z_0. Thus, if Γ is the unit circle with its usual positive (counterclockwise) orientation, we expect $n(\Gamma; z_0) = 0$ or 1, according as z_0 is outside or inside the circle Γ. If Γ is more complicated, as in Figure 7.5, then the winding number may be larger or smaller (even negative for negative orientation) and may be different for different points inside Γ.

How shall we define the winding number $n(\Gamma; z_0)$? We cannot rely on pictures for our definition, and we realize the desirability of an analytic definition, that is, one involving derivatives or integrals, the stuff of calculus. Of course any analytic definition should agree with our geometric intuition.

Our definition of winding number is motivated by the well-known fact that if Γ is a simple closed positively oriented loop such as a circle, then the integral

$$\frac{1}{2\pi i} \int_\Gamma \frac{dz}{z - z_0}$$

equals 0 or 1, according as the point z_0 is outside or inside the loop Γ. Thus, this integral gives the "correct" $n(\Gamma; z_0)$ in the case of a simple (nonself-intersecting) closed curve. It is a happy fact that even more is true.

The following lemma will enable us to define an integer $n(\Gamma; z_0)$ for an arbitrary closed curve Γ, even if the curve crosses itself or is not positively oriented.

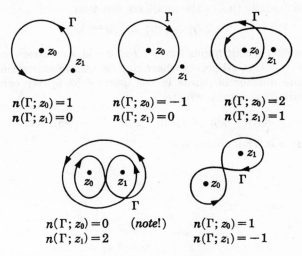

$$n(\Gamma; z_0) = 1 \qquad n(\Gamma; z_0) = -1 \qquad n(\Gamma; z_0) = 2$$
$$n(\Gamma; z_1) = 0 \qquad n(\Gamma; z_1) = 0 \qquad n(\Gamma; z_1) = 1$$

$$n(\Gamma; z_0) = 0 \quad (note!) \qquad n(\Gamma; z_0) = 1$$
$$n(\Gamma; z_1) = 2 \qquad\qquad\quad n(\Gamma; z_1) = -1$$

Winding numbers

Figure   7.5

LEMMA 2

Let Γ be a piecewise-smooth closed curve that does not pass through the point z_0. Then the integral

$$\frac{1}{2\pi i} \int_\Gamma \frac{dz}{z - z_0}$$

is an integer.

Proof: Let Γ be parametrized by $\gamma: [a, b] \to \Gamma$, a piecewise-smooth complex-valued function of a real variable. Thus, for z on Γ we have $z = \gamma(t)$ and $dz = \gamma'(t)\, dt$. Define

$$\varphi(\tau) = \int_a^\tau \frac{\gamma'(t)\, dt}{\gamma(t) - z_0}$$

and note that $\varphi(b)$ is $2\pi i$ times the value of the integral in the statement of the lemma. We will compute $\varphi(b)$.

Since $\varphi(\tau)$ is an integral, it is continuous for $a \leq \tau \leq b$. Moreover, the Fundamental Theorem of Calculus (applied to the real and imaginary parts of the integrand separately) assures us that

$$\varphi'(\tau) = \frac{\gamma'(\tau)}{\gamma(\tau) - z_0}$$

at all points where the integrand is continuous, that is, where $\gamma'(\tau)$ is continuous.

We will apply this to the auxiliary function

$$\Phi(t) = (\gamma(t) - z_0)e^{-\varphi(t)}.$$

Note that $\Phi(t)$ is continuous for $a \leq t \leq b$ and, as a direct computation shows, $\Phi'(t) = 0$ at all points where this derivative exists, that is, at all but a finite number of points in the interval $[a, b]$. We conclude that $\Phi(t)$ is actually constant, whence

$$\Phi(t) = \Phi(a) = (\gamma(a) - z_0)e^{-\varphi(a)} = \gamma(a) - z_0,$$

since $\varphi(a) = 0$. This proves that

$$e^{\varphi(t)} = \frac{\gamma(t) - z_0}{\gamma(a) - z_0},$$

whence, in particular,

$$e^{\varphi(b)} = \frac{\gamma(b) - z_0}{\gamma(a) - z_0}.$$

But since Γ is a closed curve, $\gamma(b) = \gamma(a)$, so that $e^{\varphi(b)} = 1$.

Finally, we know that the exponential function satisfies $e^z = 1$ only if z is an integer multiple of $2\pi i$. This proves the lemma.

It is worth noting that the integrand $1/(z - z_0)$ in Lemma 2 is the derivative of $\log(z - z_0)$. However, the logarithm is often rather difficult to handle, whereas $1/(z - z_0)$ is nicely behaved except at $z = z_0$. Recall that

$$\log(z - z_0) = \log|z - z_0| + i \arg(z - z_0),$$

and this is "many-valued" precisely because the angle $\arg(z - z_0)$ is defined only up to an integer multiple of 2π. By restricting ourselves to a closed curve Γ and integrating the derivative of $\log(z - z_0)$, we come up with an integer that is valuable geometrically, keeping count as it does of the number of times that the curve Γ has wound around the point z_0.

Motivated by Lemma 2, we define the *winding number* of the piecewise-smooth closed curve Γ, with respect to the point z_0 not on Γ, to be the integer

$$n(\Gamma; z_0) = \frac{1}{2\pi i} \int_\Gamma \frac{dz}{z - z_0}.$$

Some authors call $n(\Gamma; z_0)$ the *index* of Γ with respect to z_0.

Let's discuss $n(\Gamma; z_0)$ as a function (integer-valued!) of the moving point z_0. It is defined for all z_0 in $\mathbb{C} - \Gamma$, the points of the complex plane not on the curve. The set $\mathbb{C} - \Gamma$ obtained by removing Γ from the z-plane is an open set (why?), but it is not connected; it breaks up into "pieces" or "components," each of which is open and disjoint from the other pieces. $\mathbb{C} - \Gamma$ has one large unbounded "outside" component and one or more bounded "inside" components. If Γ is simple (a circle, say), then there is one outside and exactly one inside component. A "figure-eight" has two inside components, and so on.

What is $n(\Gamma; z_0)$ in the case that the point z_0 is *outside* the curve Γ? Our geometric intuition tells us that it is zero. We verify that $n(\Gamma; z_0) = 0$ in this case by actually computing the integral of $1/(z - z_0)$ around Γ, keeping the Cauchy Integral Theorem in mind. You should convince yourself of this.

Here is a somewhat more subtle question. What happens if we vary the point z_0 slightly in the formula

$$n(\Gamma; z_0) = \frac{1}{2\pi i} \int_\Gamma \frac{dz}{z - z_0} \, ?$$

Of course we do not allow z_0 to cross the curve Γ.

We agree that if z_0 is very close to z_1, then $n(\Gamma; z_0)$ is very close to $n(\Gamma; z_1)$ because there is only a slight change in the integrand. However, since the integral formula has the remarkable property that it is always an integer, and since the distance between distinct integers is not small (it is ≥ 1), we conclude that $n(\Gamma; z_0)$ actually equals $n(\Gamma; z_1)$. From the geometric point of view, of course this is reasonable; all nearby points should have the same winding number with respect to a given curve. See Figure 7.6.

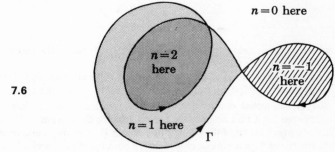

Figure   7.6

$n = 2$ here

$n = 0$ here

$n = -1$ here

$n = 1$ here

Γ

$n =$ winding number

Let us cast these observations into a lemma. There are some interesting details to be checked.

LEMMA 3

If Γ is a closed curve and z_0, z_1 are points in the same component of $\mathbb{C} - \Gamma$, then they have the same winding number, $n(\Gamma; z_0) = n(\Gamma; z_1)$. Moreover, if z_0 is outside Γ, then $n(\Gamma; z_0) = 0$.

Now that the winding number is a well-defined concept, we may return to our main business, the Argument Principle.

Exercises to Paragraph 7.2.2

1. Sketch closed curves Γ with the following winding properties. Note that it is necessary to indicate the direction of parametrization, for this affects the sign of the winding number.
   (a) All points z inside Γ satisfy $n(\Gamma; z) = 1$.
   (b) All points z inside Γ satisfy $n(\Gamma; z) = -1$.
   (c) Γ encloses various points with winding numbers 1, 2, 3.
   (d) Some points enclosed by Γ have winding number 1, the others have winding number -1.
   (e) All points enclosed by Γ have winding number 4.

2. Let Γ denote the following back-and-forth parametrization of the unit circle $|z| = 1$: The point $z(t)$ traverses the circle once in a counterclockwise direction from the point $(1, 0)$, and then it reverses direction and backtracks around the circle in a clockwise sense, returning to the starting point $(1, 0)$. Argue without computing that if $|z_0| < 1$, then $n(\Gamma; z_0) = 0$.
   To compute $n(\Gamma; z_0)$ explicitly, using the integral formula, it is necessary, of course, to write down the parametrization $z(t)$ described above as an explicit function of t.

3. Let $\beta(t) = e^{3it}$ for $0 \le t \le 2\pi$.
   (a) Observe that $\beta: [0, 2\pi] \to C$ parametrizes a path that traverses the unit circle C three times in the counterclockwise direction. We comment that e^{3it} is sometimes a more useful form than $\cos 3t + i \sin 3t$ or $(\cos 3t, \sin 3t)$.
   (b) Calculate directly that

$$\frac{1}{2\pi i} \int_\beta \frac{dz}{z} = 3,$$

where β is the parametrization in (a). This is, of course, the integral formula for $n(\Gamma; 0)$. Compare with (a).
   (c) Reverse the sense of the parametrization by altering $\beta(t)$ slightly and then verify that the winding number integral yields -3 for the clockwise path $-\Gamma$.

4. *Simply connected domains revisited.* We called a plane domain Ω "simply connected" if it had no holes or punctures. We did not define "hole," however.
   (a) Meditate on the following: The domain Ω is simply connected if and only if (whenever Γ is a simple closed curve contained in Ω and z is a point of the complement $\mathbb{C} - \Omega$) $n(\Gamma; z) = 0$. Thus, Γ encloses only points of Ω.
   (b) Prove, using (a), that the plane $\mathbb{C}$ and the disc $|z| < 1$ are simply connected.
   (c) Prove that every annulus $0 \le R_1 < |z| < R_2 \le \infty$ is multiply connected ($\equiv$ not simply connected).

5. *The proof of Lemma 3.* Prove that the winding number $n(\Gamma; z)$ is constant on components of $\mathbb{C} - \Gamma$, as follows:
   (a) The components of $\mathbb{C} - \Gamma$ are domains.
   (b) The set $Y_k = \{z \in \mathbb{C} - \Gamma \mid n(\Gamma; z) = k\}$ is open for each integer k.
   (c) Every point of $\mathbb{C} - \Gamma$ is in precisely one of the sets Y_k.
   (d) Use the connectedness of each component to argue that only one of the sets Y_k can have a nonempty intersection with that component. By (c), therefore, the set Y_k contains the component. The lemma follows.

   The style of connectedness argument in (d) appeared also in Section 1.2, and in the final result of Chapter 5.

6. Let Γ be a simple closed curve in $\mathbb{C}$. True or false?
   (a) The point $z \in \mathbb{C} - \Gamma$ is enclosed by Γ (inside Γ) if and only if $n(\Gamma; z) \neq 0$.
   (b) The point $z \in \mathbb{C} - \Gamma$ is enclosed by Γ if and only if the connected component of $\mathbb{C} - \Gamma$ that contains z is bounded.
   (c) If $z \in \mathbb{C} - \Gamma$ is enclosed by Γ, then $n(\Gamma; z) = \pm 1$. *Note:* The Jordan Curve Theorem (Chapter 1) is implicit here.

7. Convince yourself that the existence of the winding number does not immediately yield a proof of the Jordan Curve Theorem of Chapter 1; the theorem lies deeper.

7.2.3 The Complete Statement

Here is the question. If Γ is a nice closed loop in the z-plane and $w = f(z)$ is an analytic function, what does the image $f(\Gamma)$ look like?

Let's make this precise. Let Γ be a simple closed curve in the z-plane. If Γ is contained in the domain of an analytic function $w = f(z)$ which is not identically constant, then the image set

$$f(\Gamma) = \{ f(z) \mid z \in \Gamma \}$$

is a curve in the w-plane. In fact, if Γ is parametrized by the mapping $z = \gamma(t)$, $\gamma \colon [a, b] \to \Gamma$, then $f(\Gamma)$ is given by $w = f(\gamma(t))$, the composition of f and γ.

To be technically precise, we should require also that $f'(z) \neq 0$ for z on the curve Γ. See Figure 7.7. This is required because

(i) the velocity vector to the curve $w = f(\gamma(t))$ is given by $f'(\gamma(t))\gamma'(t)$, a complex number interpreted as a vector (this is the Chain Rule, of course);

(ii) in our official definition of a parametrized curve, the tangent vector is required to be nonzero at all points.

On the other hand, we know that $f'(z) = 0$ only at isolated points (since f' is analytic), so nothing very much can go wrong (we might have a finite number of zero tangents) and none of our theorems about curves is endangered.

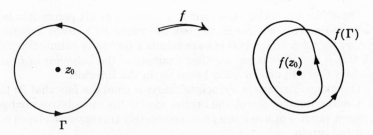

Figure   7.7

So we see that $f(\Gamma)$ is a curve in the w-plane. Suppose also that w_0 is some point in the w-plane but not on the curve $f(\Gamma)$. We ask: "How many times does $f(\Gamma)$ wrap around the point w_0?" Thus we want $n(f(\Gamma); w_0)$.

Applying the definition of winding number to the curve $f(\Gamma)$, we obtain immediately

$$n(f(\Gamma); w_0) = \frac{1}{2\pi i} \int_{f(\Gamma)} \frac{dw}{w - w_0},$$

where w is the variable on the curve $f(\Gamma)$. But since $w = f(z)$, $dw = f'(z)\, dz$, the preceding expression becomes

$$n(f(\Gamma); w_0) = \frac{1}{2\pi i} \int_\Gamma \frac{f'(z)\, dz}{f(z) - w_0}$$

where now the variable z of integration is on the original curve Γ.

What of this new integral? Actually, it is quite familiar. Notice that the integrand is the logarithmic derivative of the function $f(z) - w_0$. This is because w_0 is a constant, whence $(f(z) - w_0)' = f'(z)$.

Lemma 1 tells us that this integral is therefore equal to the number of zeros of $f(z) - w_0$ minus the number of poles of $f(z) - w_0$ occurring in the interior of the loop Γ (positively oriented). But $f(z) - w_0 = 0$ if and only if $f(z) = w_0$, and $f(z) - w_0$ has a pole at some point z if and only if $f(z)$ has a pole there also. Let us write $N_{w_0}(\Gamma)$ for the number of points z in the interior of Γ such that $f(z) = w_0$. We have proved the very useful

ARGUMENT PRINCIPLE

Let Γ be a piecewise-smooth positively oriented simple closed curve in the z-plane, and let f be analytic and nonconstant on a domain containing Γ and its interior, except perhaps for a finite number of poles strictly inside Γ. Let w_0 be a point of the w-plane not on the curve $f(\Gamma)$. Then we have

$$N_{w_0}(\Gamma) - N_\infty(\Gamma) = \frac{1}{2\pi i} \int_\Gamma \frac{f'(z)\, dz}{f(z) - w_0} = n(f(\Gamma); w_0).$$

Thus, in the special case that f is analytic at all points inside the loop Γ (no poles!), we conclude that the curve $f(\Gamma)$ wraps around the point w_0 as many times as there are points z (properly counted) inside Γ with $f(z) = w_0$. This is yet another example of the behavior of f on the boundary (here Γ) determining behavior in the interior.

The name "Argument Principle" derives from the fact that 2π times the winding number about the origin, say, is the cumulative change in argument (angle) of a moving point completely traversing a closed curve about the origin.

Some Illustrations of the Argument Principle

1. Let Γ be the usual unit circle and $f(z) = z^k$, where k is some positive integer. Let $w_0 = 0$. Without integrating anything, we see that for the function f,

$$N_0(\Gamma) = k, \qquad N_\infty(\Gamma) = 0.$$

This is because the origin $z_0 = 0$ is a k-fold root of $z^k = 0$ and because z^k has no poles. Now consider the winding number. As z traverses the unit circle once, $w = z^k$ traverses the unit circle in the w-plane exactly k times (a net change in argument of $2\pi k$), and hence winds about the origin $w_0 = 0$ the same number of times. In agreement with the Argument Principle, we have two interpretations of k:

$$N_0(\Gamma) = k = n(f(\Gamma); 0).$$

2. Let $f(z) = 1/z$, Γ = the unit circle again. Let $w_0 = 0$. Clearly, $N_0(\Gamma) = 0$, $N_\infty(\Gamma) = 1$. By the Argument Principle, we must conclude that

$$n(f(\Gamma); 0) = 0 - 1 = -1.$$

The geometric meaning of the minus sign here is that as z traverses the unit circle once in the counterclockwise sense, the point $w = f(z) = 1/z = \bar{z}$ traverses the unit circle $|w| = 1$ once in the *clockwise* sense. Draw a picture! The unit disc $|z| \leq 1$ has been turned inside-out by the mapping $w = f(z) = 1/z$, for now $|1/z| \geq 1$.

Exercises to Paragraph 7.2.3

1. Suppose f is analytic in a domain Ω that contains a disc D and the circle $C = \partial D$ with the standard parametrization.
   (a) If the image curve $f(C)$ in the w-plane is a simple (nonself-intersecting) positively oriented loop and if w is any point inside $f(C)$, what is $n(f(C); w)$? (What does "simple" mean?)
   (b) With $f(C)$ and w as in (a), how many (if any) points z in the disc D satisfy $f(z) = w$?
   (c) This time suppose there are distinct points z_1, z_2, z_3 in D, each of which is a simple (order 1) zero of f. How many times does the parametrized curve $f(C)$ wind about the origin in the w-plane if f has no other zeros in D?

2. (a) Describe the image, under the mapping $w = f(z) = z^2$, of the unit circle C with its standard parametrization $z(t) = e^{it}$, $0 \leq t \leq 2\pi$. Sketch.
   (b) What is $n(f(C); 0)$ in this case?
   (c) Clearly, $f(z) = 0$ implies $z = 0$. Doesn't the Argument Principle guarantee *two* solutions z? Explain.

3. Suppose that f is analytic in a domain Ω containing the set $0 < |z| \leq r$, and that f has a pole at $z = 0$.
   (a) Given that the origin is a simple pole and that $f(z) \neq 0$ for $|z| \leq r$, what is $n(f(C); w)$ if w is a point inside the image curve $f(C)$ and C is the circle $z(t) = re^{it}$, $0 \leq t \leq 2\pi$?

(b) Argue that the image curve $f(C)$, which is given by $w(t) = f(re^{it})$, is a simple closed loop oriented negatively.

(c) Suppose the mapping $w = f(z)$ is one-to-one in the set $0 < |z| \leq r$. Prove that every point z of this set is mapped to a point $f(z)$ *exterior* to the curve $f(C)$; that is, $n(f(C); f(z)) = 0$. *Thus, the mapping turns a neighborhood of the pole "inside-out."*

(d) Prove that the point $w = 0$ is enclosed by the image curve $f(C)$.

(e) The converse of (c). Prove that if w is a point not enclosed by the curve $f(C)$, then $w \in f(D)$.

Note: It might be helpful to visualize the mapping properties of $f(z) = 1/z$ near the origin $z = 0$.

4. Let f be analytic in a domain that contains a simple closed curve Γ and the domain of all points inside Γ. Suppose w_1, w_2 are points that lie in the same component of the complement of the image curve (a loop) $f(\Gamma)$. Argue that the two equations $f(z) = w_1, f(z) = w_2$ have the same number of solutions z (properly counted) lying inside Γ.

5. *One-to-oneness.* Let f be analytic in a domain Ω that contains a disc D and its circular boundary C. Prove that f gives a one-to-one mapping of the disc D if and only if it gives a one-to-one mapping of the circle C. *Hint:* See Exercise 1.

    Note that we do not deduce f one-to-one on the full domain Ω. The result here is fundamental in Chapter 8.

6. Give an example of an analytic function which maps the unit circle $|z| = 1$ onto the unit circle $|w| = 1$ in a 17-to-1 manner; that is, seventeen points z, properly counted, are mapped to each point w.

7. Call your example from Exercise 6 $w = f(z)$. Let $z(\theta) = e^{i\theta}$ parametrize the circle $|z| = 1$ as usual $(0 \leq \theta \leq 2\pi)$.

    (a) As θ increases from $\theta = 0$ to $\theta = 2\pi$, how many times does the point $w(\theta) = f(z(\theta))$ traverse the circle $|w| = 1$?

    (b) Check that, if your example $f(z)$ is analytic throughout the open disc $|z| < 1$ (at least), then the point $w(\theta)$ traverses the circle $|w| = 1$ in the counterclockwise (positive) direction as θ increases from $\theta = 0$ to $\theta = 2\pi$. Thus, your function preserves the orientation of circles from the z-plane to the w-plane.

8. If you did not do this in Exercise 6, give a function $w = g(z)$ which maps $|z| = 1$ onto $|w| = 1$ in a 17-to-1 manner and which reverses orientation. (By Exercise 7(b), $g(z)$ must have a singularity inside $|z| < 1$.)

Appendix: REAL INTEGRALS EVALUATED BY RESIDUES

We will show here how the Residue Theorem may be used to evaluate some types of real integral. This discussion will not be applied elsewhere in this book. Nor does it involve the Argument Principle.

Roughly, the method is this: To calculate the real value

$$I = \int_a^b f(x) \, dx,$$

(i) relate the integral I to a well-chosen complex line integral;
(ii) evaluate the complex integral by means of the Residue Theorem;
(iii) extract the value of I from the complex value found in (ii).

Examples will clarify this outline. We discuss three types.

First Type: Integrating from $-\infty$ to ∞

We define

$$\int_{-\infty}^{\infty} f(x)\, dx = \lim_{R \to \infty} \int_{-R}^{R} f(x)\, dx,$$

provided the limit exists. *Caution:* This definition, sometimes called the *principle value* of the integral, is not so restrictive as

$$\lim_{\substack{B \to \infty \\ A \to -\infty}} \int_{A}^{B} f(x)\, dx,$$

where the limits may be taken independently and yet are required to yield the same value. In our definition (principal value), every odd function (such as x^3 or arctan x) clearly has zero integral over the entire x-axis. On the other hand, an asymmetric limit such as

$$\lim_{A \to -\infty} \int_{A}^{\infty} x^3\, dx = \lim_{A \to -\infty} (\infty)$$

clearly does not converge to a finite value. Nonetheless, our definition of the integral does serve to illustrate the method of residues and, of course, gives the same value as the stronger integral when the latter exists.

Here is the key to evaluating the principal value of certain integrals over the entire x-axis.

LEMMA

Let $f(z)$ be defined and analytic in a domain containing the closed upper half-plane $y \geq 0$ except for a finite number of poles in the open upper half-plane $y > 0$. Suppose further that $f(z)$ vanishes so rapidly at infinity that

$$\lim_{|z| \to \infty} zf(z) = 0$$

uniformly for z in the closed upper half-plane. Then

$$\int_{-\infty}^{\infty} f(x)\, dx = 2\pi i \text{ (sum of residues of } f \text{ at poles in upper half-plane).}$$

Proof: We denote by $C^+(0; R)$ that part of the circle that lies in the upper half-plane. See Figure 7.8. Thus, for $z = x + iy$,

$$C^+(0; R) = \{z \in \mathbb{C} \mid |z| = R, y \geq 0\}.$$

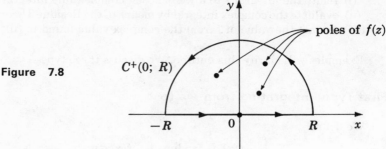

Figure 7.8

Now we observe that

$$\int_{-R}^{R} f(x)\, dx + \int_{C^+(0;R)} f(z)\, dz$$

is equal to $2\pi i$ times the sum of the residues of f inside the half-disc bounded by the segment $[-R, R]$ and the semicircle $C^+(0; R)$ (why?).

If R is large enough, *all* the poles of f with $y > 0$ will be contained inside this half-disc. In this case we know that

$$\lim_{R \to \infty} \int_{-R}^{R} f(x)\, dx + \lim_{R \to \infty} \int_{C^+(0;R)} f(z)\, dz$$

equals $2\pi i$ times the sum of the residues at all poles of $f(z)$ in the upper half-plane. Since the first term of this sum is the integral to be evaluated, it suffices to prove that the second limit is zero. This is the key observation. We will, of course, use the hypothesis that $f(z)$ vanishes rapidly for $|z|$ large.

For $|z| = R$, we have $z = Re^{i\theta}$, $dz = iRe^{i\theta}\, d\theta$. Thus,

$$\int_{C^+(0;R)} f(z)\, dz = \int_0^\pi f(Re^{i\theta}) iRe^{i\theta}\, d\theta.$$

We now apply a variant of the ML-inequality to show that the right-hand integral tends to zero as $R \to \infty$. The hypothesis on $zf(z)$ assures us that for each $\varepsilon > 0$, there is a radius R_ε such that if $R > R_\varepsilon$, then $|f(Re^{i\theta})R| < \varepsilon$. Hence, the integral in θ is bounded in absolute value by $\pi\varepsilon$. Since ε may be taken arbitrarily small, we conclude that

$$\lim_{R \to \infty} \int_{C^+(0;R)} f(z)\, dz = 0.$$

This proves the lemma.

Here is an application. We compute the improper integral

$$\int_{-\infty}^{\infty} \frac{dx}{1 + x^2}.$$

1. We are given

$$f(x) = \frac{1}{1 + x^2} \quad \text{so} \quad f(z) = \frac{1}{1 + z^2}.$$

2. We verify that $zf(z) \to 0$ as $|z| \to \infty$. This is true because $zf(z)$ behaves like $1/z$ for large $|z|$. Hence the lemma applies.

3. $f(z)$ has one pole in the upper half-plane, namely, $z_0 = i$.

4. We calculate $\text{res}(f; i) = -i/2$, as done in Section 7.1.

5. We conclude immediately from the lemma that

$$\int_{-\infty}^{\infty} \frac{dx}{1 + x^2} = 2\pi i \left(\frac{-i}{2}\right) = \pi.$$

6. Happily, we may check this particular integral if we recall from calculus that $(d/dx) \arctan x = 1/(1 + x^2)$. This means that

$$\int_{-\infty}^{\infty} \frac{dx}{1 + x^2} = \lim_{R \to \infty} \{\arctan x]_{-R}^{R}\}$$

$$= \frac{\pi}{2} - \left(-\frac{\pi}{2}\right)$$

$$= \pi,$$

as found above in (5) using residues.

Here is a timely question. To which real functions $f(x)$ does the lemma apply? We give a partial answer.

Let $f(x) = A(x)/B(x)$ be a quotient of polynomials such that $B(x) = 0$ has no real roots and also

$$\text{degree } B(x) \geq \text{degree } A(x) + 2.$$

In this case, $f(x)$ is defined for all real x, so we might ask for

$$\int_{-\infty}^{\infty} f(x)\, dx.$$

Moreover, $f(z) = A(z)/B(z)$ has only a finite number of poles in the z-plane and vanishes rapidly as $|z| \to \infty$. For $A(z)/B(z)$ behaves like $1/z^k$ with $k \geq 2$ when $|z|$ is large, thanks to the hypothesis on degrees. Thus, $zf(z)$ behaves like $1/z^{k-1}$, which vanishes as $|z|$ gets very large. Perhaps the simplest example of this is $f(x) = 1/(x^2 + 1)$ done above.

Some Exercises

You are invited to apply the lemma to verify

$$\int_{-\infty}^{\infty} \frac{dx}{x^2 + x + 1} = \frac{2\pi}{\sqrt{3}}, \quad \int_{-\infty}^{\infty} \frac{x - 2^{1/3}}{x^3 - 2}\, dx = \frac{2\pi}{3^{1/2} 2^{1/3}}.$$

Here, $2^{1/3}$ is the real cube root of 2. Check first that the lemma applies in each problem.

Second Type: Integrals Involving Exponentials; Jordan's Lemma

We wish to deal with improper integrals of the forms

$$\int_{-\infty}^{\infty} g(x)e^{ikx}\, dx, \qquad \int_{-\infty}^{\infty} g(x)\cos kx\, dx, \qquad \int_{-\infty}^{\infty} g(x)\sin kx\, dx.$$

The method is similar to that used above, but a somewhat different lemma is called for. It is this.

JORDAN'S LEMMA

Let $g(z)$ be analytic in a domain containing the closed upper half-plane $y \geq 0$ except perhaps for a finite number of poles in the open upper half-plane $y > 0$. Suppose further that

  (i) *$g(z) \to 0$ uniformly as $|z| \to \infty$ in the closed upper half-plane, and*

  (ii) *$k > 0$.*

Then

$$\lim_{R \to \infty} \int_{C^+(0;R)} g(z)e^{ikz}\, dz = 0,$$

where the integral is taken over the semicircle in the upper half-plane.

COROLLARY

$$\int_{-\infty}^{\infty} g(x)e^{ikx}\, dx = 2\pi i \text{ (sum of residues of } g(z)e^{ikz} \text{ in upper half-plane).}$$

Proof of Jordan's Lemma: If R is taken large enough, we have $|g(z)| < \varepsilon$ for $|z| = R$. Now we deal with the exponential. We have $|\exp ikz| = |\exp\{ikR(\cos\theta + i\sin\theta)\}| = \exp\{-kR\sin\theta\}$, since $|\exp it| = 1$ for t real. Hence,

$$\left| \int_{C^+(0;R)} g(z)e^{ikz}\, dz \right| = \left| \int_0^\pi g(z)e^{ikz}Re^{i\theta}\, d\theta \right|$$

$$< \varepsilon \int_0^\pi e^{-kR\sin\theta}\, R\, d\theta$$

$$= 2R\varepsilon \int_0^{\pi/2} e^{-kR\sin\theta}\, d\theta.$$

We used the symmetry of $\sin\theta$ about $\theta = \pi/2$ for the last equality.

Now let us replace the $\sin\theta$ in the exponent by something simpler. We assert that if $0 \leq \theta \leq \pi/2$, then $\sin\theta/\theta \geq 2/\pi$. You may prove this by evaluating $\sin\theta/\theta$ at $\theta = 0$ and $\theta = \pi/2$, and considering its first derivative on the interval $0 < \theta < \pi/2$. Or graph the curve $y = \sin\theta$ and the line $y = 2\theta/\pi$ and observe $\sin\theta \geq 2\theta/\pi$ when $0 \leq \theta \leq \pi/2$.

Since $-kR \sin \theta \le -2kR\theta/\pi$, we have now that

$$\left| \int_{C^+(0;R)} g(z)e^{ikz}\, dz \right| < 2R\varepsilon \int_0^{\pi/2} e^{-2kR\theta/\pi}\, d\theta$$

$$= \frac{\pi}{k}\, \varepsilon(1 - e^{-kR}) < \frac{\pi\varepsilon}{k}\,.$$

Thus, our original integral may be bounded by an arbitrarily small quantity if R is large enough. Done.

Now we single out a class of integrals to which Jordan's Lemma applies. Suppose $g(x) = A(x)/B(x)$ is a rational function with real coefficients such that the polynomial $B(x)$ has no roots on the x-axis, and also the degree of $B(x)$ is ≥ 1 + the degree of $A(x)$. Then, as we have asserted previously, $g(z) = A(z)/B(z)$ tends to zero uniformly as $|z|$ becomes large. Thus, such rational functions $g(z)$ satisfy the conditions of Jordan's Lemma.

By using the same method as with the first type of integral we discussed, we may show readily (thanks to Jordan's Lemma) that

$$\int_{-\infty}^{\infty} g(x)e^{ikx}\, dx = 2\pi i \text{ (sum of residues of } g(z)e^{ikz} \text{ in upper half-plane)}$$

Note that the integrand $g(z)e^{ikz}$ has poles at the poles of $g(z)$ (why?). However, the residues of the integrand may, of course, differ from the residues of $g(z)$, thanks to e^{ikz}.

Note also that the integrand is a complex-valued function of x. By taking real and imaginary parts of both sides of the last equation, we may evaluate integrals of the type

$$\int_{-\infty}^{\infty} g(x) \cos kx\, dx, \qquad \int_{-\infty}^{\infty} g(x) \sin kx\, dx.$$

Example

Here is a typical problem. Compute

$$\int_{-\infty}^{\infty} \frac{\cos kx}{x^2 + a^2}\, dx \qquad (a > 0,\, k > 0).$$

1. First we observe that this is twice the integral from $x = 0$ to $x = \infty$ because the integrand is an even function of x. Likewise, if we replace $\cos kx$ by $\sin kx$ in the integrand, the integral will vanish (why?).

2. Of course our integral is the real part of

$$\int_{-\infty}^{\infty} \frac{e^{ikx}}{x^2 + a^2}\, dx.$$

3. We leave it to you to check that the rational function $g(z) = (z^2 + a^2)^{-1}$ satisfies the conditions outlined above, whence our labors are reduced to computing the residues of $g(z)e^{ikz}$ in the upper half-plane.

4. Since $g(z) = 1/(z - ia)(z + ia)$, the only pole of $g(z)e^{ikz}$ satisfying $y > 0$ is at $z = ia$ (recall $a > 0$).

5. We compute

$$\text{res}(g(z)e^{ikz}; ia) = \lim_{z \to ia} (z - ia)g(z)e^{ikz}$$

$$= \frac{e^{-ka}}{2ai}.$$

6. Thus, the original integral has a real value

$$\int_{-\infty}^{\infty} \frac{e^{ikx}}{x^2 + a^2} \, dx = 2\pi i \left(\frac{e^{-ka}}{2ai} \right) = \frac{\pi}{a} e^{-ka}.$$

7. The desired integral is the real part of this value. We conclude that

$$\int_{-\infty}^{\infty} \frac{\cos kx}{x^2 + a^2} \, dx = \frac{\pi}{a} e^{-ka},$$

$$\int_{-\infty}^{\infty} \frac{\sin kx}{x^2 + a^2} \, dx = 0.$$

Some Exercises Involving Jordan's Lemma

Verify for $k > 0$ that

$$\int_{-\infty}^{\infty} \frac{\cos kx \, dx}{x^2 + x + 1} = 2\pi e^{-\sqrt{3}k/2} \sin \left(\frac{k}{2} \right),$$

$$\int_{-\infty}^{\infty} \frac{\sin kx \, dx}{x^2 + x + 1} = 2\pi e^{-\sqrt{3}k/2} \cos \left(\frac{k}{2} \right).$$

Third Type: Integrals of Trig Functions

We now consider real integrals of the type

$$I = \int_0^{2\pi} \Phi(\cos \theta, \sin \theta) \, d\theta,$$

where Φ is a rational function (with real coefficients) in two variables. The method now will be rather different. The integral I can be shown equal to a certain complex contour integral around the circle $|z| = 1$, as follows:

Let $z = e^{i\theta} = \cos \theta + i \sin \theta$. In this case, $\bar{z} = 1/z = \cos \theta - i \sin \theta$. Now we may solve for sine and cosine, obtaining

$$\cos \theta = \frac{1}{2} \left(z + \frac{1}{z} \right), \qquad \sin \theta = \frac{1}{2i} \left(z - \frac{1}{z} \right).$$

Note also that $dz = de^{i\theta} = ie^{i\theta}\,d\theta$, so that

$$d\theta = \frac{dz}{iz}.$$

Making these substitutions yields a line integral around the positively oriented unit circle C given by $|z| = 1$:

$$I = \int_0^{2\pi} \Phi(\cos\theta, \sin\theta)\,d\theta = \int_C \varphi(z)\,dz,$$

where $\varphi(z)$ is easily seen to be a rational function in z with complex coefficients. But this last integral may be calculated by the Residue Theorem (provided $\varphi(z)$ has no poles on C or—same thing—the original integral converged).

Example

Here is an application of this method of substitution. We examine the integrals (real a, b)

$$I = \int_0^{2\pi} \frac{d\theta}{a + b\cos\theta}.$$

1. Note first that if $|a| \le |b|$, then the denominator here has a zero for some θ between 0 and 2π. Hence, we require $|a| > |b| > 0$. There is no loss now in assuming $a > 0$; if not, multiply the integral by -1.

2. Substitution for $\cos\theta$ and $d\theta$ yields an integral in z:

$$I = \frac{-2i}{b} \int_C \frac{dz}{z^2 + (2az/b) + 1}.$$

3. To obtain residues, we must first calculate the roots of the denominator in (2). Using the familiar quadratic formula, we obtain the roots

$$z_1 = \frac{-a + \sqrt{a^2 - b^2}}{b}, \qquad z_2 = \frac{-a - \sqrt{a^2 - b^2}}{b}.$$

Note that they are real and unequal because $a^2 > b^2$.

4. Which of these roots (poles of the integrand) lies inside the unit circle $|z| = 1$? We note $z_1 z_2 = $ constant term of the polynomial in the denominator $= 1$, so that $|z_1 z_2| = 1$. But $|z_1| \ne |z_2|$, so exactly one of z_1, z_2 must satisfy $|z| < 1$.

Clearly, $|z_1| < 1$ because $a > 0$ and $\sqrt{a^2 - b^2} > 0$, whence the numerator of z_1 has smaller absolute value than the numerator of z_2.

5. We may now write the integrand as

$$\varphi(z) = \frac{1}{(z - z_1)(z - z_2)}.$$

The residue at $z = z_1$ is obtained in the usual way:

$$\operatorname{res}(\varphi; z_1) = \lim_{z \to z_1} (z - z_1)\varphi(z) = \frac{1}{z_1 - z_2}$$

$$= \frac{b}{2\sqrt{a^2 - b^2}}.$$

6. The Residue Theorem now tells us that

$$I = \frac{-2i}{b} \int_C \varphi(z)\, dz = \frac{-2i}{b} \left\{ 2\pi i \left(\frac{b}{2\sqrt{a^2 - b^2}} \right) \right\}.$$

That is,

$$\int_0^{2\pi} \frac{d\theta}{a + b \cos \theta} = \frac{2\pi}{\sqrt{a^2 - b^2}} \qquad (a > |b|)$$

An Exercise

Verify

$$\int_0^{2\pi} \frac{d\theta}{a + b \sin \theta} = \frac{2\pi}{\sqrt{a^2 - b^2}} \qquad (a > |b|)$$

You may do this in two ways: either by carrying out the method described above *or* by using the result obtained in the example above; note that we got this same value when cos θ appeared in the integrand instead of sin θ. Why is this?

8

Analytic Functions as Conformal Mappings

Section 8.1 MAPPING BY ANALYTIC FUNCTIONS

8.1.0 Introduction

Now we will be considering an analytic function $w = f(z)$ as a geometric mapping or transformation of a domain Ω in the z-plane into a subset of the w-plane. This is a huge subject, one that is highly technical. Therefore, let us organize our discussion (most of it, at any rate) around two general problems.

1. *Description of the Image.* If S is a subset of the z-plane with certain properties or a certain structure, and if f is given, does the image set $f(S)$—that is, $\{f(z) \mid z \in S\}$—also have these properties? For example, if Ω is a domain, is $f(\Omega)$ a domain?

2. *The Search for Mappings.* Given subsets S_1, S_2 of the z- and w-planes, respectively, does there exist an analytic function f mapping S_1 onto S_2 in a nice way? For example, does there exist a one-to-one analytic mapping of the upper half of the z-plane onto the open unit disc $|w| < 1$? Can we write down all such mappings?

Most of what we shall say in Section 8.1 relates to problem (1). We will use the Argument Principle crucially here. The second problem will be emphasized in Sections 8.2 and 8.3.

8.1.1 Local Approximation to *f*

We adopt the following principle, very much in the spirit of calculus: Before we can hope to discuss *global* questions such as the nature of the image $f(\Omega)$ for a given (possibly enormous) domain Ω, we must first have some idea of the *local* behavior of f, that is, behavior near a point z_0 of Ω. We ask: "If f is analytic in Ω and z_0 is a point of Ω, what does f do to a small enough open neighborhood around z_0?" We rule out the trivial case where f is identically constant.

A great deal is known about this local problem. In summary, the answer is this: First expand f in Taylor series centered at z_0, the point of interest:

$$f(z) = f(z_0) + \frac{f^{(k)}(z_0)}{k!} (z - z_0)^k + \text{higher powers of } z - z_0,$$

with $f'(z_0) = \cdots = f^{(k-1)}(z_0) = 0$, but $f^{(k)}(z_0) \neq 0$. Then the following statements hold (proofs later):

(i) For z near z_0, $f(z)$ behaves like a simple polynomial $P(z)$, namely,

$$w = P(z) = f(z_0) + \frac{f^{(k)}(z_0)}{k!} (z - z_0)^k.$$

(ii) If $k = 1$, $f'(z_0) \neq 0$, then $f(z)$ behaves very well near z_0 because $P(z)$ is a linear function. But if $k \geq 2$, then f behaves less well.

(iii) Happily, a point z_0 at which $f'(z_0) = 0$ is isolated from the collection of all other points with this property.

This prompts some definitions. If $f'(z_0) = 0$, then we say that z_0 is a *critical point* for f. We refine this notion as follows: If the first $k - 1$ derivatives of f at z_0 vanish, $f'(z_0) = \cdots = f^{(k-1)}(z_0) = 0$, but $f^{(k)}(z_0) \neq 0$; then we say that f has *order* k at z_0. Note that z_0 is critical if and only if f has order $k \geq 2$ at z_0.

The standard example is $f(z) = z^k$, which has order k at $z_0 = 0$ and order 1 at all other points.

Since critical points are zeros of the analytic function f', they are isolated by Theorem 16, Paragraph 5.3.4. This is statement (iii) above.

You may recall how important the condition $f'(x) = 0$ was in ordinary real calculus. We shall see many similarities now, as well as some significant differences.

In order to emphasize the good behavior of analytic functions near noncritical points, and also because these points are more abundant (nonisolated), we will first devote ourselves to their study.

Exercises to Paragraph 8.1.1

1. (a) Verify that $f(z) = \sin z$ has order 1 (is noncritical) at $z_0 = 0$.
   (b) Verify that statement (i) of the text implies that the mapping $w = f(z) = \sin z$ behaves near the origin like the identity mapping $w = P(z) = z$.
   (c) Relate (b) to the Maclaurin expansion $\sin z = z - z^3/3! + \cdots$.

2. Locate all critical points of the following functions. Determine the order in each case.
   (a) $\cos z$.
   (b) z^2.
   (c) e^z.
   (d) $z + (1/z)$.

3. Let $f(z) = (az + b)/(cz + d)$ with a, b, c, d complex and $ad - bc \neq 0$. Prove that f has no critical points.

4. *Basic terminology.* Let f map the plane set S_1 into the plane set S_2 (written $f: S_1 \to S_2$ or $f(S_1) \subset S_2$). What does it mean to say that f is *one-to-one* from S_1 into S_2? that f maps S_1 *onto* S_2?

8.1.2 Behavior Near Noncritical Points; Conformality

We will be dealing with two sorts of theorems. One is algebraic, concerned with the number of solutions z to the equation $f(z) = w$, where w is fixed. The other is geometric, concerned with distortion of a small disc centered at z_0 under mapping by f.

THEOREM 1

Let f be analytic in a domain Ω containing the point z_0. If z_0 is a noncritical point for f, $f'(z_0) \neq 0$, then f is locally one-to-one near z_0. That is, there is an open neighborhood Ω_0 of z_0 inside Ω such that if $z_1 \neq z_2$ in Ω_0, then $f(z_1) \neq f(z_2)$.

Proof: We first choose $D = D(z_0; r)$ with $\partial D = C = C(z_0; r)$ such that

(i) the closure $\overline{D}$ is contained in Ω;
(ii) $\overline{D}$ contains no point z except z_0 such that $f(z) = f(z_0)$.
This is possible because points z with $f(z) = f(z_0)$ are isolated.

By (ii), the curve $f(C)$ does not pass through $w_0 = f(z_0)$. Let w_0 be the connected component of $\mathbb{C} - f(C)$, which contains the point w_0. We have, in the notation of Chapter 7, that

$$N_w(C) = n(f(C); w) = n(f(C); w_0) = N_{w_0}(C) = 1,$$

provided w is in W_0. The first and third equal signs here follow from the Argument Principle, the second follows from the constancy of n on components of $\mathbb{C} - f(C)$, and the fourth equal sign follows from (ii) above.

Now we consider $f^{-1}(W_0) \cap D$. See Figure 8.1. This set is open. The reason is standard: W_0 is open (a component) and f is continuous. We note that z_0 is in the open set $f^{-1}(W_0) \cap D$ and also that f maps this set in a one-to-one fashion. Thus, we let $\Omega_0 = f^{-1}(W_0) \cap D$. Done.

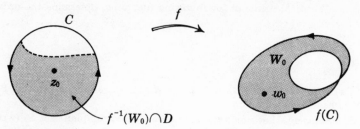

Figure   8.1

Comments

1. We do not estimate the size of Ω_0. This is a more difficult problem.

2. The proof depended heavily on the Argument Principle, which in turn depended on power series (in our treatment).

3. In real calculus, if $f'(x_0) \neq 0$, then the graph $y = f(x)$ is increasing or decreasing above x_0, and so the function is locally one-to-one, just as in Theorem 1 here. However, it is possible for *real* functions to be one-to-one in the vicinity of a critical point; an example is given by $y = f(x) = x^3$, $x_0 = 0$. See Figure 8.2. We will see that this does *not* happen with analytic functions. In fact, contrast $w = f(z) = z^3$. This is locally three-to-one near $z_0 = 0$, not one-to-one, thanks to the existence of complex cube roots!

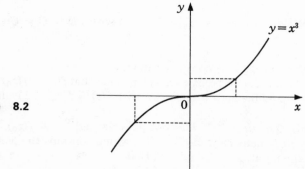

Figure   8.2

This function is one-to-one
for real x and y.

A Geometric Result

Now we will see what the analytic mapping $w = f(z)$ does to straight lines drawn through the noncritical point z_0 in Ω. This will give us a picture of how a small disc about z_0 is distorted by the mapping f as it is carried onto a subset of the w-plane.

It is easy to check that a straight line through z_0 making an angle φ with the horizontal is given by the set of points $\sigma(s)$ of the form (here s is real)

$$\sigma(s) = z_0 + se^{i\varphi}.$$

Since $|e^{i\varphi}| = 1$, s is the arc length parameter, in fact.

The line $z = \sigma(s)$ is carried by f to the curve

$$w = f(\sigma(s)) = f(z_0 + se^{i\varphi}).$$

The velocity vector to this curve at the image point $w_0 = f(z_0)$ is, when written as a complex number,

$$\frac{d}{ds} f(\sigma(s)) \bigg|_{s=0} = f'(z_0)\sigma'(0) = f'(z_0)e^{i\varphi}.$$

What angle does this vector make with the horizontal? It is the argument of the corresponding complex number; that is,

$$\arg\{f'(z_0)e^{i\varphi}\} = \arg f'(z_0) + \varphi.$$

Thus, the tangent line to the new curve $w = f(\sigma(s))$ makes an angle with the horizontal that differs from the original tangent angle by $\arg f'(z_0)$. This is true for all φ. Hence, near z_0, f tends to rotate lines radiating from z_0 through a constant angle $\arg f'(z_0)$. See Figure 8.3. In particular,

z-plane w-plane

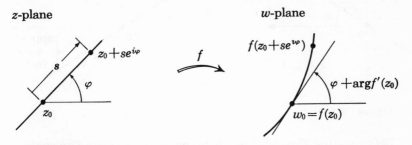

Figure 8.3

if two lines L_1 and L_2 through z_0 are mapped to two curves through w_0, then the counterclockwise angle between the image curves (that is, between their tangents) at w_0 equals the counterclockwise angle ψ between the original lines. See Figure 8.4.

This calls for a definition. It is classical. A mapping $f: \Omega \to \mathbb{C}$ is *conformal at z_0* in Ω if and only if the mapping preserves both magnitude and sense of angles at z_0; that is, the counterclockwise angle between two straight lines through z_0 is equal to the counterclockwise angle between their respective image curves through the point $w_0 = f(z_0)$.

Note that the mapping $f(z) = \bar{z}$, complex conjugation, is *not* conformal; reflection across the x-axis reverses the sense of angles. See

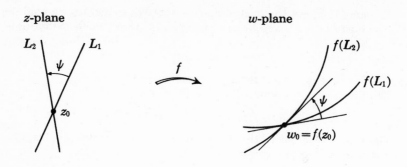

Conformality at z_0

Figure   8.4

Figure 8.5, where the angle from L_1 to L_2 is ψ and the angle from $f(L_1)$ to $f(L_2)$ is $-\psi$.

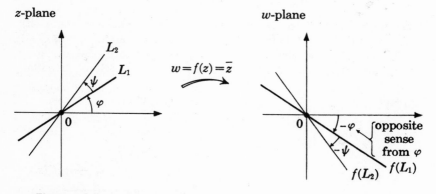

Figure   8.5

Before stating Theorem 2 on the geometric properties of an analytic mapping at a noncritical point, we make another observation. Suppose z is near z_0. How far is the image point $f(z)$ from $f(z_0)$? An approximate answer is indicated by the elementary observation that

$$|f'(z_0)| = \lim_{z \to z_0} \frac{|f(z) - f(z_0)|}{|z - z_0|}.$$

For z close to z_0, therefore, $|f'(z_0)|$ is approximately equal to the quotient on the right. Cross-multiplying gives

$$|f(z) - f(z_0)| \approx |f'(z_0)| \, |z - z_0|.$$

Thus, mapping by f tends to magnify all distances near z_0 in the z-plane by a factor equal to $|f'(z_0)|$. This positive number is termed the *local magnification factor*.

In summary, we have seen that both the absolute value and the argument of $f'(z_0)$ carry geometric information about the local mapping accomplished by f. This is essentially the content of the following classical result.

THEOREM 2

Let $f: \Omega \to \mathbb{C}$ be analytic. If z_0 in Ω is a noncritical point, $f'(z_0) \neq 0$, then

    (i) *f is conformal at z_0;*
    (ii) *in fact, mapping by f rotates angles at z_0 through a constant amount $\arg f'(z_0)$;*
    (iii) *moreover, mapping by f magnifies distances near z_0 by a factor approximately equal to $|f'(z_0)|$.*

Moral

If $f(z) = f(z_0) + f'(z_0)(z - z_0) + \cdots$ with $f'(z_0) \neq 0$, then f behaves near z_0 very much like the linear approximation

$$w = P(z) = f(z_0) + f'(z_0)(z - z_0)$$

obtained by chopping off the power series after the first-order term. For $P(z)$ is one-to-one, is conformal, and—because $P'(z) = f'(z_0)$ for all z— $P(z)$ rotates lines at z_0 through a constant angle $\arg f'(z_0)$ and has the local magnification factor $|f'(z_0)|$. Of course this particular $P(z)$ gives a good approximation to $f(z)$ only if z is near z_0.

The reason behind all this is, of course, that higher-order terms in $(z - z_0)^2, (z - z_0)^3, \ldots$ are negligible compared with $z - z_0$ when the latter is small.

Caution—Local vs. Global: A mapping $w = f(z)$ may have no critical points in Ω and hence be locally one-to-one near each point z_0. Yet f may fail to be globally one-to-one in Ω! In fact, there may be infinitely many solutions z to $f(z) = w$ for most w. Consider $f(z) = e^z$, which has no critical points in $\Omega = \mathbb{C}$. If $w \neq 0$, then $e^z = w$ has infinitely many solutions z. Locally one-to-one, globally infinite-to-one!

Exercises to Paragraph 8.1.2

1. (a) Verify that $f(z) = e^z$ has a noncritical point at the origin.
   (b) Locate all critical points of the exponential function.
   (c) Find the largest radius $R > 0$ such that $w = e^z$ gives a one-to-one mapping of the disc $D(0; R)$ into the w-plane. Note that Theorem 1 guarantees the existence of such a disc about the origin.

2. (a) Verify that $g(z) = z^2$ has a critical point (order 2) at the origin $z_0 = 0$.
   (b) Show that the mapping $w = g(z)$ is not one-to-one when defined on any disc of the type $D(0; R)$ in the z-plane.
   (c) Reinterpret (b) in terms of existence of square roots of w.

3. (a) Verify that the mapping $w = g(z) = z^2$ is locally one-to-one near any point $z_1 \neq 0$. Contrast Exercise 2.

(b) In fact, if $z_1 \neq 0$ and $R_1 = |z_1|$, then $w = g(z)$ gives a one-to-one mapping when restricted to the disc $D(z_1; R_1)$, which does not contain the origin.

(c) Is R_1 the largest such radius for discs centered at z_1? Proof or counter-example.

4. (a) Prove there exists no domain Ω in the z-plane such that the mapping $w = g(z) = z^2$ is one-to-one when restricted to Ω and also $g(\Omega)$ equals the entire w-plane. Note that this would give us a nice unique $z = \sqrt{w}$ for all w.

(b) Is (a) possible if we delete the origin from the w-plane?

(c) Let S be the subset of the z-plane consisting of the upper half-plane $y > 0$ together with the nonnegative x-axis, $y = 0$, $x \geq 0$ (origin included). Verify that the squaring function $g: S \to \mathbb{C}$ is one-to-one and onto. Does this contradict (a)?

(d) Deduce that every complex number w that is not zero or a positive real number has precisely one square root $z = \sqrt{w}$ in the upper half-plane $y > 0$.

(e) Let the relation $z = \sqrt{w}$ in (d) be defined on the entire w-plane and give a one-to-one mapping of the w-plane onto the set S of (c). Prove that this mapping is analytic at w only if w does not lie on the ray $u \geq 0$, $v = 0$. *Note:* $w = 0$ is a branch point for $z = \sqrt{w}$.

5. (a) Verify that the x-axis and y-axis near $z = 0$ are mapped by $w = e^z$ to a pair of curves that are perpendicular at $w = 1$, as predicted by Theorem 2.

(b) Verify that every vertical line $x = x_0$ is wrapped by $w = e^z$ around the circle of radius $R = e^{x_0}$ in the w-plane. Likewise, the horizontal line $y = y_0$ is mapped onto the ray arg $w = y_0$. Draw a picture. *Hint:* Review our discussion of exponential mapping in Paragraph 3.4.2.

(c) Let $f(z) = e^z$. What is arg $f'(z_0)$ in terms of $z_0 = x_0 + iy_0$?

(d) Verify in your picture (b) that, near their point of intersection, the lines $x = x_0$ and $y = y_0$ have each been rotated by the mapping $w = f(z) = e^z$ through an angle equal to arg $f'(z_0)$.

6. True or false?

(a) The mapping $w = f(z) = e^z$ is conformal at each point z_0.

(b) The exponential function gives a one-to-one mapping of the z-plane onto the punctured w-plane $w \neq 0$.

(c) If $w = f(z)$ is conformal at z_0, then straight-line segments through z_0 are mapped to straight-line segments through $w_0 = f(z_0)$.

(d) If $w = f(z)$ is conformal at z_0, then it is one-to-one near z_0.

(e) The mapping $w = f(z) = az$, $|a| = 1$, rotates the z-plane through an angle arg a.

7. Let $f(z) = z^2$, $z_0 = i$.

(a) Compute $|f'(z)|$.

(b) Let $z = z_0 + (1/10)$. Compute $|f(z) - f(z_0)|$.

(c) Verify that $|f(z) - f(z_0)| \approx |f'(z_0)| \, |z - z_0|$, as claimed in Theorem 2.

8. Let $w = f(z)$ be conformal at z_0. Prove that a sufficiently small rectangle with one corner at z_0 and area equal to A is mapped by f to a "curvilinear rectangle" whose area is approximately $|f'(z_0)|^2 A$.

9. Let f be analytic on and throughout the interior of a simple closed curve Γ in the z-plane. Suppose the mapping $w = f(z)$ is one-to-one when restricted

to Γ. Prove that it is one-to-one in the interior of Γ as well. *Hint:* Winding number. See Exercise 5 to Paragraph 7.2.3.

10. *An elementary proof of local one-to-oneness.* Try Exercise 9 to Paragraph 4.1.2. It involves only basic facts about integrals. No power series, no residues.

8.1.3 Behavior Near Critical Points; Local Coverings

Now we will prove analogs of the algebraic Theorem 1 and the geometric Theorem 2 which will show that, near a critical point, the mapping function f is still describable, though no longer one-to-one and conformal.

We begin with a definition. Let $f: \Omega \to \mathbb{C}$ and let z_0 be in Ω. Then f gives a *local k-to-one covering near* z_0, provided there exists an open neighborhood Ω_0 of z_0 in Ω such that (writing $w_0 = f(z_0)$)

    (i) $z = z_0$ is a k-fold root of the equation $f(z) - w_0 = 0$, in the sense that $f(z) - w_0 = (z - z_0)^k f_1(z)$ with $f_1(z_0) \neq 0$;

    (ii) if $z \in \Omega_0$ satisfies $f(z) = w_0$, then $z = z_0$;

    (iii) if $w \in f(\Omega_0) - \{w_0\}$, then there exist precisely k distinct points $z_1, \ldots, z_k \in \Omega_0$ such that $f(z_j) = w$ for $j = 1, \ldots, k$.

Note: We do not require that Ω_0 be a disc.

Examples

Let $f(z) = z^2$. Then f gives a two-to-one cover near $z_0 = 0$; define Ω_0 to be any disc $|z| < R$. Near any point different from the origin, f gives a one-to-one covering, *not* two-to-one (why?).

Now we generalize Theorem 1 to the case z_0 is critical of order k.

THEOREM 3

Let $f: \Omega \to \mathbb{C}$ be analytic. If f has order k at the point z_0 of Ω, then f gives a local k-to-one covering near z_0.

Proof: This is similar to the proof of Theorem 1. Define the disc D with boundary C and no critical point but z_0. Define W_0 as before. For w in W_0, then, we have $N_w(C) = N_{w_0}(C)$. But this equals k because $z = z_0$ is a k-fold solution to

$$f(z) - w_0 = (z - z_0)^k f_1(z) = 0.$$

We now define Ω_0 to be the intersection $f^{-1}(W_0) \cap D$. We check (ii) and (iii) in the definition of local covering. Since Ω_0 is inside D, (ii) is immediate. For (iii), let $w \in f(\Omega_0) \subset W_0$. Then $N_w(C) = k$, as noted before, giving points $z_1, \ldots, z_k$ inside $f^{-1}(W_0) \cap D$ which map to w.

Moreover, these points z_j are all distinct, for if $z_1 = z_2$, say, then $f(z) - w = (z - z_1)^2 f_1(z)$ with f_1 analytic, so that $f'(z_1) = 0$. This is impossible in D, whose only critical point is z_0. Done.

We turn once more to geometry. Again we examine what the mapping $w = f(z)$ does to lines through the point z_0 (now critical). In this case it is preferable to consider *rays* of the form $z = \sigma(s) = z_0 + se^{i\varphi}$ with the angle φ fixed and $s \geq 0$. Hence, the ray starts at the point z_0.

Such a ray is mapped by f to the curve consisting of points $w = f(\sigma(s)) = f(z_0 + se^{i\varphi})$. We ask for the tangent to this curve at the point $w_0 = f(z_0)$.

Suppose f has order k at z_0. Thus, the Taylor series is

$$f(z) = f(z_0) + \frac{f^{(k)}(z_0)}{k!}(z - z_0)^k + \frac{f^{(k+1)}(z_0)}{(k+1)!}(z - z_0)^{k+1} + \cdots,$$

with $f^{(k)}(z_0) \neq 0$. From $z = \sigma(s)$, we obtain the curve

$$w = f(\sigma(s)) = w_0 + \frac{f^{(k)}(z_0)}{k!}s^k e^{ik\varphi} + \frac{f^{(k+1)}(z_0)}{(k+1)!}s^{k+1}e^{i(k+1)\varphi} + \cdots.$$

To compute the tangent, we reparametrize by means of the substitution $t = s^k$. Thus, the curve is given by

$$w = w_0 + \frac{f^{(k)}(z_0)}{k!}te^{ik\varphi} + \frac{f^{(k+1)}(z_0)}{(k+1)!}t^{(k+1)/k}e^{i(k+1)\varphi} + \cdots.$$

Differentiating with respect to t and setting $t = 0$ gives a velocity vector to the curve, namely,

$$\frac{f^{(k)}(z_0)}{k!}e^{ik\varphi}.$$

As with Theorem 2, we have written this as a complex number rather than as a vector (pair of real numbers).

This vector (complex number) makes an angle with the horizontal equal to

$$\arg\left\{\frac{f^{(k)}(z_0)}{k!}e^{ik\varphi}\right\} = \arg f^{(k)}(z_0) + k\varphi.$$

Since $k \geq 2$ at a critical point, the action of f is not merely a simple rotation through the constant angle $\arg f^{(k)}(z_0)$.

We have generalized Theorem 2 as follows:

THEOREM 4

Let $f: \Omega \to \mathbb{C}$, analytic, have order $k \geq 1$ at a point z_0 of Ω. Then a ray extending from z_0 and making an angle φ with the horizontal is mapped by f to a curve extending from $w_0 = f(z_0)$ and making (at w_0) an angle with the horizontal equal to $\arg f^{(k)}(z_0) + k\varphi$.

See Figure 8.6.

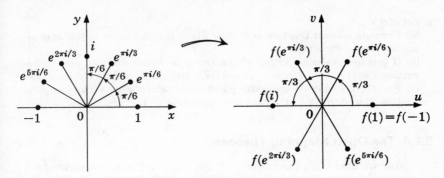

$f(z) = z^2$ doubles angles at the origin;
$\arg f(z) = \arg f''(0) + 2 \arg z = 2 \arg z$.

Figure   8.6

Moral

Theorems 3 and 4 shows that if

$$w = f(z)$$

$$= f(z_0) + \frac{f^{(k)}(z_0)}{k!}(z - z_0)^k + \text{higher powers of } z - z_0,$$

then, for z near z_0, the mapping f behaves like the polynomial

$$P(z) = f(z_0) + \frac{f^{(k)}(z_0)}{k!}(z - z_0)^k.$$

You should convince yourself that the mapping $w = P(z)$ gives a local k-to-one covering near z_0, wrapping a disc centered at z_0 exactly k times around a disc centered at w_0.

We comment that using the differentiability of a complicated function f to approximate it by a simpler polynomial is very much in the spirit of calculus.

Exercises to Paragraph 8.1.3

1. Verify that $f(z) = z^2$ gives a local two-to-one covering near $z_0 = 0$, provided we take $\Omega_0 = D(0; r)$ for any $r > 0$. Is it necessary that z_0 be the origin for two-to-oneness?

2. Describe the local covering property of $g(z) = (z - i)^3 e^z$ at the point $z_0 = i$. Is it locally one-to-one?

3. Let $f(z) = z^2$ and let $z(t) = (t, t)$, $t \geq 0$, parametrize the ray that makes an angle of $\pi/4$ with the x-axis. That is, $y = x$ with $w \geq 0$.
   (a) Sketch the image curve $w(t) = f(z(t))$, $t \geq 0$.
   (b) What angle does $w(t)$ make with the horizontal u-axis at $t = 0$? Compare Theorem 4.

4. Let $g(z) = z^3$.

(a) Compute a linear function $w = A + B(z - i)$ which gives a close approximation to $g(z)$ for z near $z_0 = i$.

(b) If you used calculus in (a), obtain the same linear function now by rewriting z as $(z - i) + i$ in $w = z^3$ and forgetting terms of degree ≥ 2.

(c) Are there linear or quadratic polynomials which give a close approximation to $w = g(z)$ for all z near the origin $z_0 = 0$? Discuss.

8.1.4 The Open Mapping Theorem

Now we deal with the question "Is the image under a nonconstant analytic mapping of a domain again a domain?" Since the image of a connected set under a continuous mapping is again connected (proof?), we need only consider the openness of the image.

Thus, let us say that a mapping $w = f(z)$ is *open*, provided $f(\Omega)$ is open in the w-plane whenever Ω is open in the z-plane.

The following result settles these questions. All the labors of the proof were essentially carried out in Theorems 1 and 3.

OPEN MAPPING THEOREM

A nonconstant analytic function is an open mapping.

Proof: Let $w_0 \in f(\Omega)$, where Ω is open. Say $f(z_0) = w_0$ with $z_0 \in \Omega$. Then, since f is nonconstant, we may choose a disc D with boundary C just as in the proof of Theorem 1. Then the set W_0 defined there is an open neighborhood of w_0 and is contained in $f(D)$; hence, in $f(\Omega)$. Thus, w_0 is an interior point and so $f(\Omega)$ is open. Done.

Comment

This result leads to a geometric proof of the assertion that an analytic function (on a disc, say) that assumes real boundary values must be a (real) constant. For the image of the disc would be squashed onto a portion of the real u-axis in the image plane and hence could not be an open subset of the plane.

There is a simpler nongeometric proof of this assertion which utilizes the Cauchy–Riemann equations.

The Open Mapping Theorem also yields a simple intuitive proof of the Maximum Modulus Principle. See Exercise 4 below. (Of course, a simple proof is now possible because we have done some hard work with series and the Argument Principle. Another proof of the maximum Modulus Principle, using Parseval's identity, was offered in Exercise 4 to Paragraph 5.3.2.)

Exercises to Paragraph 8.1.4

1. Verify that the image under the mapping $w = f(z) = z^2$ of each of the following domains is again open; the disc $|z| < 1$; the first quadrant ($x > 0, y > 0$);

the entire z-plane. These are, of course, special cases of the Open Mapping Theorem.

2. Suppose f is an entire function with a removable singularity at $z = \infty$. That is, $\lim_{z \to \infty} f(z)$ exists and is finite. It follows that f is bounded in the entire z-plane and hence constant by Liouville's Theorem. Here is another approach to this principle. This method generalizes to analytic functions defined on Riemann surfaces.

   (a) Observe that if f is entire and has either a removable singularity or a pole at $z = \infty$, then it may be regarded as a mapping $f : \Sigma \to \Sigma$, where Σ is the Riemann sphere constructed in Chapter 6.

   (b) Argue that the image $f(\Sigma)$ is open unless f is a constant. *Hint:* The image of every open disc, including discs containing ∞, must be open because f is analytic!

   (c) Argue that $f(\Sigma)$ is also compact (closed and bounded). *Hint: f* is continuous, Σ is compact. A basic result.

   (d) Conclude that if f is not constant, then f maps *onto* Σ, $f(\Sigma) = \Sigma$, and so f must have a pole; it cannot be everywhere finite-valued on Σ. *Hint:* Which subsets of Σ are both open and closed?

   Note: The general theorem here is this: If $f : X \to \mathbb{C}$ is analytic and finite-valued on the compact Riemann surface X, then f is constant. Thus, the natural "analytic" functions on X are meromorphic, $f : X \to \Sigma$, and must assume the value ∞. Compare the rational functions on Σ and the exercises to Section 6.3.

3. Compare the proofs (via the Cauchy–Riemann equations or the Open Mapping Theorem) of the fact that an analytic function which assumes only real boundary values on the rim of the disc must be a constant.

4. Show how Open Mapping implies the Maximum Modulus Principle.

5. Prove that a one-to-one analytic mapping f of the z-plane onto the w-plane is linear, $f(z) = az + b$, $a \neq 0$. *Hint:* How does f behave at infinity?

6. Is it true that a one-to-one analytic mapping defined on the entire z-plane *must* map onto the full w-plane (and hence be linear by Exercise 5)? *Hint:* See hint for preceding exercise.

7. *A nonopen mapping.* Suppose f maps the entire z-plane onto the closed upper half w-plane, given by $v > 0$.

   (a) Prove (one line) that f is not an open mapping.

   (b) Give an explicit formula for such a mapping.

   (c) Is your example in (b) analytic?

8.1.5 Rouché's Theorem

Now we study "perturbations" of a given analytic mapping $f(z)$, that is, functions of the form $F(z) = f(z) + h(z)$. We will see that if $h(z)$ is small in a certain sense, then F and f assume the same number of zeros inside a closed curve Γ. The exercises offer an amusing interpretation.

THEOREM (ROUCHÉ)

Let Ω be a domain and Γ a Jordan curve in Ω whose interior is also contained in Ω. Let f be analytic in Ω and nonzero on Γ. Suppose $h(z)$

is analytic in Ω such that $|h(\zeta)| < |f(\zeta)|$ for all ζ on Γ. Then the functions $f(z)$ and $F(z) = f(z) + h(z)$ have the same number of zeros inside Γ.

Proof: We have

$$F(z) = f(z) \left\{ 1 + \frac{h(z)}{f(z)} \right\} = f(z)g(z),$$

where $g(z)$ is analytic except for possible poles at the zeros of $f(z)$. Also, for ζ on Γ, we note $|h(\zeta)/f(\zeta)| < 1$, whence $g(\zeta) \neq 0$. Likewise, F has no zeros on Γ.

By the Argument Principle, the number of zeros of F inside Γ is given by

$$\frac{1}{2\pi i} \int_\Gamma \frac{F'(\zeta)}{F(\zeta)}\, d\zeta = \frac{1}{2\pi i} \int_\Gamma \frac{f'(\zeta)g(\zeta) + f(\zeta)g'(\zeta)}{f(\zeta)g(\zeta)}\, d\zeta$$

$$= \frac{1}{2\pi i} \int_\Gamma \frac{f'(\zeta)}{f(\zeta)}\, d\zeta + \frac{1}{2\pi i} \int_\Gamma \frac{g'(\zeta)}{g(\zeta)}\, d\zeta$$

Since the first term here equals the number of zeros of $f(z)$ inside Γ, we are done if we can show that the second integral vanishes.

Now the second integral equals the winding number $n(g(\Gamma); 0)$. But $g(z) = 1 + (h(z)/f(z))$ and since $|h(\zeta)/f(\zeta)| < 1$, the curve $g(\Gamma)$ consisting of points $g(\zeta)$ stays inside the open disc $D(1; 1)$ of radius 1 centered at $w_0 = 1$. Hence, $g(\Gamma)$ cannot wind around the origin in the w-plane. Thus, $n(g(\Gamma); 0) = 0$. By the remarks above, we are done.

Comments

1. The Argument Principle was crucial in the proof here.
2. It might be helpful to draw the picture of $g(\Gamma)$ and the disc $D(1; 1)$ indicated by the reasoning of the last paragraph of the proof.

Algebraic Applications of Rouché's Theorem

The Fundamental Theorem of Algebra. You may construct a proof as follows: Given

$$F(z) = a_n z^n + a_{n-1} z^{n-1} + \cdots + a_0,$$

define $f(z) = a_n z^n$. Thus, $h(z) = F(z) - f(z)$ has degree $n - 1$ or less. Now construct a circle Γ of radius so large that $|h(\zeta)| < |a_n \zeta^n|$ for all ζ on Γ (work!). Conclude that $F(z)$ has n complex zeros inside Γ.

Bounding the Roots of a Polynomial. We carry out the procedure described just above in a special case. Given

$$F(z) = z^4 + z^3 + 1,$$

whence $f(z) = z^4$, $h(z) = z^3 + 1$. Now we observe easily that if $|\zeta| = 3/2$, then

$$|\zeta^4| = \frac{81}{61} > \frac{35}{8} = |\zeta^3| + 1 \geq |\zeta^3 + 1|.$$

We conclude that all four roots of $F(z) = 0$ satisfy $|z| < 3/2$. This is not the smallest bound, of course.

Exercises to Paragraph 8.1.5

1. *The Dog-Walking Theorem.* Relate the theorem of Rouché to the following statement: A man $f(\zeta)$ walks a dog $h(\zeta)$ around a fire hydrant $w = 0$. If the leash is sufficiently short, then man $f(\zeta)$ and dog-on-the-end-of-the-leash $f(\zeta) + h(\zeta)$ circle the hydrant the same number of times.

2. Sketch the curve $g(\Gamma)$ inside the disc $D(1; 1)$ as indicated in the proof of Rouché's Theorem.

3. (a) Verify that all roots of $z^5 + z + 1 = 0$ satisfy $|z| < 5/4$.
   (b) Argue that there is a real root satisfying $-1 < x < 0$.
   (c) Argue that there is only one real root. *Hint:* Calculus.

8.1.6 The Brouwer Fixed-Point Theorem

A version of this famous theorem of topology states that if $g \colon \overline{D} \to \overline{D}$ is a continuous function mapping a closed disc into itself, then g has at least one fixed point z_0; that is, $g(z_0) = z_0$. For example, a rotation of the disc about its center point must leave the center point fixed.

We will prove a less general version of this theorem, which deals only with analytic functions. It follows from the theorem of Rouché.

THEOREM

Let g be an analytic function that maps the closed disc $\overline{D}(0; r)$ into the open disc $D(0; r)$. Then g has exactly one fixed point in $D(0; r)$.

Proof: Let $f(z) = -z$, $F(z) = g(z) - z$. Then $h(z) = F(z) - f(z) = g(z)$ and $|g(\zeta)| < |\zeta|$ for ζ on $C(0; r)$. Also, $f(\zeta) \neq 0$, clearly. By Rouché, f and F have the same number of zeros in $D(0; r)$. But f has only one zero, the origin. Finally, the one solution of $g(z) - z = 0$ is a fixed point of g. Done.

The key idea here was to realize the given function $g(z)$ as a perturbation (in the sense of Paragraph 8.1.5) of the simple function $f(z) = -z$.

Exercises to Paragraph 8.1.6

1. Let $g(z)$ az^k with $|a| < 1$.
   (a) Verify that g maps the closed unit disc into the open disc $|z| < 1$.
   (b) Verify that the origin $z = 0$ is the unique fixed point of the mapping g that satisfies $|z| \leq 1$.
   (c) Does $g(z)$ have any other fixed points in $\mathbb{C}$?

2. *The one-dimensional Brouwer Fixed-Point Theorem.* Let $f(x)$ be a continuous function, $0 \leq x \leq 1$, such that $0 \leq f(x) \leq 1$. Prove that $f(x_0) = x_0$ for some x_0. *Hint:* Look at the graphs $y = f(x)$ and $y = x$. Of course this case of the theorem has nothing to do with analyticity.

3. Ask a friendly topologist to prove the Brouwer Theorem for you in the case of continuous functions from $|z| \leq 1$ into $|z| \leq 1$.

Section 8.2 LINEAR FRACTIONAL TRANSFORMATIONS

8.2.0 Introduction

This important class of analytic mappings consists of all rational functions $T(z)$ of the form

$$w = T(z) = \frac{az + b}{cz + d} \quad (ad - bc \neq 0),$$

where a, b, c, d are complex. You should convince yourself that if the "determinant" $ad - bc$ were zero, then $T(z)$ would reduce to $0/0$, or a complex constant. We wish to rule out these occurrences.

Linear fractional transformations are also called *bilinear transformations* or *Möbius transformations*.

It is easy to see that if $c \neq 0$ in the above expression for $T(z)$, then $T(z)$ has a simple pole at the point $z_0 = -d/c$. We might say $T(z_0) = \infty$. This encourages us to define $T(z)$ on the *extended complex plane* or *Riemann sphere* Σ rather than merely on the ordinary complex plane $\mathbb{C}$. As a set, the Riemann sphere consists of the complex numbers with infinity adjoined:

$$\Sigma = \mathbb{C} \cup \{\infty\}.$$

In addition to the usual algebraic operations among the finite complex numbers, we define in the set Σ:

$$a + \infty = \infty \ (a \in \Sigma), \qquad a \cdot \infty = \infty \ (a \neq 0),$$

$$\frac{a}{\infty} = 0 \ (a \neq \infty), \qquad \frac{a}{0} = \infty \ (a \neq 0).$$

Note that we do not use both $+\infty$ and $-\infty$, which sometimes occur in real calculus.

Where is the point ∞? For us, it suffices to say that a sequence $\{z_n\}$ of points in $\mathbb{C}$ converges to infinity, provided, in the familiar sense,

$$\lim_{n \to \infty} |z_n| = \infty.$$

However, the following convention is more important. If we take any straight line in the plane and travel toward either end, then we come to the point ∞. Thus, straight lines in $\mathbb{C}$ become circles in Σ, with opposite ends meeting at ∞.

It is possible to put the elements of the Riemann sphere $\Sigma = \mathbb{C} \cup \{\infty\}$ into one-to-one correspondence with points of the ordinary sphere (hollow globe) in three-dimensional space. The method (Figure 8.7) is

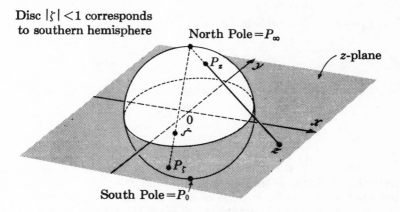

Disc $|\zeta| < 1$ corresponds to southern hemisphere

North Pole $= P_\infty$

z-plane

Stereographic projection
Point z in the plane corresponds to P_z on sphere.

Figure   8.7

called *stereographic projection*. In this case, ∞ corresponds to the North Pole of the globe and points of the "finite" complex plane correspond to the remaining points of the globe. In particular, the origin $z = 0$ of $\mathbb{C}$ corresponds to the South Pole of the globe.

However, we do not stress this hollow-globe interpretation. For conformal mapping purposes, we choose to concentrate on the flat plane $\mathbb{C}$ and employ ∞ without fear when necessary.

We are about to see that linear fractional transformation gives a nice mapping of Σ onto Σ. Moreover, given reasonable subsets S_1, S_2 of Σ, it is often possible to write down a linear fractional transformation that maps S_1 onto S_2 conformally. This is the second general problem mentioned at the start of this chapter.

8.2.1 Examples and Basic Properties

Here are some types of linear fractional transformations:

Identity: $T(z) = z$
Translation by b: $T(z) = z + b$
Multiplication: $T(z) = az$
Dilation (stretching): $T(z) = rz$ (r real > 0)
Rotation by φ: $T(z) = e^{i\varphi}z$, (φ real)
Linear or Affine: $T(z) = az + b$
Inversion: $T(z) = 1/z$

Now we make some basic observations about these transformations.

Property 1. Invertibility: Given

$$w = T(z) = \frac{az + b}{cz + d},$$

we may solve for z in terms of w, obtaining the inverse transformation

$$z = T^{-1}(w) = \frac{dw - b}{-cw + a}.$$

Property 2. $T: \Sigma \to \Sigma$ Is One-to-one and Onto: This statement follows from the existence of the inverse just found. Given w_0, we obtain the unique z_0 such that $w_0 = T(z_0)$ as $z_0 = T^{-1}(w_0)$.

At this point we discuss the value $T(\infty)$. Note that if $T(z) = az + b$ in linear, then $T(\infty) = \infty$. From this it follows that if $T(z)$ has the form $1/(cz + d)$, then $T(\infty) = 0$.

What if

$$T(z) = \frac{az + b}{cz + d} \quad \text{with } ac \neq 0?$$

In this case, $T(\infty)$ is found by employing the convention that the behavior of $T(z)$ at $z = \infty$ is the same as that of $T(1/\zeta)$ at $\zeta = 0$. Thus, we compute

$$T(z) = T\left(\frac{1}{\zeta}\right) = \frac{a(1/\zeta) + b}{c(1/\zeta) + d} = \frac{a + b\zeta}{c + d\zeta}.$$

Setting $\zeta = 0$ shows $T(\infty) = a/c$, a finite complex number.

Property 3. Composition: Given the transformations

$$w = T(z) = \frac{az + b}{cz + d}, \qquad z = T_1(\zeta) = \frac{a_1\zeta + b_1}{c_1\zeta + d_1}$$

then the composition $w = T(T_1(\zeta))$ is readily seen to be

$$w = T(T_1(\zeta)) = \frac{(aa_1 + bc_1)\zeta + (ab_1 + bd_1)}{(ca_1 + dc_1)\zeta + (cb_1 + dd_1)}.$$

You may check that the determinant here is nonzero (in fact, it equals $(ad - bc)(a_1 d_1 - b_1 c_1)$) and hence this composite function is again a linear fractional transformation.

Culture: We have shown that the family of complex linear fractional transformations forms a group in the sense of algebra. The group operation is composition of functions, as in Property 3. This is readily seen to be associative. The identity element of the group is the identity mapping (see the examples above), while inverses exist as in Property 1.

Property 4. Structure of $T(z)$: We claim that every linear fractional transformation may be obtained by composing three simple types: multiplication by a constant, translation, and inversion
　　To see this, let $T(z) = (az + b)/(cz + d)$. If $c = 0$, then we are done: Without loss, $d = 1$ and T is the same as $z \to az \to az + b$, a multiplication followed by a translation.
　　If $c \neq 0$ then we divide the denominator into the numerator, obtaining

$$T(z) = \frac{a}{c} + \frac{b - ad/c}{cz + d}.$$

Thus, $T(z)$ is obtained by the process

$$z \to cz \to cz + d \to \frac{1}{cz + d} \to \frac{b - ad/c}{cz + d} \to \frac{a}{c} + \frac{b - ad/c}{cz + d}.$$

At each step we have one of the three types mentioned. Done.

Property 5. Conformality: We claim that the linear fractional transformation $w = T(z) = (az + b)/(cz + d)$ is conformal (except at $z = -d/c, \infty$, where a special definition is required).
　　Here are two methods of proof: First, $T(z)$ is obtained by a succession of translations, multiplications by constants, and inversions. Each of these simpler transformations is conformal—except, of course, for the pole of the inversion. Since a sequence of conformal transformations is conformal, $T(z)$ is conformal as stated.
　　For a second proof, you may compute the derivative

$$T'(z) = \frac{bc - ad}{(cz + d)^2}$$

and observe that it is finite and nonzero except at $z = -d/c, \infty$. Conformality follows from Theorem 2 of this chapter.

Actually, you may deduce that $T'(z) \neq 0$ from the global one-to-oneness of $T(z)$ (Property 2) and Theorem 3.

Property 6. $T(z)$ Preserves Circles in Σ: By a "circle in Σ" we mean an ordinary circle in $\mathbb{C}$ or a straight line in $\mathbb{C}$ that we consider a circle (of large radius)! with both ends meeting at ∞. We claim that the image of a circle in Σ under a linear fractional transformation is again a circle in Σ. This will be important in some explicit mapping problems.

To prove this, we use Property 4. It suffices to prove that circles are preserved under multiplication by a complex scalar, under translation and under inversion. The first two of these are clear. Hence, we must examine the effect of the inversion $w = T(z) = 1/z$ on circles and lines.

The equation (with B, C, D real and not all zero)

$$A(x^2 + y^2) + Bx + Cy + D = 0$$

determines any circle or line ($A = 0$) in the xy-plane. Writing

$$w = u + iv = T(x + iy) = \frac{x}{x^2 + y^2} - \frac{yi}{x^2 + y^2},$$

we note $|w|^2 = |z|^{-2}$; that is, $u^2 + v^2 = 1/(x^2 + y^2)$. It follows that

$$x = \frac{u}{u^2 + v^2}, \qquad y = \frac{-v}{u^2 + v^2}.$$

Under these substitutions the equation for the image of the circle given above is (check!)

$$A + Bu - Cv + D(u^2 + v^2) = 0.$$

This is again a line or circle. Property 6 is established.

Property 7. One or Two Fixed Points. Now we argue that if $T(z)$ is not the identity mapping, then $T(z)$ has either one or two fixed points in Σ, that is, points z satisfying $T(z) = z$. Of course the identity mapping fixes all points.

To see this, let $T(z) = (az + b)/(cz + d)$ be different from the identity. If $c = 0$, then without loss $d = 1$ and $T(z) = az + b$. Clearly, then, $T(\infty) = \infty$. Also, $T(z) = z$ has one more solution z if and only if $a \neq 1$ (geometrically, $w = T(z)$ is not "parallel" to $w = z$). Thus, there are one or two fixed points if $c = 0$.

In case $c \neq 0$, we note first that $T(\infty) = a/c \neq \infty$. Thus, ∞ is not a fixed point. Also, multiplying both sides of the fixed-point equation $T(z) = z$ by the denominator $cz + d$ yields the quadratic equation

$$cz^2 + (d - a)z - b = 0,$$

which has one (repeated) or two (distinct) roots. These are all the fixed points of T.

Hence, *if a linear fractional transformation is given which fixes three or more points, then it must be the identity.*

Property 8. T Is Determined by Three Values: We claim that if z_0, z_1, z_2 are distinct points of Σ, and w_0, w_1, w_2 another triple of distinct points of Σ, then there exists a unique linear fractional transformation T such that $T(z_k) = w_k$ for $k = 0, 1, 2$.

Let us prove the uniqueness first. If both T and T_1 have the stated property, then $T^{-1}(T_1(z_k)) = z_k$, $k = 0, 1, 2$. Since the composition of T^{-1} and T_1 has three fixed points, it must be the identity (Property 7) whence $T = T_1$.

To prove existence, we first observe that it suffices to prove that there is a transformation T_1 that takes any triple z_0, z_1, z_2 of distinct points to the three points $0, 1, \infty$. For if we can do this, then we can also construct a transformation T_2 that takes the given w_0, w_1, w_2 to $0, 1, \infty$. But it is easy to see that the composite transformation $T(z) = T_2^{-1}(T_1(z))$ satisfies $T(z_k) = w_k$ for $k = 0, 1, 2$, as desired.

We therefore conclude the proof by noting that if z_0, z_1, z_2 are distinct complex numbers, then

$$T_1(z) = \frac{z_1 - z_2}{z_1 - z_0} \cdot \frac{z - z_0}{z - z_2}$$

takes z_0, z_1, z_2 to $0, 1, \infty$, respectively. If, however, one of the given z_0, z_1, z_2 happens to be ∞, then we alter the argument slightly as follows: If $z_0 = \infty$, then $T_1(z) = 1/(cz + d)$ will map ∞, z_1, z_2 to $0, 1, \infty$, provided we choose c, d to satisfy $cz_1 + d = 1$, $cz_2 + d = 0$. If $z_1 = \infty$, then the transformation $T_1(z) = (z - z_0)/(z - z_2)$ does the job. If $z_2 = \infty$, then $T_1(z)$ must have the form $az + b$ (with suitable a, b), for this fixes ∞.

Comment

Except for our mention of conformality, the eight properties we developed above involved elementary algebra and geometry only.

Exercises to Paragraph 8.2.1

1. Suppose $ad - bc = 0$ in $(az + b)/(cz + d)$. Show that this fraction is either $0/0$ or a complex constant.

2. Let $T(z) = (z + i)/(z - i)$. Compute $T(z)$ for these values of z:
   (a) $z = 1$,
   (b) $z = i$,
   (c) $z = -i$,
   (d) $z = \infty$.

3. Given $w = T(z) = (z + i)/(z - i)$, compute $z = T^{-1}(w)$ as a linear fractional transformation in the variable w. *Hint:* See Property 1 in the text.

4. Let $w = T(z)$ be as in Exercise 3. Find z_0 such that $T(z_0) = i$.

5. (a) Verify that the composition $T(T_1(\zeta))$ is given by the formula in the text, Property 3.
   (b) Write $w = T(T_1(\zeta))$ as a linear fractional transformation, where T is as in Exercise 3 and $z = T_1(\zeta) = \zeta/(\zeta - 1)$.

6. Decompose $w = T(z)$ of Exercise 3 into a composite of simpler types (multiplication by a constant, translation, inversion) as described in Property 4 of the text.

7. *The inversion mapping $w = 1/z$*
   (a) Write $w = 1/z$ in polar form, where $z = re^{i\theta}$ as usual.
   (b) Where in the w-plane does this mapping send the ray $\theta = \theta_0$ in the z-plane?
   (c) Likewise for the circle $|z| = r_0$.
   (d) Sketch the disc $|z| \le 1$, adding some concentric circles centered at the origin and some segments ("spokes") from the origin to the rim of the disc. Then sketch the image of this configuration under the inversion $w = 1/z$.
   (e) Determine the image of the circle $|z - \frac{1}{2}| = \frac{1}{2}$ under the mapping $w = 1/z$. Is this circle preserved?

8. *Some fixed points*
   (a) Locate the fixed points of the inversion $w = 1/z$.
   (b) Likewise for $T(z) = (z + i)/(z - i)$.

9. (a) Write down a linear fractional transformation that maps z_0, z_1, z_2 to $0, 1, \infty$, where $z_0 = 1, z_1 = i, z_2 = -1$.
   (b) Where does your transformation send the origin $z = 0$?
   (c) Where does your transformation send the circle $|z| = 1$? *Hint:* Circles in Σ are preserved.
   (d) Where does your transformation send the disc $|z| < 1$?

10. *A useful convention.* Suppose we wish to map z_0, z_1, z_2 in the z-plane to w_0, w_1, w_2 in the w-plane. Let $\zeta = (az + b)/(cz + d)$ transform z_0, z_1, z_2 to $0, 1, \infty$, respectively (refer to Property 8 of the text) and likewise let $\zeta = (\alpha w + \beta)/(\gamma w + \delta)$ send w_0, w_1, w_2 to $0, 1, \infty$. Meditate upon the following: The equation (suppressing ζ)

$$\frac{\alpha w + \beta}{\gamma w + \delta} = \frac{az + b}{cz + d}$$

specifies a mapping of the extended z-plane onto the extended w-plane, which sends z_0, z_1, z_2 to w_0, w_1, w_2 as desired. Note that it is simpler to write this equation than to solve for w in terms of ζ and then compose two transformations.

11. Let C and C' be circles in the z- and w-planes, respectively. Does there exist a linear fractional transformation $w = T(z)$ that maps C onto C'? If so, is it unique?

12. Is $w = f(z) = z + (1/z)$ a linear fractional transformation?

13. *The Cross Ratio.* Let z_0, z_1, z_2, z_3 be distinct points of Σ. Their *cross ratio* is the number

$$[z_0, z_1, z_2, z_3] = \frac{z_0 - z_1}{z_0 - z_3} \cdot \frac{z_2 - z_3}{z_2 - z_1}.$$

Show that a linear fractional transformation T preserves cross ratio

$$[T(z_0), T(z_1), T(z_2), T(z_3)] = [z_0, z_1, z_2, z_3].$$

Hint: Prove it separately for multiplications, translations, and inversion.

14. Prove that a linear fractional transformation $w = T(z)$ which sends given distinct z_1, z_2, z_3 to distinct w_1, w_2, w_3 is uniquely determined by the requirement

$$[w, w_1, w_2, w_3] = [z, z_1, z_2, z_3].$$

Thus, the map T is obtained by solving for w in

$$\frac{w - w_1}{w - w_3} \cdot \frac{w_2 - w_3}{w_2 - w_1} = \frac{z - z_1}{z - z_3} \cdot \frac{z_2 - z_3}{z_2 - z_1}.$$

But see Exercise 10 also.

8.2.2 Some Explicit Mapping Problems

Now we ask for transformations $w = T(z)$, which map a given domain onto another one in a one-to-one manner. As usual, we write $z = x + iy, w = u + iv$.

Problem 1

Map the right half-plane $x > 0$ onto the upper half-plane $v > 0$. Refer to Figure 8.8.

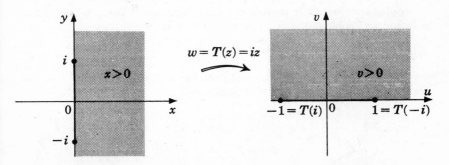

Figure   8.8

Solution: We need only rotate the entire plane through one right angle. The mapping $w = T(z) = iz$ accomplishes this. More generally, any mapping $w = iz + r$ where the translation term r is real will map the right half-plane onto the upper half-plane. And these are not the only linear fractional transformations with this property.

Problem 2

Map the entire z-plane onto the open unit disc $|w| < 1$. Refer to Figure 8.9.

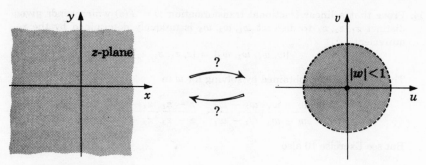

Figure 8.9

Solution: No such *analytic* mapping exists, for it would be bounded and therefore constant by Liouville's Theorem. And, even more strongly, no such linear fractional transformation exists because these must map $\Sigma = \mathbb{C} \cup \{\infty\}$ onto Σ. A solution to Problem 2 could not do this (why?).

Problem 3

Map the open unit disc onto the entire plane by a linear fractional transformation.

Solution: No such mapping exists. For its inverse would be a solution to Problem 2!

Problem 4

Map the open disc $|z| < 1$ onto the upper half w-plane by a linear fractional transformation. Refer to Figure 8.10.

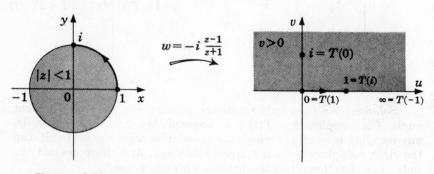

Figure 8.10

Solution: Such a mapping must send the circle $|z| = 1$ onto the horizontal u-axis. (Proof?)

We note that the u-axis is the "circle" determined by the three points $0, 1, \infty$. Since three values determine the transformation (see

Property 8), we choose $z_0 = 1, z_1 = i, z_2 = -1$ on the unit circle and will send these to $0, 1, \infty$, respectively. Then we will check that this indeed maps $|z| < 1$ onto the *upper* half-plane, not the lower half-plane.

As in the proof of Property 8, the transformation

$$w = T(z) = \frac{i+1}{i-1} \cdot \frac{z-1}{z+1} = -i \cdot \frac{z-1}{z+1}$$

sends $1, i, -1$ to $0, 1, \infty$, respectively. To verify that $T(z)$ maps the unit disc onto the upper half-plane, we note first that $T(0) = i$, which is in the upper half-plane. You should convince yourself that this implies the full result. Figure 8.10 might help. Note, too, that T was constructed to preserve orientations: counterclockwise on the circle, positive on the u-axis.

Problem 5

Find all linear fractional transformations that map the upper half-plane onto the unit disc. Refer to Figure 8.11.

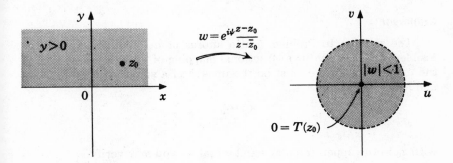

Figure   8.11

Solution: We proceed as in Problem 4, first finding the transformations that map the x-axis and ∞ (boundary of the upper half-plane) onto the unit circle (boundary of the disc) and then singling out those that map the upper half-plane onto the interior (rather than exterior) of the disc.

Suppose $T(z) = (az + b)/(cz + d)$ maps the x-axis onto the unit disc. Then $|T(0)| = |b/d| = 1$, so $|b| = |d|$. Likewise, $|T(\infty)| = |a/c| = 1$, so $|a| = |c| \neq 0$. We write $a/c = e^{i\psi}$ for some angle ψ, whence $T(z)$ has the form

$$T(z) = e^{i\psi} \frac{z - \beta}{z - \delta},$$

with $|\beta| = |\delta|$. We must now specify β, δ.

If $z = x$ is on the x-axis, then $|x - \beta| = |x - \delta|$ because $|T(x)| = |e^{i\psi}| = 1$. It follows that for all x,

$$(x - \beta)(x - \bar{\beta}) = |x - \beta|^2 = |x - \delta|^2 = (x - \delta)(x - \bar{\delta});$$

that is,

$$x^2 - (\beta + \bar{\beta})x + \beta\bar{\beta} = x^2 - (\delta + \bar{\delta})x + \delta\bar{\delta}.$$

Equating coefficients, we see that $\mathrm{Re}(\beta) = \frac{1}{2}(\beta + \bar{\beta}) = \frac{1}{2}(\delta + \bar{\delta}) = \mathrm{Re}(\delta)$. Since $|\beta| = |\delta|$, we conclude that either $\delta = \beta$ or $\delta = \bar{\beta}$. (Picture!) But $\delta = \beta$ is impossible for linear fractional transformations (why?). Thus, $\delta = \bar{\beta}$.

Writing $z_0 = \beta$, we assert finally that $T(z)$ must have the form

$$T(z) = e^{i\psi}\frac{z - z_0}{z - \bar{z}_0},$$

with $z_0 = x_0 + iy_0$ in the upper half-plane, $y_0 > 0$. For $T(z_0) = 0$, which is inside the unit disc. We must require, therefore, that z_0 be above the x-axis.

Comment

The solution to Problem 5 also affords us a complete solution to Problem 4. For every linear fractional mapping of the disc $|w| < 1$ onto the upper half z-plane must be the inverse of a solution to Problem 5 and hence has the form

$$z = \frac{\bar{z}_0 e^{i\varphi}w - z_0}{e^{i\varphi}w - 1},$$

with z_0 in the upper half-plane and φ real, as you may verify.

Problem 6

Find all linear fractional transformations that map the unit disc onto itself.

Solution: There are some obvious answers: The rotations

$$w = e^{i\psi}z.$$

There are others, however. And we have essentially found them already, as we will make clear.

Choose any linear fractional transformation of the disc $|z| < 1$ onto the upper half ζ-plane; say, $\zeta = U(z)$. See Figure 8.12. Now we claim that every linear fractional transformation $w = T(z)$ of the disc $|z| < 1$ onto the disc $|w| < 1$ has a unique factorization of the form

$$w = T(z) = T^*(U(z)),$$

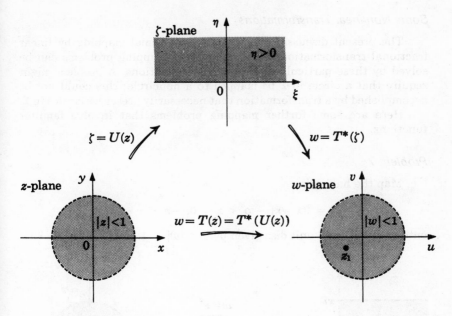

Figure 8.12

where $w = T^*(\zeta)$ is one of the mappings of the upper half-plane onto the unit disc that we characterized in Problem 5. In fact, $T^*(\zeta) = T(U^{-1}(\zeta))$. Hence, T^* is determined uniquely by T (once U has been selected and fixed, of course).

What will $T(z)$ look like? We know that every T^* has the form (see Problem 5)

$$w = T^*(\zeta) = e^{i\psi} \frac{\zeta - \zeta_0}{\zeta - \overline{\zeta}_0},$$

with ψ real and ζ_0 in the upper half-plane. Also, for U we choose a nice mapping of the disc $|z| < 1$ onto the upper half-plane (see Problem 4), say

$$\zeta = U(z) = -i \cdot \frac{z - 1}{z + 1}.$$

It follows (with a little bit of arithmetic) that T has the form

$$w = T(z) = T^*(U(z)) = e^{i\varphi} \frac{z - z_1}{\overline{z}_1 z - 1},$$

where φ is real and $|z_1| < 1$. Note that $T(0) = e^{i\varphi} z_1$ *is* inside the unit disc.

Some Nonlinear Transformations

The present discussion has stressed conformal mapping by linear fractional transformation. Of course not all mapping problems can be solved by these particularly simple transformations. A problem might require that a circle in Σ be mapped to a noncircle; this could not be accomplished by a transformation that necessarily preserves circles in Σ.

Here are some further mapping problems that involve familiar functions.

Problem 7

Map the half-strip

$$S = \{(x, y) \mid -\infty < x < 0, -\pi < y \leq \pi\}$$

onto the punctured unit disc given by $0 < |w| < 1$. Refer to Figure 8.13.

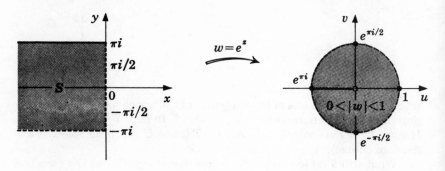

Figure   8.13

Solution: We saw in Chapter 3 that the exponential mapping $w = e^z$ is a solution.

Note here that S is not an open set, although the punctured disc is open. Does this contradict the Open Mapping Theorem for the inverse $z = \log w$?

Problem 8

Map the open half-disc

$$\Omega = \{w = u + iv \mid |w| < 1, v > 0\}$$

onto the half-strip (open)

$$S_1 = \{z = x + iy \mid -\infty < x < 0, 0 < y < \pi\}$$

Refer to Figure 8.14.

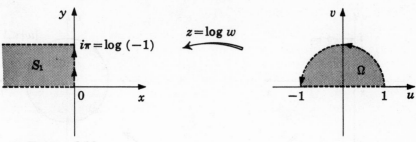

Figure 8.14

Solution: $z = \log w$.

Problem 9

Map the open quarter-disc

$$Q = \{z = x + iy \mid |z| < 1, x > 0, y > 0\}$$

onto the half-disc Ω of Problem 8. Refer to Figure 8.15.

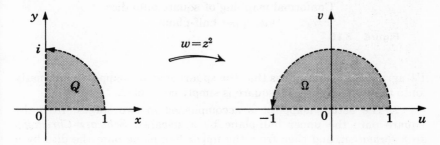

Figure 8.15

Solution: $w = f(z) = z^2$. Note that $f(z)$ here does have an analytic inverse mapping Ω onto Q. It is, of course, a branch of the square root function $z = \sqrt{w}$.

Note also that $f(z)$, conformal in Q, is *not* conformal at its critical point $z_0 = 0$ on ∂Q.

Problem 10

Map the open unit square

$$S = \{z = x + iy \mid 0 < x, y < 1\}$$

conformally onto the open unit disc $|w| < 1$, with the boundary of the square mapping onto the circular boundary of the disc. See Figure 8.16.

We do not solve this problem here, but make the following comments.

1. The problem is solvable. The Riemann Mapping Theorem of

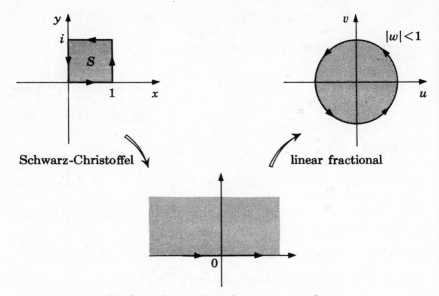

Conformal mapping of square onto disc
via upper half-plane

Figure   8.16

Paragraph 8.2.3 guarantees that the spuare may be mapped conformally onto the disc because the square is simply connected.

2. The actual mapping is accomplished in two stages: from the square onto the upper half-plane by a so-called *Schwarz–Christoffel transformation*, and then from the upper half-plane onto the disc by a standard linear fractional transformation.

3. We do not develop the Schwarz–Christoffel transformations here. Suffice it to say that one uses these transformations to map polygons (simply connected, such as squares, triangles, the interior of any simple closed polygonal path) onto the upper half-plane. Not surprisingly, the proper Schwarz–Christoffel transformation is constructed by mapping the boundary (a path made of line segments) onto the real axis with ∞. Thus, one must "unbend" the corners of the square, for instance, to map its boundary onto the real axis. This involves functions that are not conformal (not angle-preserving) at the corners of the polygon. Of course the mapping function must be conformal *inside* the polygonal domain.

4. You might check that the mapping functions we have stressed so far (linear fractional maps, sine, cosine, exponential) will not suffice to map the square onto the disc. Something new is needed. Several books listed in "Suggestions for Further Reading" contain discussions of the Schwartz–Christoffel transformation.

Exercises to Paragraph 8.2.2

1. Prove that a linear fractional transformation that maps the disc $|z| < 1$ onto the upper half w-plane must send the circle $|z| = 1$ to the horizontal u-axis with ∞ adjoined. Refer to the fourth mapping problem in the text.

2. (a) Write down a linear fractional transformation that maps the upper half-plane onto the lower half-plane.
   (b) Does there exist such a transformation that fixes the horizontal x-axis, $T(x) = x$ for all x?

3. *Mapping a wedge*
   (a) Does there exist a linear fractional transformation that maps the open wedge $0 < \arg z < \pi/3$ onto the open upper half w-plane?
   (b) Can this mapping be effected by any conformal transformation, not necessarily linear?

4. Let distinct points w_0, w_1, w_2 lie on the circle $|w| = 1$. Let T be the unique linear fractional transformation that sends 1, i, -1 on $|z| = 1$ to w_0, w_1, w_2, respectively. What further condition should w_0, w_1, w_2 satisfy to guarantee that T maps the interior $|z| < 1$ onto the interior $|w| < 1$? What happens if your condition is not satisfied?

5. Let the linear fractional transformation $w = T(z)$ map the disc $|z| < 1$ onto the disc $|w| < 1$ and also suppose $T(1) = 1$. Give proof or counterexample for the following statements:
   (a) T is the identity mapping $T(z) = z$ for all z.
   (b) T maps the x-axis onto a straight line or circle in the extended w-plane.
   (c) T is conformal at $z = 1$.
   (d) If T maps the x-axis to a straight line in the (extended) w-plane, then this straight line is in fact the u-axis.
   (e) If T maps the x-axis to a straight line in the (extended) w-plane, then $T(-1) = -1$.
   (f) If $T(-1) = -1$, then T must have a third fixed point, $-1 < x < 1$, $T(x) = x$.
   (g) If T maps the x-axis onto a straight line in the extended w-plane, then T is the identity mapping.
   (h) If $T(0)$ is real, then T is the identity mapping.
   (i) If $T(0) = w_0$, $|w_0| < 1$, is not real, then T maps the x-axis with infinity adjoined onto a circle in the finite w-plane.
   (j) There is only one mapping T of the given type such that $T(0) = w_0$ (= some preassigned image point).

6. *Another derivation of the Poisson Integral Formula.* Let $u(z)$ be harmonic in a domain containing $|z| \le 1$. Let ζ be a typical point of $|z| < 1$. Choose a transformation $z = T(w)$ that maps $|w| < 1$ onto $|z| < 1$ with $T(0) = \zeta$.
   (a) Verify

$$u(T(0)) = \frac{1}{2\pi} \int_0^{2\pi} u(T(e^{i\omega}))\, d\omega$$

where $w = e^{i\omega}$ if $|w| = 1$. *Hint:* Circumferential mean of $u(T(w))$.
   (b) Let $\zeta = \rho e^{i\varphi}$, $|\rho| < 1$. Then for $z = e^{i\theta}$ we have

$$e^{i\omega} = T^{-1}(e^{i\theta}) = \kappa\, \frac{e^{i\theta} - \rho e^{i\varphi}}{\rho e^{i(\theta - \varphi)} - 1} \qquad \text{with } |\kappa| = 1.$$

Hint: See the sixth mapping problem of the text.

(c) $i\omega = \log \kappa + \log(e^{i\theta} - \rho e^{i\varphi}) - \log(\rho e^{i(\theta - \varphi)} - 1)$.

(d) Check

$$d\omega = \frac{d\omega}{d\theta}\, d\theta = \frac{1 - \rho^2}{1 + \rho^2 - 2\rho \cos(\theta - \varphi)}\, d\theta.$$

Hint: Differentiate.

(e) Conclude that

$$u(\rho e^{i\varphi}) = \frac{1}{2\pi} \int_0^{2\pi} \frac{u(e^{i\theta}) \cdot (1 - \rho^2)}{1 + \rho^2 - 2\rho \cos(\theta - \varphi)}\, d\theta.$$

Summary: The Poisson Integral is obtained from the Circumferential Mean-Value Theorem by transforming points inside the unit disc to the origin.

7. *Mapping an infinite plank.* Let Ω be the open plank

$$\Omega = \{z = x + iy \mid 0 < x < \pi, y > 0\}.$$

Prove that $w = \cos z$ maps Ω onto the lower half-plane $v < 0$, as follows:

(a) The segment $0 < x < \pi, y = 0$, is mapped to a segment on the real u-axis.

(b) The vertical sides $x = 0, y > 0$, and $x = \pi, y > 0$, are mapped to infinite intervals on the real u-axis.

(c) $\cos(\partial\Omega) =$ the u-axis.

(d) $\cos(\Omega) =$ the lower half-plane $v < 0$, as claimed.

8. *A boundary-value problem.* Find a function $\varphi(z) \not\equiv 0$ harmonic on the plank Ω of Exercise 7, continuous on $\overline{\Omega}$, such that φ vanishes identically on $\partial\Omega$. *Hint:* Use the result of Exercise 7.

9. The function that is identically zero is also a solution to the boundary-value problem of Exercise 8. Does this contradict the principle of unique determination by boundary values? Explain.

10. *Mapping the infinite wedge.* Let Ω consist of those points $z = re^{i\theta}$ with $r > 0, 0 < \theta < \varepsilon\pi$, where $0 < \varepsilon < 1$.

(a) Sketch Ω as a "wedge" in the upper half z-plane.

(b) Verify that Ω is mapped one-to-one onto the upper half of the w-plane (compare $u > 0$) by a branch of the mapping $w = z^\varepsilon$. Note that ε may be irrational.

(c) Concoct a boundary-value problem on the wedge Ω which is solved by means of this mapping.

11. *Univalent mappings.* Let $w = f(z)$ be a nonconstant analytic function defined on a domain Ω_1 in the z-plane. By the Open Mapping Theorem, the image $\Omega_2 = f(\Omega_1)$ is a domain in the w-plane. The mapping is *univalent* or *schlict* if f is one-to-one on Ω_1. True or false?

(a) If f is univalent in Ω_1, then $f'(z) \neq 0$ for z in Ω_1.

(b) If $f'(z) \neq 0$ in Ω_1, then f is univalent in Ω_1.

(c) If f is univalent in Ω_1, then it is conformal at each point of Ω_1.

(d) If f is conformal at each point of Ω_1, then f is univalent in Ω_1.

(e) There exists a univalent analytic mapping of the open disc $|z| < 1$ onto the w-plane.

(f) There exists a univalent analytic mapping of the open disc $|z| < 1$ onto the right half w-plane $u > 0$.

(g) There exists a univalent analytic mapping of the open disc $|z| < 1$ onto the annulus $1 < |z| < 2$.

12. *The inverse of a univalent mapping.* See Exercise 11. Let f be a univalent analytic mapping of Ω_1 onto Ω_2.
    (a) Prove that f has an inverse function $g \colon \Omega_2 \to \Omega_1$ defined by $g(f(z)) = z$ *Hint:* See Exercise 12 to Paragraph 3.2.3.
    (b) Prove that the inverse function g is analytic on Ω_2.
    (c) Prove in fact that $g'(w) = 1/f'(z)$, where $w = f(z)$. Note that $f'(z) \neq 0$ throughout Ω_1.

13. *Mapping the unit disc.* We know there exist univalent analytic mappings (linear fractional transformations, in fact) of the disc $|z| < 1$ onto the upper half w-plane $v > 0$. These may be composed with other mappings.
    (a) Find a univalent mapping of $|z| < 1$ onto a slit ζ-plane, that is, a plane with the nonnegative real axis removed. *Hint:* $\zeta = (T(z))^2$, where $T(z)$ maps $|z| < 1$ onto $v > 0$.
    (b) Find a univalent mapping of $|z| < 1$ onto an infinite horizontal strip of height 2π. *Hint:* Use (a) and an appropriate branch of the logarithm.
    (c) Is it possible to map the disc $|z| < 1$ in a univalent fashion onto an infinite "plank" Ω in the w-plane, say,

    $$\Omega = \{w = u + iv \mid 0 < u < \pi, v > 0\}?$$

    Hint: See Exercise 7.

14. Let z_0 be a noncritical point of the analytic function $w = f(z)$. Then this function is approximated near z_0 by a linear fractional transformation of a particularly well-behaved type. To which transformation do we refer? (See Section 1.) Is this true if z_0 is a critical point for f?

15. *Mapping a triangle.* Suppose z_0, z_1, z_2 are the vertices of an ordinary triangle in the z-plane and we map them by a linear fractional transformation $w = T(z)$ to $0, 1, \infty$, respectively. Is the triangle mapped by T onto the infinite plank (with straight sides) in the w-plane given by $0 < u < 1$, $v > 0$? *Hint:* T is conformal at z_0, z_1.

16. *Mapping discs to discs.* Let $D(z_0; r)$ and $D(w_0; R)$ be open discs in the z- and w-planes, respectively. Write down a linear fractional transformation $w = T(z)$ that maps the one disc onto the other, with $w_0 = T(z_0)$. *Hint:* Let $T(z)$ be affine, of the form $T(z) = w_0 + b(z - z_0)$.

17. *Exponential mapping of a rectangle*
    (a) Let $R = \{z = x + iy \mid -1 < x < 0, -\pi < y \le \pi\}$. Describe the image of the rectangle R under the mapping $w = e^z$.
    (b) Locate a rectangle in the z-plane that the exponential function maps onto the annulus $1 < |w| < 2$.
    (c) Same question for the punctured disc $0 < |w| < 1$.

8.2.3 A Dirichlet Problem Solved by Conformal Mapping

Let Ω be the open upper half-plane $y > 0$; $\partial\Omega$, the x-axis. Given the "boundary values" $B(x) = 1/(1 + x^2)$, we are to find $H(z) = H(x, y)$

continuous in $\overline{\Omega}$, harmonic in Ω, and equal to B on $\partial\Omega$; that is, $H(x, 0) = B(x)$. We proceed as follows:

1. We begin with a doomed attempt. Define

$$B(z) = \frac{1}{1 + z^2}.$$

Then the real part of the analytic function $B(z)$ is harmonic and surely agrees with $1/(1 + x^2)$ when $z = x$, that is, on $\partial\Omega$. Unfortunately, $B(z)$ is not analytic at $z = i$, so that its real part is not harmonic there. We desire $H(z)$ to be harmonic at *every* point of Ω. Thus, we must look elsewhere for $H(z)$. Refer to exercises to Section 4.5.

2. Now we apply conformal mapping. We will map the circle $|\zeta| = 1$ in the ζ-plane onto the x-axis (with ∞ adjoined) by means of a transformation $z = T(\zeta)$. Thus, $B(x)$ becomes $B(T(\zeta))$, a function on $|\zeta| = 1$. Next we will solve the Dirichlet problem on $|\zeta| \le 1$ for the boundary values $b(\zeta) = B(T(\zeta))$. This will give a harmonic function $h(\zeta)$ on $|\zeta| < 1$. The function $H(z) = h(T^{-1}(z))$ will then be seen to solve the original problem on Ω. Here are the details:

(a) The transformation

$$z = T(\zeta) = -i \cdot \frac{\zeta - 1}{\zeta + 1}$$

maps $|\zeta| < 1$ onto Ω and $|\zeta| = 1$ onto the x-axis with ∞ $(= T(-1))$. Refer to the fourth mapping problem of Paragraph 8.2.2.

(b) Write $\zeta = e^{i\varphi}$ on $|\zeta| = 1$. Then $b(e^{i\varphi})$ is defined as $B(T(e^{i\varphi}))$, and since $B(x) = 1/(1 + x^2)$,

$$b(e^{i\varphi}) = \frac{1}{1 + T(e^{i\varphi})^2} = \frac{1}{1 - \left(\dfrac{e^{i\varphi} - 1}{e^{i\varphi} + 1}\right)^2}$$

$$= \frac{2 + e^{i\varphi} + e^{-i\varphi}}{4} = \frac{1}{2} + \frac{1}{2}\cos\varphi.$$

(c) Now we extend $b(e^{i\varphi})$ into the disc $|\zeta| = \rho < 1$ by defining (this takes a bit of insight!)

$$h(\zeta) = h(\rho e^{i\varphi}) = \tfrac{1}{2} + \tfrac{1}{2}\rho\cos\varphi.$$

Note that h and b are equal on the circle $\rho = 1$.

(d) Observe now that $h(\zeta) = h(\rho e^{i\varphi})$ is harmonic (in the entire ζ-plane, in fact). For if we write $\zeta = \xi + i\eta$ as usual, we have

$$h(\zeta) = h(\xi, \eta) = \tfrac{1}{2} + \tfrac{1}{2}\xi,$$

so surely $h_{\xi\xi} + h_{\eta\eta} = 0$. Thus, the Dirichlet problem is solved on $|\zeta| \le 1$, and without recourse to the Poisson formula. The problem was simple on $|\zeta| \le 1$.

(e) Now we have

$$\zeta = T^{-1}(z) = \frac{-iz - 1}{iz - 1},$$

as you may check. We obtain a function of z by defining

$$H(z) = h(T^{-1}(z)) = \frac{1}{2} + \frac{1}{2}\,\text{Re}\left(\frac{-iz - 1}{iz - 1}\right),$$

where Re denotes the real part, so that $\zeta = \text{Re}\,\zeta$. You may solve for the real part here to obtain

$$H(z) = H(x, y) = \frac{y + 1}{x^2 + (y + 1)^2}.$$

This is harmonic because h is harmonic and T^{-1} is analytic. Clearly also,

$$H(x, 0) = \frac{1}{1 + x^2} = B(x),$$

so that $H(x, y)$ has the correct boundary values. The Dirichlet problem is solved.

Exercises to Paragraph 8.2.3

1. Graph the boundary value function

$$B(x) = \frac{1}{1 + x^2}$$

above the x-axis. Include the points $x = -\infty$ and $x = \infty$ as "end points" of your x-axis, and note that $B(-\infty) = B(\infty) = 0$. Now graph the function $b(e^{i\varphi}) = \frac{1}{2} + \frac{1}{2}\cos\varphi$ above the φ-axis for $-\pi \le \varphi \le \pi$. Compare the two graphs to see how the conformal transformation $x = T(e^{i\varphi})$ has preserved the nature of the boundary values in carrying $B(x)$ to the unit circle in the ζ-plane.

2. We solved the Dirichlet problem in the disc $\rho \le 1$ by taking $b(e^{i\varphi}) = \frac{1}{2} + \frac{1}{2}\cos\varphi$ on $\rho = 1$ and defining $h(\rho e^{i\varphi}) = \frac{1}{2} + \frac{1}{2}\rho\cos\varphi$. Was this the only satisfactory definition? Could we have defined a different function $h(\rho e^{i\varphi})$ that would also have solved this problem for $\rho \le 1$?

3. Use the familiar formula

$$\frac{1}{a + bi} = \frac{a}{a^2 + b^2} - \frac{b}{a^2 + b^2}\,i$$

to verify that the solution $H(x, y)$ found in the text is the imaginary part of the analytic function

$$f(z) = \frac{-1}{x + (y + 1)i} = \frac{-1}{z + i}.$$

Deduce that $H(x, y)$ is harmonic.

4. Given the boundary value function

$$B_1(x) = \frac{x}{1 + x^2}$$

on the x-axis, solve the Dirichlet problem in the upper half-plane $y \geq 0$ in two ways: by conformal mapping as in the text, and by adapting the insight of Exercise 3. Of course you should obtain the same harmonic function from each method.

Section 8.3 WHAT IS THE RIEMANN MAPPING THEOREM?

This celebrated theorem deals with the question: "Which domains can be mapped onto the open unit disc by a one-to-one analytic function?" Thus, we saw in Section 8.2 that the entire complex plane cannot be mapped onto the disc, whereas the upper half-plane can (by linear fractional transformations, in fact).

We first raised this question in the Appendix to Section 4.5. It was seen to be crucial in the solution of boundary-value problems, notably the Dirichlet problem, on a domain Ω. For (as we saw) if $F: \Omega \to D$ is a one-to-one analytic mapping onto the disc D, then a Dirichlet problem on Ω (that is, the search for a harmonic function that assumes prescribed boundary values) may be "translated" by the mapping F into a Dirichlet problem on the disc D. And Dirichlet problems on D may be solved using the Poisson Integral Formula (Appendix to Chapter 2). Thus, the Dirichlet problem on certain domains reduces to a question about analytic mappings of those domains onto the disc.

Now we state

RIEMANN MAPPING THEOREM

Let Ω be a domain in $\mathbb{C}$. Then there exists a one-to-one analytic (hence conformal) mapping F of Ω onto the open unit disc $\Leftrightarrow$ the domain Ω is simply connected but not equal to the entire plane $\mathbb{C}$.

Recall that a simply connected domain is one with no holes or punctures; every closed curve lying in a simply connected domain Ω must have its interior contained entirely in Ω also. It is not hard to see that if Ω can be mapped one-to-one onto the disc by a continuous mapping F (analyticity not required), then Ω must be simply connected. And if F is analytic as well, then Ω cannot equal $\mathbb{C}$ because of Liouville's Theorem. (Details?)

The converse, however, the existence of the conformal mapping F, is not at all obvious. We will not give a proof here. It is something for you to look forward to in an advanced course in complex function theory.

Further Comments

1. The Riemann Mapping Theorem is an existence theorem. It does not give an explicit method of writing down the mapping function F.

2. In particular, the theorem does not assert that the analytic mapping can be accomplished by a linear fractional transformation in all cases. We know this is false.

3. You may have observed the necessity of describing the behavior of the mapping function F on the boundary $\partial\Omega$ as well as on Ω. Does F extend to a continuous mapping of $\partial\Omega$ onto the circle ∂D? What properties of $\partial\Omega$ are preserved by the extended mapping F? "Behavior on the boundary" is a study in itself. Suffice it to say now that if $\partial\Omega$ is reasonably nice (a piecewise-smooth loop, say), then F has a continuous extension to the closure $\overline{\Omega} = \Omega \cup \partial\Omega$, and all is well.

4. It is now clear that if Ω_1, Ω_2 are simply connected subdomains of $\mathbb{C}$, then there is a conformal mapping F of Ω_1 onto Ω_2. For the Riemann Mapping Theorem gives conformal $F_k: \Omega_k \to D$ $(k = 1, 2)$ and the composite mapping $F(z) = F_2^{-1}(F_1(z))$ is conformal from Ω_1 to Ω_2. See Figure 8.17.

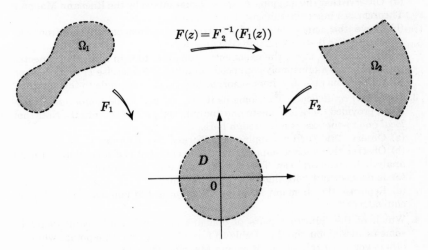

Figure 8.17

5. What can we say about the mapping of nonsimply connected domains? For instance, let Ω_1 and Ω_2 be annuli (one hole)

$$\Omega_1 = \{z \mid r_1 < |z| < R_1\}, \qquad \Omega_2 = \{w \mid r_2 < |w| < R_2\}.$$

Is there a conformal mapping of one onto the other? Are every two annuli "conformally equivalent"? The answer is "No." In fact, a conformal mapping $F: \Omega_1 \to \Omega_2$ exists if and only if the ratios R_1/r_1 and R_2/r_2 are equal (if so, it is easy to write down such a mapping). We omit the argument.

6. You may have found the Riemann Mapping Theorem surprising: a topological property (simple connectedness) guaranteeing analytic equivalence. Well, paragraph (5) shows that the surprises have not ceased. For any two annuli such as Ω_1, Ω_2 surely are *topologically* equivalent (surely there exists a one-to-one *continuous* mapping of Ω_1 onto Ω_2 with continuous inverse), yet now the mapping cannot be *analytic* unless $R_1/r_1 = R_2/r_2$. Why? And what if the domain Ω has more than one hole? We are peering into vast realms of inquiry.

Exercises to Section 8.3

1. Let $F: \Omega \to D$ be one-to-one, onto, and analytic, as in the statement of the Riemann Mapping Theorem. Prove that F is conformal at each point of Ω.

2. *Homeomorphisms.* Let S_1, S_2 be subsets of the plane. A function f mapping S_1 onto S_2 is a *homeomorphism*, provided it is continuous, one-to-one, and its inverse, which maps S_2 onto S_1, is also continuous. Another name is *topological* mapping. This definition makes no mention of analyticity.
   (a) Observe that the mapping $F: \Omega \to D$ guaranteed by the Riemann Mapping Theorem is a homeomorphism.
   (b) Prove the impossibility of constructing a homeomorphism from the punctured disc $0 < |z| < 1$ onto the disc $|w| < 1$.
   (c) It is possible, using the exponential function, to map a simply-connected rectangle onto a nonsimply-connected annulus. See Exercise 17 to Paragraph 8.2.2. Does this give us a homeomorphism from the rectangle to the annulus?

3. *Conformal equivalence.* The domains Ω_1, Ω_2 are said to be *conformally equivalent*, provided there is a one-to-one analytic mapping of Ω_1 onto Ω_2. Note that *local* one-to-oneness is not enough here.
   (a) Observe that such a mapping *is* conformal. See Exercise 1.
   (b) Observe that such a mapping has an inverse that is also analytic; it is an analytic homeomorphism. This justifies the term "equivalent."
   (c) Is it important here that Ω_1, Ω_2 be domains?
   (d) Rephrase the Riemann Mapping Theorem in the language of conformal equivalence.

4. Which of the following pairs of domains are conformally equivalent? In some of these, you should be able to furnish the explicit mapping, while in others you must rely on the Riemann Mapping Theorem.
   (a) $\Omega_1 =$ the entire z-plane, $\Omega_2 =$ the unit disc.
   (b) $\Omega_1 =$ the disc $|z| < 1$, $\Omega_2 =$ the upper half-plane.
   (c) $\Omega_1 =$ the disc $|z| < 1$, $\Omega_2 =$ the open first quadrant $0 < \theta < \pi/2$.
   (d) $\Omega_1 =$ the interior of any simple closed curve, $\Omega_2 =$ the disc $|z| < 1$.
   (e) $\Omega_1 =$ the interior of any simple closed curve, $\Omega_2 =$ the interior of any other simple closed curve.
   (f) $\Omega_1 =$ the open wedge $0 < \theta < \pi/4$, $\Omega_2 =$ the entire w-plane.

5. According to the Riemann Mapping Theorem, it is possible to map $\Omega =$ the upper half of the unit disc (that is, $x > 0$, $|z| < 1$) onto the unit disc. How is this accomplished?
   (a) Observe that this is *not* accomplished by squaring, which maps Ω onto a slit disc. We are not done yet.

(b) Map the slit disc onto an infinite plank, using a branch of the logarithm. You may have to slide things about.

(c) Map the infinite plank onto a half-plane, using a cosine or sine. See Exercise 7 to Paragraph 8.2.2.

(d) Map the half-plane onto the open disc.

(e) Observe that the composition of mappings (b)–(d) will send the half-disc Ω onto the unit disc, as desired.

6. Let Ω be given by $|z| > 1$ and let $f(z) = 1/z$. Note that Ω is *not* simply connected. Also, if $z \in \Omega$, then $f(z) \in D(0; 1)$. Why doesn't this contradict the Riemann Mapping Theorem? For f maps Ω conformally into $D(0; 1)$.

7. *The mapping of annuli.* Let Ω_k, with $k = 1, 2$, be the annuli $r_k < |z| < R_k$. We allow $r_k = 0$ or $R_k = \infty$.

   (a) Prove that there is a homeomorphism from Ω_1 onto Ω_2. *Hint:* Construct it. See Exercise 2 for definitions.

   (b) Prove that if $R_1/r_1 = R_2/r_2$, then a homeomorphism may be constructed that is in fact a linear fractional transformation (and hence analytic) of a complex variable.

   (c) Suppose that $f: \Omega_1 \to \Omega_2$ is a conformal equivalence (see Exercise 3), that Ω_1 is a punctured disc $0 < |z| < R_1$, and that $R_2 \neq \infty$, so that Ω_2 is bounded. Prove that Ω_2 is also a punctured disc, $r_2 = 0$. *Hint:* Show that $z = 0$ must be a removable singularity for f and that, in fact, $f(0) = 0$.

8. *Mapping an infinite strip.* Consider the infinite horizontal strip $\Omega = \{z = x + iy \mid -\pi < y < \pi\}$. According to the Riemann Mapping Theorem, Ω is conformally equivalent to the unit disc $|w| < 1$. Verify this by constructing an equivalence. See Exercise 3 for definitions.

9. Suppose that Ω is a plane domain that has no points in common with some ray (= a straight-line path from some finite point out to infinity). Prove that Ω is conformally equivalent to a subdomain of the unit disc.

   Note that this result is a very crude approximation to the Riemann Mapping Theorem and not very useful in solving the Dirichlet problem on Ω.

10. Give a proof or a counterexample for each of the following statements. You may refer to a standard theorem if convenient. As usual, Ω is a plane domain.

    (a) If f is a nonconstant analytic function defined on the domain Ω, then the image $f(\Omega)$ is a domain.

    (b) If f is a nonconstant analytic function on Ω, then $f(\Omega)$ is conformally equivalent to Ω.

    (c) If Ω is conformally equivalent to the unit disc, then Ω is bounded.

    (d) The entire plane and the slit plane (nonpositive x-axis removed) are conformally equivalent.

    (e) The punctured disc $0 < |z| < 1$ and the unbounded annulus $1 < |w| < \infty$ are conformally equivalent.

    (f) If the analytic function $f: \Omega_1 \to \Omega_2$ maps Ω_1 onto Ω_2 and is conformal at each point of Ω_1, then it gives a conformal equivalence of Ω_1 with Ω_2.

Suggestions for Further Reading

I. SOME OTHER INTRODUCTORY TEXTS

These begin with complex numbers and analytic functions, and only later deduce results about real harmonic functions.

CHURCHILL, R. V. *Complex Variables and Applications.* 2nd ed. New York: McGraw-Hill, 1960.

KAPLAN, W. *Introduction to Analytic Functions.* Reading, Mass.: Addison-Wesley, 1966.

LEVINSON, N. and REDHEFFER, R. *Complex Variables.* San Francisco: Holden-Day, 1970.

SPIEGEL, M. R. *Theory and Problems of Complex Variables.* New York: Schaum, 1964.

II. SOME CLASSICS IN ENGLISH

For those who have acquired (or wish to) the taste in: careful topology, the Goursat theorem, proof of the Riemann Mapping Theorem, higher transcendental functions, the power series approach (Cartan), three proofs of the Maximum Modulus Principle (Titchmarsh).

AHLFORS, L. V. *Complex Analysis.* New York: McGraw-Hill, 1953.

CARTAN, H. *Elementary Theory of Analytic Functions of One or Several Complex Variables.* Reading, Mass.: Addison-Wesley, 1963.

HILLE, E. *Analytic Function Theory.* 2 vols. New York: Blaisdell, 1959, 1962.

KNOPP, K. *Theory of Functions.* 2 vols. New York: Dover, 1945.

SAKS, S. and ZYGMUND, A. *Analytic Functions.* 2nd ed. Warsaw: PWN, 1965.

TITCHMARSH, E. C. *The Theory of Functions.* 2nd ed. London: Oxford University Press, 1939.

III. SOME RELATED BOOKS

APOSTOL, T. M. *Mathematical Analysis*. Reading, Mass.: Addison-Wesley, 1957.

KNOPP, K. *Theory and Application of Infinite Series*. Glasgow: Blackie and Sons, 1928.

KOBER, H. *Dictionary of Conformal Representation*. New York: Dover, 1952.

NEHARI, Z. *Conformal Mapping*. New York: McGraw-Hill, 1952.

NEWMAN, M. *Elements of the Topology of Plane Sets of Points*. Cambridge: Cambridge University Press, 1954.

Solutions to Selected
Exercises

Chapter 1

PARAGRAPH 1.1.1

1. (a) (1, 3), (b) $(-3, 7)$, (c) $(5, -9)$, (d) $(-2, 18)$.
2. (a) $w = (1/3)(-z - 2\zeta) = (0, -8/3)$.
    (b) $w = -2z - \zeta = (-3, -1)$.

PARAGRAPH 1.1.2

1. (a) $\sqrt{17}$, (b) $2\sqrt{2}$, (c) $\sqrt{13}$, (d) same as (c).
2. Same as 1(c), (d).

PARAGRAPH 1.1.3

1. (c) Yes.
2. (b) The annulus is connected.
4. The key here is to observe that $\Omega - S$ must be empty (no points) or else Ω would not be connected.
6. No, not in the terminology of this book.

PARAGRAPH 1.1.4

1. (b) and (c) are the only bounded sets here.
2. (a) All points $(0, y)$ with $y \geq 0$ (the nonnegative y-axis) together with all points $(x, 0)$ with $x \geq 0$.
    (b) The circle $|z - z_0| = 2$. Note that this is contained in the given set.
    (c) The circle $|z - z_0| = 2$ together with the point $(0, 0)$.
    (d) The vertical lines $x = 0$, $x = 1$.

Paragraph 1.2.1

1. $\alpha(t) = (\cos 5t, \sin 5t)$.
2. (a) $\beta(t)$ traverses the unit circle $|k|$ times, starting at the point $(1, 0)$, in a counterclockwise or clockwise fashion according as k is positive or negative. (b) $\beta'(t) = \langle -k \sin kt, k \cos kt \rangle$. Its length is $|k|$; its direction is reversed if k is negative instead of positive.
3. Let $\gamma(t) = (t, 0)$ for $0 \le t \le 1$, $\gamma(t) = (1, t - 1)$ for $1 \le t \le 2$, $\gamma(t) = (3 - t, 1)$ for $2 \le t \le 3$, $\gamma(t) = (0, 4 - t)$ for $3 \le t \le 4$.
4. (b) $\alpha'(x) = \langle 1, f'(x) \rangle$.

Paragraph 1.2.2

1. $3\pi R$.
2. Length $= \int_0^1 \sqrt{1 + 4x^2}\, dx = (2\sqrt{5} + \ln(2 + \sqrt{5}))/4$. A table of integrals might help here.

Paragraph 1.2.3

1. (a) $\alpha(\theta) = (R \cos \theta, R \sin \theta)$ with $0 \le \theta \le \pi$.
   (b) $\sigma(s) = (R \cos(s/R), R \sin(s/R))$. Here $\sigma(s) = \alpha(E(s))$ with $\theta = E(s) = s/R$.
2. (a) $L = k(b - a)$, (b) $t = E(s) = a + s/k$.

Paragraph 1.2.4

1. Only (a).
2. Here the inside is given by $|z| < r$, the outside by $|z| > r$.
3. See the concept of *winding number* in Chapter 7.
4. (a) False (unbounded), (b) True, (c) True, (d) False; the "outside" domain will not be bounded. (e) True, (f) False.

Paragraph 1.2.5

1. (a) 47,   (b) $|U| = \sqrt{58}$, $|V| = \sqrt{73}$,   (c) $\cos \theta = U \cdot V / |U|\,|V| = 47/\sqrt{4234}$,
   (d) $\theta = \arccos(47/\sqrt{4234})$. Use tables.
2. (a) $V = \langle 7, -3 \rangle$, (b) $|V|^{-1}V = \langle 7/\sqrt{58}, -3/\sqrt{58} \rangle$ has unit length and is a positive multiple of V.

Paragraph 1.2.6

2. (c) $V \cdot N = 0$ means that V points in the same direction as T or in the opposite direction. But $V \cdot T > 0$ means that V points in the same direction as T.
   (d) V points in the opposite direction to T.
   (e) $V \cdot T = 0$ means that V points in the normal direction or its opposite. But $V \cdot N > 0$ means that V points in the normal direction.

Paragraph 1.3.1

1. $L(z; z_0) = 12 - (x - 1) + 12(y - 2)$.
2. $1 + 8 \sin t \cos t$.

4. *Mean Value Theorem.* Given a continuously differentiable real function $g(x)$ and points x, x_0, then there is a point x_1 between x and x_0 such that $g(x) - g(x_0) = g'(x_1) \cdot (x - x_0)$.

In the proof of Theorem 3, we let $u(x, y)$ play the role of $g(x)$ (note y is fixed here). Thus $u(x, y) - u(x_0, y) = u_x(x_1, y) \cdot (x - x_0)$ as claimed, and likewise for $u(x_0, y)$ as a function of y.

PARAGRAPH 1.3.2

1. (a) 1, (c) -12, (e) $11/\sqrt{2}$.
2. (a) $U = \langle 3/5, -4/5 \rangle$. Note that we simply divided the given vector by its length.
   (b) $4e/5$.
3. (b) Since r is constantly equal to 1 on the unit circle, we expect its tangential derivative (rate of change with respect to change in the angle θ) to be 0. Also, we expect the tangential derivative of θ to be 1.

PARAGRAPH 1.3.3

1. (a) $\langle 1 - 2x, 6y \rangle$, (c) $\langle -1, 12 \rangle$.
3. (c) $\nabla\theta(z_0)$ points in the direction of the tangent vector to the circle $C(0; r_0)$ at the point z_0, where $r_0 = |z_0|$.

PARAGRAPH 1.3.4

3. (a) $x = r \cos \theta$, so that $(\partial x/\partial n)(1, \theta) = (\partial x/\partial r)(1, \theta) = \cos \theta$.
   (b) $u(z) = |z|^2 = r^2$, so that $(\partial u/\partial n)(1, \theta) = 2r = 2$.
6. (a) One example is $u(z) = -\ln |z|$.
   (b) $v(z) = 1/(1 + x^2 + y^2)$.
7. (a), (d), (g) are false; the others are true.

PARAGRAPH 1.3.5

1. Harmonic: (a), (b), (d), (f).

PARAGRAPH 1.4.1

1. (a) 0, (b) π, (c) π, (d) 0.
2. (b) The converse is false if the domain Ω has a "hole," for example, if Ω is an annulus or a punctured disc. See Exercise 8 to Paragraph 1.4.3, as well as Exercise 5 below.
3. (a) Exact, $u(x, y) = xy$, (b) Exact, $u(x, y) = x$, (c) Not exact, (d) Exact, $u(x, y) = e^x \cos y$.
5. (d) One point of the loop Γ doesn't lie in the domain Ω of the function θ. Thus the theorem of Exercise 4(a) does not apply to the present situation.
6. (a) $u(x, y) = ye^x$, (b) $u(\pi/2, 1) - u(0, 0) = e^{\pi/2}$.

PARAGRAPH 1.4.2

1. (a) 1 ($=$ area of the rectangle), (b) 1/4.
2. (a) True, (b) True, (c) True, (d) False, (e) True.
5. $2\pi \int_0^1 r^3 \, dr = \pi/2$.

PARAGRAPH 1.4.3

2. Key: observe that if we translate $\int_\Gamma du$ into a double integral by means of Green's Theorem, then we obtain an integrand which is identically zero.
4. Note $p_y(x) = q_x(y) = 0$.
7. (a) The key here is Theorem 8 of Paragraph 1.4.1. Note that, unless we know that all points enclosed by the loop Γ are also in Ω, we cannot apply Green's Theorem.
8. (a) Conditions (2) and (4) each allow us to construct a function $u(x, y)$ in Ω whose differential is $p\,dx + q\,dy$. Note that (2) and (4) say the same thing.
   (b) Each of the four conditions guarantees exactness now.

PARAGRAPH 1.4.4

3. Both $u(\zeta)$ and $v(\zeta)$ have the same integral respresentation.
7. Write Green's III, but use $\Delta u \equiv 0$.
8. See Section 3 of Chapter 2.

Chapter 2

SECTION 2.1

1. Harmonic: (a), (d), (f).
2. First show $\Delta u(x, y) = e^x(g''(y) + g(y))$. Thus u is harmonic precisely when $g(y)$ is a solution to the famous second order linear differential equation $g'' + g = 0$. It is easy to check that $A \cos y + B \sin y$ is a solution to this equation (plug it in!), and somewhat more difficult to prove that *all* solutions are of the form $A \cos y + B \sin y$ for certain real A, B.
3. Note $u_{xx} = v_{yx}$ and $u_{yy} = -v_{xy}$. But $v_{xy} = v_{yx}$.
5. The double integral, which involves the Laplacian.
6. Note that $\ln |\zeta|$ is not defined at one point of the disc.
7. Both are harmonic. For instance, $u_{xx} + u_{yy} = 0$ implies $u_{xxx} + u_{yyx} = 0$. But this equals $(u_x)_{xx} + (u_x)_{yy}$, so that u_x is harmonic.

SECTION 2.2.1

3. (c) Zero temperature occurs only on the unit circle.
   (d) Integrate.
   (f) The origin is an infinite source of heat.
   (g) The same amount of heat energy must flow across each circle centered at the origin, independent of its radius R.
4. See Exercise 3(g).
5. (b) The source is ζ_1 (temperature $= +\infty$), the sink ζ_0.
   (c) $u(\zeta) = 0$ if and only if $\ln (|\zeta - \zeta_0|/|\zeta - \zeta_1|) = 0$ if and only if $|\zeta - \zeta_0| = |\zeta - \zeta_1|$. This determines a straight line of points equidistant from ζ_0 and ζ_1.
7. You might observe that heat would tend to flow away from any point with highest positive temperature towards colder points. It follows that this point would tend to become cooler, contradicting the time-independence of temperature. Thus there can be no highest positive (or lowest negative) temperature, so the temperature is everywhere zero.
8. If $u(x, y, t)$ is constant in time, then $u_t = 0$.

PARAGRAPH 2.2.2

1. (a) False (note that q was not necessarily harmonic), (b) False, (c) True, (d) True (continuity), (e) True (continuity); compare (c).
2. Prove there is a point x_1 with $g(x_1) > 0$ if and only if there is a subinterval $c \le x \le d$ with $\int_c^d g(x)\,dx > 0$.

   On the other hand, if all we know is that $\int_a^b g(x)\,dx = 0$ over the full interval, we cannot conclude whether $g(x)$ is identically zero or not.

PARAGRAPH 2.2.3

1. (a) True, (b) True, (c) True; in fact, $u(z) = u(0)$ for all z with $|z| \le 1$. (d) False; in fact $A \ln |z|$ is not harmonic at $z = 0$ in Ω. (e) True, (f) True.
2. We've used $\Delta u = 0$ in the Inside-Outside Theorem and in Green's III, and then used these theorems to examine harmonic functions.

SECTION 2.3

2. (a) $2(= u(1, 2))$, (c) 0.
5. (a) $32\pi\ (= \pi R^2 u(\zeta))$, (b) 0.
7. Yes.

PARAGRAPH 2.4.1

1. The Circumferential Mean Value Theorem.
2. A simple example would be: Let $u(z) = 1$ for all z in Ω_1 and $u(z) = 2$ for all z in Ω_2. You must prove that this is continuous provided Ω_1 and Ω_2 are disjoint.

PARAGRAPH 2.4.2

2. (a) Maxima at $(\sqrt{2}/2, \sqrt{2}/2)$ and $(-\sqrt{2}/2, -\sqrt{2}/2)$; minima at $(\sqrt{2}/2, -\sqrt{2}/2)$ and $(-\sqrt{2}/2, \sqrt{2}/2)$. All these points are on the unit circle, as guaranteed by the Weak Maximum Principle.
   (c) Because $\ln |z|$ is an increasing function (calculus!) of $r = |z|$, its minima on the annulus occur precisely on the inner rim $|z| = 1$ and its maxima occur on the outer rim $|z| = 2$.

PARAGRAPH 2.4.3

1. (b) The only example is $v(x, y) = x^2 - y^2$. (c) Yes.
2. (a) The upper half plane is unbounded.
   (b) $\ln |z|$ "blows up" (is not finite-valued, not defined) at the point $z = 0$.
3. (a) Theorem 4 allows us to omit consideration of $(\partial u/\partial n)(z)$ for harmonic functions.
   (b) If Ω is a disc, then the Poisson Integral Formula (see Appendix 2 to this chapter) does this.

PARAGRAPH 2.5.1

4. The infinite limit at a point $(z = 1)$ in the plane should make us think of the natural logarithm. The function $u(z) = - \ln |(z - 1)/2|$ will satisfy the requirements.

PARAGRAPH 2.5.2

1. (b) In applying Harnack's Inequality, the student is assuming that the function is everywhere nonnegative, and so is bounded below.
2. (a) $\sin xy$ is bounded (in fact, $|\sin xy| \le 1$) but not constant, contradicting Liouville's Theorem.
   (b) $u(x, y)$ is identically 0. For $|u(0, 0)| \le |\sin 0| = 0$, so $u(0, 0) = 0$. And u is constant by Liouville's Theorem.

Chapter 3

PARAGRAPH 3.1.1

1. (a) $3 + 2i$, (b) $-1 - 4i$, (c) $5 + 5i$.
2. $(-1/5) + (7/5)i$. You might obtain this by writing $\tau = a + bi$ and solving for a, b in the two linear equations implicit in $z\tau = \zeta$. But see Paragraph 3.1.3.
3. (a) $\pm i$, (b) $\pm 3i$.
4. i^n is real if and only if n is even, positive (real) if and only if n is divisible by 4 (that is, $n = 0, \pm 4, \pm 8, \dots$), and negative if n is of the form $4k + 2$ (that is, $n = \pm 2, \pm 6, \dots$).
5. Multiply both sides of the given equation by z, so as to introduce z^3 into the discussion. This gives $z^3 = -z^2 - z$. Now use $z^2 + z = -1$.
6. Re $z\zeta = 5$, Im $z\zeta = 5$.

PARAGRAPH 3.1.2

2. 5.
3. This is the circle $|z| = 2$, that is, $C(0; 2)$.
4. $|z| = \sqrt{5}$, arg $z = \arctan(-1/2) = -\arctan(1/2)$ radians. Now $iz = 1 + 2i$, so that $|iz| = \sqrt{5} = |z|$ and arg $iz = \arctan(2)$. Thus multiplication by i has rotated the point z through one right angle counterclockwise to iz. (Check that $\arctan(2) = -\arctan(1/2) + \pi/2$.) Keep your eyes open for geometric gifts coming from complex arithmetic.
5. Let $\tau = -i$.

PARAGRAPH 3.1.3

1. (a) $(2/5) + (1/5)i$, (b) $(-1/5) + (7/5)i$.
3. The statement is true. Use $|z|^2 = x^2 + y^2$.
4. $-i$.

PARAGRAPH 3.2.1

1. $u(x, y) = x^3 - 3xy^2$, $v(x, y) = 3x^2y - y^3$.
2. $g(z) = 1/z = (x/(x^2 + y^2)) + i(-y/(x^2 + y^2))$, so that $u(x, y) = x/(x^2 + y^2)$, $v(x, y) = -y/(x^2 + y^2)$.
4. $|f(z_0) - w_0| < \varepsilon$.
7. Continuous.

PARAGRAPH 3.2.2

1. (a) $6z^2 + i$, (c) $-2/(\zeta - 1)^2$, (e) $H'(z) = g'(f(z))f'(z) = 2/(z + 1)^2$.
3. The key is that $(f(z) - f(z_0))/(z - z_0)$ approaches $f'(z_0)$ as z approaches z_0.
4. The key is that $f(z) - f(z_0)$ approaches $f'(z_0)(z - z_0)$ as z approaches z_0. It follows that $f(z)$ approaches $f(z_0)$, whence continuity.
5. See Exercise 4 to Paragraph 1.3.1. One difficulty with a complex analog of the Mean Value Theorem is that we cannot hope to say "there is a point z_1 *between* z and z_0" due to the fact that the domain of a complex function is 2-dimensional, not a line segment.
6. Look at the difference quotient for the derivative.

PARAGRAPH 3.2.3

1. (a) All points in the plane, (b) At all points except $z = \pm i$, (c) At all points except $z = 0$.
2. For analyticity at a point it is required that the function be continuously complex differentiable *in an open neighborhood of the point*.
3. Nowhere.
6. Analyticity of all derivatives follows from the Cauchy Integral Formula in Chapter 4.
7. Yes. See Exercise 4 to Paragraph 3.2.2.
8. Yes. This follows from the Chain Rule together with some elementary considerations of continuity. See Exercise 13.
9. See Section 3.3.
10. In elementary calculus this result is either accepted as obvious or proved using the Mean Value Theorem. In the complex case we have no Mean Value Theorem; one method of proof is to reduce things to the real two-variable case by means of the Cauchy–Riemann equations of the preceding exercise.
11. (a) It is the quotient of analytic functions.
    (b) $f'(z_0)$.
    (c), (d) The answer is yes in each case. See Chapter 5.
12. (a) $f(\Omega)$ is the open upper half plane $v > 0$ in the w-plane. If w has polar coordinates (ρ, φ), then its inverse image $z = g(w)$ has polar coordinates $(r, \theta) = (\sqrt{\rho}, \varphi/2)$.
    (c) The best characterization of continuity to employ here is the topologist's, Exercise 8 to Paragraph 3.2.1.
14. (a) True, (b) False.

SECTION 3.3

3. Note that this is true only on a *connected* domain.
   One method of proof: $f'(z) = 0$ everywhere implies $u_x(z) = u_y(z) = v_x(z) = v_y(z) = 0$, where $f(z) = u(z) + iv(z)$, which implies that $u(z)$ and $v(z)$ are both constant.
   A second method, which works for polynomials and rational functions: Compute $f'(z)$ explicitly and then argue that $f'(z)$ is identically zero only if $f(z)$ is constant.
   A third method, involving Theorem 1 of Chapter 4: This says that $\int_a^b f'(z) \, dz = f(b) - f(a)$. Thus if $f'(z) = 0$ we have $f(b) = f(a)$ for all points a, b in the domain.

Note that none of these methods involve the difference quotient $(f(z) - f(z_0))/(z - z_0)$ used to define the derivative. Such a proof is also possible.

4. Use the Cauchy–Riemann equations to prove that the real part of the complex function is constant. (Clearly the imaginary part is identically zero.)

PARAGRAPH 3.4.1

1. (a) $-e$, (c) 1, (e) 1.
5. Yes. In fact, $w = \log z$. See Paragraph 3.4.2.
6. This is the set of all points $w \neq 0$. See Paragraph 3.4.2.
7. (a) $e^{i\pi/4}$, (b) $e^{3\pi i/2}$, (c) $e^{\pi i}$.
9. $e^{2\pi i} - 1 = 0$.
11. (a) $e^{\pi i/8}$, $e^{9\pi i/8}$ $(= -e^{\pi i/8})$, (b) $e^{5i\pi/8}$, $e^{13\pi i/8}$ $(= -e^{5\pi i/8})$.

PARAGRAPH 3.4.2

1. (a) $\ln \sqrt{2} + i(3\pi/4)$, (c) $\ln 4 + i(\pi/7)$.
2. (a) $\ln \sqrt{2} + i(11\pi/4)$, (c) $\ln 4 + i(15\pi/7)$.
3. Yes; a nonprincipal branch of the logarithm *must* assume a nonreal value at a real point $z = x$. For example, $\log 1 = 2k\pi i, k = \pm 1, \pm 2, \ldots$, for the various nonprincipal branches.
4. Ω_1 is mapped by exp onto the punctured disc $0 < |w| < 1$. Ω_2 is mapped onto the annulus $1 < |w| < 2$. Ω_3 is mapped onto the open upper half disc $|w| < 1$, $|v| > 0$. Ω_4 is mapped to the exterior $|w| \geq 1$.
5. See the answers to Exercise 4 above.
6. Onto the punctured plane $w \neq 0$. Since $f(z)$ may be decomposed as $z \mapsto 2\pi z \mapsto \exp(2\pi z)$, its inverse is $w \to \log w \mapsto (1/2\pi) \log w$, provided we choose the principal branch of the logarithm.
8. (a) The composition of analytic functions is analytic.
   (b) $h(z)$ constant implies $\log h(z)$ constant.
9. (a) Construct $\log z = \ln |z| + i\theta$ where $-\varphi < \theta < \varphi$ (where it is reasonable to suppose that the ray is specified by an an angle satisfying $0 \leq \varphi < 2\pi$).

PARAGRAPH 3.4.3

2. This quickly reduces to solving $e^{iz} = \pm 1$.
3. $z = x = ((2k + 1)/2)\pi, k = 0, \pm 1, \pm 2, \ldots$.
4. All points z for which $\cos z \neq 0$.
6. (b) $\sin(z_1 + z_2) = \sin z_1 \cos z_2 + \cos z_1 \sin z_2$
      $\cos(z_1 + z_2) = \cos z_1 \cos z_2 - \sin z_1 \sin z_2$.
7. True. Use Exercise 6(b).
9. (b) $\sin z = -5$ if and only if $e^{iz} - e^{-iz} = -10i$ if and only if $(e^{iz})^2 + 10ie^{iz} - 1 = 0$ if and only if $e^{iz} = (-5 \pm \sqrt{24})i$ if and only if $z = -i \log((-5 \pm \sqrt{24})i)$. Note that this gives a "doubly infinite" family of solutions, $z = (5k\pi/2) - i \ln |-5 \pm \sqrt{24}|$ with $k = 0, \pm 1, \ldots$.
11. The entire w-plane.
12. $\arcsin z = -i \log(iz \pm \sqrt{1 - z^2})$; the logarithm is no surprise!
13. (b) $(d/dz) \sinh z = \cosh z$, $(d/dz) \cosh z = \sinh z$.
    (c) $\sinh^2 z - \cosh^2 z = 1$.
14. (a) ce^z, c arbitrary complex, (b) $c_1 \cos z + c_2 \sin z$, (c) $c_1 \cosh z + c_2 \sinh z$.

Paragraph 3.4.4

1. (a) $e^{-2k\pi}e^{i\ln\pi}$, $k = 0, \pm 1, \dots$, though you may wish to absorb the minus sign in the exponent into the k.
   (c) $e^{-(8k+1)\pi/4}e^{i\ln\sqrt{2}}$, $k = 0, \pm 1, \dots$.
2. Three choices z^b, z^c, z^{b+c} will satisfy $z^{b+c} = z^b z^c e^{2\pi i(kb+mc)}$ for certain integers k, m.

Section 3.5

1. $y^2 - x^2 + c$, with c an arbitrary real number.
2. Two nonintersecting paths from z_0 to z readily yield a closed loop. The difference of the two line integrals from z_0 to z_1 (over the two paths) is seen to be the integral of $-u_y\,dx + u_x\,dy$ around the closed loop. Since Ω is simply connected, Green's Theorem may be applied to convert this line integral into a double integral. But u harmonic readily implies that the new integrand is identically zero. This proves that both paths from z_0 to z yielded the same value of the line integral.
3. Because the natural candidate $\theta = v(z) = \arg z$ is a harmonic function only if we select a continuous branch, and continuous branches of $\arg z$ live on domains such as the slit plane (with $-\pi < \theta < \pi$), but *not* on the punctured plane.
4. See 3 above.

Chapter 4

Paragraph 4.1.1

1. (a) $(-2 + 2i)/3$, (b) $(2 - 2i)/3$, (c) $(-2 + 2i)/3$.
2. $2\pi i$.

Paragraph 4.1.2

1. (a), (b) See Exercise 1 to Paragraph 4.1.1. (c) No; the integral is independent of the path from z_0 to z_1.
3. (a) 0 (by Theorem 1), (b) All these integrals are zero.
6. $e^\pi - 1$.
8. A most useful example: Let $\Omega = \mathbb{C} - \{0\}$, the punctured plane, and $f(z) = 1/z$. Then $f(z)$ is not the derivative of an analytic function on all of Ω (although it is the derivative of any branch of $\log z$ on various slit planes).

Paragraph 4.1.3

1. (a) Note $M = 1/r$, so that the absolute value of the integral is not greater than $ML = (1/r)(2\pi r) = 2\pi$.
2. Given any $\varepsilon > 0$, there exist radii $r > 0$ such that $|I_r| < \varepsilon$. (In fact, simply let r satisfy $0 < r < 2\pi M$.) This means (by the second hypothesis), that all $|I_r| < \varepsilon$, for any $\varepsilon > 0$. It must therefore be the case that all $I_r = 0$.

SECTION 4.2

1. (a) 0, (c) $2\pi i$, (e) 0.
2. (a) True, (b) False (see Exercise 1e above), (c) True, (d) False (Ω might have a puncture enclosed by Γ), (e) False (consider $f(z) = 1/z^2$), (f) True, (g) True (beware!).
4. (a) 0, (b) 1, (c) $F(z) = \ln|z| + i \arg z$, $-\pi < \arg z < \pi$, (d) $1/z$.
5. Use Exercise 3b to Paragraph 4.1.2.
6. Use Exercise 2 to Paragraph 4.1.3.
13. Ω is not simply connected.

SECTION 4.3

1. (a) $2\pi i$, (c) 0, (e) 0.
2. Corollary 5 in Section 4.2.
5. a, b, c, d, e, f, j and no others.
8. (a) Note that $f(z) - g(z)$ is identically zero on the loop Γ. Now use the integral formula to get $f(z) - g(z)$ inside Γ.
   (b) Extensions from Γ into its interior do not always exist, even if Γ is a circle. It depends on the given boundary values $f(z)$. This is in contrast to the Dirichlet problem for the disc, in which a real harmonic function assuming given real boundary values is sought. The complex extension problem with given boundary values $f(z)$ may fail because the real and imaginary parts of $f(z) = u(z) + iv(z)$ might not extend to *conjugate* harmonic functions in the interior of the disc: the Cauchy–Riemann equations again.

SECTION 4.4

1. (a) Once ζ is fixed, the integrand $f(z)(z - \zeta)^{-2}$ is defined (finite-valued) and continuous for all z on the curve Γ. And it's standard that continuous (even bounded piecewise-continuous) functions on a "finite interval" have finite integrals. Note that we need ζ off the curve Γ.
   (b) Differentiation (in ζ) under the integral sign in the Cauchy Integral Formula.
2. (b) No. For if $|\zeta| > 1$, then the integral is 0.
3. (a) $\pi i/60$, (b) $2\pi i(e - 2)$.
4. Note that the integral equals $2\pi i/(n - 1)!$
5. $f'(\zeta) = \int_{|z|=1} (z - \zeta)^{-2} \sin xy \, dz$.
6. $\int_\Gamma f(z) \, dz = \int_\Gamma g(z)(z - z_0)^{-m} \, dz = (2\pi i/(m - 1)!)g^{(m-1)}(z_0)$.

SECTION 4.5

1. (a) False (though it is true if Ω is simply connected), (b) False (as in (a)), (c) True, (d) False (consider z^2 and $z^2 + 2\pi i$), (e) False, (f) True (see (d)).
2. There remained the technical issue of proving that both parts actually were twice continuously differentiable.
5. $h(z) = x$. Note that $|e^z| = e^x$.
6. Let Ω be simply connected.
7. $u(z) = \ln|z|$. We've seen this already.

SECTION 4.6

1. (a) $2\pi(= 2\pi \cos 0)$, (b) $0(= 2\pi \log 1)$.
2. (a) 0, (b) 0.
4. Corollary 5.
5. $1(= \cos 0)$.

SECTION 4.7

2. (a) The maximum modulus $|\cos z|$ for $|z| \leq 1$ must be attained only on $|z| = 1$. But $\cos 0 = 1$. Hence we must have $|\cos z| > 1$ for at least one z with $|z| = 1$.
   (b) Let z_1 satisfy $|z_1| = 1$, $|\cos z_1| > 1$, as in (a). Now let $\{\zeta_n\}$ be a sequence of points with $|\zeta_n| < 1$ and $\lim_{n \to \infty} \zeta_n = z_1$. By continuity, $\lim_{n \to \infty} \cos \zeta_n = \cos z_1$. Thus $|\cos \zeta_n|$ approaches a positive number >1, and so we can't have all $|\cos \zeta_n| \leq 1$.
   (c) Of course $|\cos x| \leq 1$ for all real x.
3. False. The "limit at ∞" might not exist. A very simple example: Let $f(z) = \sin z$. Then $\lim_{x \to +\infty} \sin x$ does not exist, so surely the general limit as $z \to \infty$ fails to exist. Others: $\cos z$, e^z.
4. (a) This is the Weak Maximum Modulus Principle.
   (b) It is true. This is Liouville's Theorem (Section 4.9).
6. If $f(\Omega)$ intersected the circle $|w| = r$, then $f(z)$ would be assuming its maximum modulus at some z in Ω.
8. The only rotations have $n = 1$, $|c| = 1$. All the others contract, that is, $|f(z)| < |z|$ if $0 < |z| < 1$.

SECTION 4.8

1. Consider the polynomial $f(z) = p(z) - c$.
3. Take the complex conjugate of both sides of the equation $p(z_1) = 0$ to obtain $p(\bar{z}_1) = 0$. Key: The coefficients of the polynomial are the same as their complex conjugates.

SECTION 4.9

1. The imaginary part $v(z)$, being harmonic and bounded below, must be constant. By the Cauchy–Riemann equations, the real part of $f(z)$ must also be constant.
2. If $f(\mathbb{C})$ missed a line then it would lie to one side or the other of this line. Rotating, we have $e^{i\varphi}f(\mathbb{C})$ above a horizontal line (here φ is some fixed angle) Thus $w = e^{i\varphi}f(z)$ is constant, by Exercise 1, so that $f(z)$ is constant. This is a contradiction.
3. The exercise is stronger; it shows $p(\mathbb{C}) = \mathbb{C}$, which is definitely unbounded.
7. Since $|f(z) - w_0| > \varepsilon$, we have $|g(z)| = 1/|f(z) - w_0|$ bounded. But $g(z)$ is entire. Contradiction.
8. (b) No branch of $\log z$ is entire (cf. slit plane).

SECTION 4.10

1. No. Morera's Theorem does not apply, because there are loops Γ which pass through the point $z_0 = 0$ for which $\int_\Gamma (1/z^2)\, dz$ is not zero.
2. (a) The integral exists because the integrand is continuous on the curve Γ.
   (b) The Cauchy Integral Theorem.

(c) This argument is standard. We used it in the proof of the Cauchy Integral Formula. It depends strongly on the fact that $f(z)$ is bounded even at the "bad" point z_0.

SECTION 4.11

2. We have $|f^{(n)}(z_0)| \le n! \, M(z_0; r)/r^n \le n! \, ar^k/r^n$. Fix $n > k$ and let $r \to \infty$. The right hand side of this inequality tends to zero. The result follows.
3. We sketch an approach. First, integrate $f^{(k+1)}(z)$ $k + 1$ times to obtain $f(z)$. But note that the $(k + 1)$st iterated antiderivative of $0 (= f^{(k+1)}(z))$ is a polynomial of degree $\le k$.

Chapter 5

PARAGRAPH 5.1.1

1. See a calculus book.

PARAGRAPH 5.1.2

1. (a) $i - (1/2) - (i/3) + (1/4) + (i/5) - (1/6)$.

PARAGRAPH 5.1.3

2. (a) Absolutely, (b) Diverge, (c) Absolutely, (d) Conditionally.

PARAGRAPH 5.1.4

1. The harmonic series is one counterexample.
2. See 1.
4. The series in z converges. The absolute value of its sum is less than 2.

PARAGRAPH 5.2.1

1. $1 + i$.
2. (c) No; it need not be true. (d) No. (e) Obviously not; see (a). (f) No.
3. (a) Yes. (b) No. (c) Possibly. (d) $R = 1$.
4. $R = 1$.
5. $R = 1$.
6. (a) $R = 1$. (b) $R = 1$.
12. (a) False; see (e). (b) True. (c) False. (d) False. (e) True. (f) True. (g) True; see Exercise 9 above. (h) True. (i) False; it might converge at $z = 0$.

PARAGRAPH 5.2.2

2. (a) True, (b) True, (c) False, (d) False, (e) True.
5. (a) $1/(1 - z)$.

PARAGRAPH 5.2.3

1. (a) True, (b) True, (c) True.
6. (a) True, (b) False, (c) True; the functions must be analytic, (d) False; not all continuous real functions can be obtained by restricting. (e) True, (f) False.
8. (b) The principal branch of $\log z$.

PARAGRAPH 5.3.1

1. (a) $a_k = 2^k/k!$, (b) $a_{2k} = 0$, $a_{2k+1} = (-1)^k/(2k+1)!$, (c) $a_k = 0$ for $k > 1$.

PARAGRAPH 5.3.2

1. (a) True, (b) False, (c) True; see Exercise 8 below, (d) False, (e) True.
2. (a) False; the series might also converge at certain points on the rim of the disc of convergence. (b) True, though not immediately obvious. (c) True. (d) True. (e) True. (f) True.
3. (a) $R = 1$, (b) $R = \sqrt{10}$, (c) $R = \infty$.

PARAGRAPH 5.3.3

1. (a) $e^{\pi z} = \sum_{k=0}^{\infty} \pi^k z^k/k!$
   (b) $\sin \pi z = \sum_{k=0}^{\infty} (-1)^k (\pi z)^{2k+1}/(2k+1)!$
   (c) $1/(2+z) = \sum_{k=0}^{\infty} (-1)^k z^k/2^{k+1}$.
   (d) $1/(2+z) = \sum_{k=0}^{\infty} (-1)^k (z+1)^k$.
   (e) $1/(1+2z) = \sum_{k=0}^{\infty} (-1)^k (2z)^k$.
   (f) $\pi i + (1/z) = \pi i + \sum_{k=0}^{\infty} i^{k-1}(z-i)^k$.
2. (a) $R = \infty$, (b) $R = \infty$, (c) $R = 2$, (d) $R = 1$, (e) $R = 1/2$, (f) $R = 1$.
3. (a) $\sinh z = z + z^3/3! + \cdots = \sum_{k=0}^{\infty} z^{2k+1}/(2k+1)!$

PARAGRAPH 5.3.4

2. Only statements (d) and (e) are necessary consequences.

PARAGRAPH 5.3.5

1. Statements (b), (d), (e), (f) each imply $f = g$ on Ω. The others do not.
3. (a) True. (b) False; consider the definition. (c) False. (d) True; see the Schwarz Reflection Principle in Exercise 6. (e) True.
7. $f(z) = e^z$ is the only one.
8. Counterexample: $\ln |z| = 0$ on $|z| = 1$, but $\ln |z| \not\equiv 0$ on the punctured plane.

Chapter 6

PARAGRAPH 6.1.1

1. (a) Poles at $z = 0$ and $z = -1$.
   (b) No singularities in the z-plane; an entire function.
   (c) Essential singularity at $z = 0$.
   (d) Pole at $z = 0$.
   (e) Poles at $z = x = 0$, $\pm \pi$, $\pm 2\pi$, $\pm 3\pi$, ...
   (f) Removable singularity at $z = -1$, pole at $z = -2$.
4. (a) $(1/z) + \sin z$ or e^z/z or $1/\sin z$.
   (b) $e^{1/(z-1)}$.
   (c) $f(z) = 1/z$, $g(z) = (1/z) + z$.
   (d) $5z^3$.
   (e) $(z^2 - 1)(z^2 + 1)$.
   (f) $z^{-1}e^{1/(z-1)}$ or $z^{-1} + e^{1/(z-1)}$.

PARAGRAPH 6.1.2

2. (a) $h = k$.
   (b) $h = k$.
   (c) If $h > k$, then $\lim_{z \to z_0} f(z)/g(z) = 0$; if $h < k$, then the limit is infinite.

PARAGRAPH 6.2.1

1. (b) It would be possible to write down a series in $(z - z_0)$ which converged only at z_0, and one in $(z - z_0)^{-1}$ which converged only at ∞. The sum of these series would converge nowhere.
   (d) Yes.
2. (a) True. (b) True. (c) True. (d) True. (e) True; this series is the one used to study $f(z)$ near z_0. (f) True. (g) True.
3. (a) The geometric series $\sum_{n=0}^{\infty} z^n$.
   (b) $1/(z - 1)$; this is a Laurent series!
   (c) $-\sum_{n=1}^{\infty} z^{-n}$.
5. No. The logarithm does not have an isolated singularity at $z = 0$.

PARAGRAPH 6.2.2

1. The origin is (a) nonsingular, (b) essential, (c) removable, (d) pole, (e) not isolated, (f) essential, (g) logarithmic. The series are (a) $z - (z^3/3!) + \cdots$, (b) $(1/z) - (1/z^3 3!) + \cdots$, (c) $1 - (z^2/3!) + \cdots$, (d) $(1/z) - (z/3!) + \cdots$, (e) no series, (f) $1 - (1/z^2 3!) + \cdots$, (g) no series.
2. (a) $1 + z^2 + z^4 + \cdots$.
   (b) $z^{-2} + z^{-4} + z^{-6} + \cdots$.
   (c) $\sum_{n=-1}^{\infty} (-1)^{n+1}(i/2)^{n+2}(z + i)^n$.

SECTION 6.3

1. (a) True. (b) False; the limit does not exist. (c) True (in the sense of removable singularities). (d) True. (e) True; see (b). (f) True. (g) False, essential singularity; in fact, you might observe that the function is $\cos(1/z)$. (h) True.
2. (a) Meromorphic: polynomials, rational functions, entire functions, $(\sin z)/z$, $1/(\sin z)$, e^z/z.
   (b) Yes.

SECTION 6.4

1. One example is $e^{1/p(z)}$ with $p(z) = z(z - 1)$.
2. The naive student used the wrong Laurent series for studying $1/(z - 1)$ at $z_0 = 0$.
3. Here $f(z) = (\sin z)/z$.
5. The statement is true. One might begin the proof by writing $f(z) = z^{-n} f_1(z)$ with $\lim_{z \to z_0} f_1(z)$ finite and nonzero.

Chapter 7

PARAGRAPH 7.1.1

1. (a) $2\pi i$, (b) 0, (c) 0, (d) $\pi i/2$, (e) 0, (f) 0.
2. $10\pi i$.
4. (i) True, (ii) False, (iii) True, (iv) False, (v) True.

PARAGRAPH 7.1.2

1. (a) res(csc z; 0) = 1, (b) z csc z has a removable singularity at $z = 0$, (c) as in (b), (d) res($f(z)$; 0) = 1/4, (e) The integrand has simple poles at $z_0 = -1/2$ and $z_1 = 1/2$ with residues -1 and 1, respectively, (f) The integrand is analytic.
   The values of the integrals are given above.
2. (a) Either a removable singularity or a simple pole, (b) $g(z_0)$, (c) $2\pi i g(z_0)$, (d) Cauchy Integral Formula.
3. (a) Simple pole at $z_0 = 0$, double pole at $z_1 = -1$; res(f; 0) = 1, res(f; -1) = -1.
   (b) 0.

PARAGRAPH 7.2.1

1. (a) -2, (b) -5.
3. The value of the integral is 1.

PARAGRAPH 7.2.2

6. (a) True, (b) True, (c) True.

PARAGRAPH 7.2.3

1. (a) 1, (b) 1, (c) 3 times.
2. (a) The image is the unit circle $|w| = 1$ traversed twice in the counterclockwise direction.
   (b) 2.
   (c) The origin *is* a double zero of $f(z) = 0$.

Chapter 8

PARAGRAPH 8.1.1

2. (a) The critical points of cos z are the zeros of $-\sin z$ which are, as we have seen several times already, all real: $z = x = 0, \pm\pi, \pm 2\pi, \ldots$. Each of these is a simple zero, that is, a critical point of order 2 for the cosine.
   (b) The origin is the only critical point (order 2).
   (c) No critical points.
   (d) Critical points at $z = \pm 1$, each of order 2.
3. Begin by computing $f'(z)$.
4. The function (mapping) f is *one-to-one* from S_1 into S_2 if $z_1 \neq z_2$ in S_1 implies that $f(z_1) \neq f(z_2)$ in S_2; in other words, distinct points are mapped by f to distinct points. And f maps S_1 *onto* S_2 if for every w in S_2 there is at least one z in S_1 such that $f(z) = w$; in other words, $f(S_1) = S_2$.

PARAGRAPH 8.1.2

1. (c) $R = \pi$. Consider the periodicity of e^z.
3. (c) Yes.
4. (b) No. Contrast (c).
6. (a) True. (b) False, although it is locally one-to-one near every point z_0. (c) False, although it is true up to a first order approximation. (d) True. (e) True.

PARAGRAPH 8.1.3

1. The key here is that $z^2 = (-z)^2$.
2. $g(z)$ is locally 3 to 1 near $z_0 = i$.

PARAGRAPH 8.1.4

5. The key here is to observe the behavior of $f(z)$ at infinity: $\lim_{z \to \infty} f(z) = \infty$ (a pole). By Exercise 11(c) to Paragraph 6.4, $f(z)$ is a polynomial. Now use one-to-oneness. Note that this has little connection with the Open Mapping Theorem.

PARAGRAPH 8.1.5

3. (a) Regard $z^5 + z + 1$ as a perturbation of z^5 and check that $|z + 1| < |z^5|$ if $|z| = 5/4$. Even sharper estimates are not hard to find.

PARAGRAPH 8.1.6

1. (c) The other fixed points are the roots of $az^{k-1} - 1 = 0$.

PARAGRAPH 8.2.1

2. (a) i, (b) ∞, (c) 0, (d) 1.
3. $z = i(w + 1)/(w - 1)$.
4. $z_0 = 1$.
6. $T(z) = 1 + (2i/(z - i))$.
7. (a) $w = r^{-1}e^{-i\theta}$, (b) Onto the ray $\varphi = -\theta_0$ in the plane $|w| > 0$, (c) Onto the circle $|w| = 1/r_0$, (e) The vertical line through $w = 1$ in the w-plane.
9. (a) $T(z) = -i(z - 1)/(z + 1)$.
   (b) $T(0) = i$.
   (c) Onto the circle in Σ determined by 0, 1, ∞, that is, the horizontal u-axis.
   (d) Onto the open upper half plane $v > 0$.
11. Such transformations always exist, but they are never unique.
12. No. It maps both $z = i$ and $z = -i$ to 0.

PARAGRAPH 8.2.2

2. (a) Perhaps the simplest is $T(z) = -z$.
   (b) No. Too many fixed points.
3. (a) No. For if $T(0)$ were finite, then T would not be conformal at $z = 0$. Thus $T(0) = \infty$. Now follow T with an inversion. This yields a linear fractional transformation of the wedge onto the lower half plane which fixes the origin. But this is not conformal at the origin.
   (b) Yes. Let $w = z^3$.
4. The points w_0, w_1, w_2 should occur *in that order* on a counterclockwise journey once around the circle $|w| = 1$.
11. (a) True, (b) False, (c) True (see (a)), (d) False (see (b)), (e) False (see Exercise 12), (f) True, (g) False.
14. The approximation is the linear map $w = f(z_0) + f'(z_0)(z - z_0)$.

PARAGRAPH 8.2.3

2. No. The harmonic solution is uniquely determined by its boundary values.
4. $H(x, y) = x/(x^2 + (y + 1)^2) = $ the real part of $1/(z + i)$.

SECTION 8.3

1. F one-to-one implies F locally one-to-one (at each point of Ω). Thus F' is never zero on Ω. Now see Section 8.1.
2. (c) No. Note that the mapping in Exercise 17 of Paragraph 8.2.2 is not one-to-one; both top and bottom of the rectangle are mapped to the same image.
4. Only pair (a) and pair (f) fail to be conformally equivalent.
6. Note that 0 is not in $f(\Omega)$.
8. See Exercise 7 to Paragraph 8.2.2.
9. We may assume that Ω does not intersect the nonpositive x-axis. Thus the principal branch of $w = \log z$ maps Ω into a strip of finite height. Now apply Exercise 8 above.
10. The true statements are (a), (e). Use $w = 1/z$ in (e).

Index

A CATALOGUE OF
SELECTED DOVER BOOKS
IN ALL FIELDS OF INTEREST

A CATALOGUE OF SELECTED DOVER
BOOKS IN ALL FIELDS OF INTEREST

CELESTIAL OBJECTS FOR COMMON TELESCOPES, T. W. Webb. The most used book in amateur astronomy: inestimable aid for locating and identifying nearly 4,000 celestial objects. Edited, updated by Margaret W. Mayall. 77 illustrations. Total of 645pp. 5⅜ x 8½.
20917-2, 20918-0 Pa., Two-vol. set $9.00

HISTORICAL STUDIES IN THE LANGUAGE OF CHEMISTRY, M. P. Crosland. The important part language has played in the development of chemistry from the symbolism of alchemy to the adoption of systematic nomenclature in 1892. ". . . wholeheartedly recommended,"—Science. 15 illustrations. 416pp. of text. 5⅝ x 8¼.
63702-6 Pa. $6.00

BURNHAM'S CELESTIAL HANDBOOK, Robert Burnham, Jr. Thorough, readable guide to the stars beyond our solar system. Exhaustive treatment, fully illustrated. Breakdown is alphabetical by constellation: Andromeda to Cetus in Vol. 1; Chamaeleon to Orion in Vol. 2; and Pavo to Vulpecula in Vol. 3. Hundreds of illustrations. Total of about 2000pp. 6⅛ x 9¼.
23567-X, 23568-8, 23673-0 Pa., Three-vol. set $27.85

THEORY OF WING SECTIONS: INCLUDING A SUMMARY OF AIR-FOIL DATA, Ira H. Abbott and A. E. von Doenhoff. Concise compilation of subatomic aerodynamic characteristics of modern NASA wing sections, plus description of theory. 350pp. of tables. 693pp. 5⅜ x 8½.
60586-8 Pa. $8.50

DE RE METALLICA, Georgius Agricola. Translated by Herbert C. Hoover and Lou H. Hoover. The famous Hoover translation of greatest treatise on technological chemistry, engineering, geology, mining of early modern times (1556). All 289 original woodcuts. 638pp. 6¾ x 11.
60006-8 Clothbd. $17.95

THE ORIGIN OF CONTINENTS AND OCEANS, Alfred Wegener. One of the most influential, most controversial books in science, the classic statement for continental drift. Full 1966 translation of Wegener's final (1929) version. 64 illustrations. 246pp. 5⅜ x 8½.
61708-4 Pa. $4.50

THE PRINCIPLES OF PSYCHOLOGY, William James. Famous long course complete, unabridged. Stream of thought, time perception, memory, experimental methods; great work decades ahead of its time. Still valid, useful; read in many classes. 94 figures. Total of 1391pp. 5⅜ x 8½.
20381-6, 20382-4 Pa., Two-vol. set $13.00

DRAWINGS OF WILLIAM BLAKE, William Blake. 92 plates from Book of Job, *Divine Comedy, Paradise Lost,* visionary heads, mythological figures, Laocoon, etc. Selection, introduction, commentary by Sir Geoffrey Keynes. 178pp. 8⅛ x 11. 22303-5 Pa. $4.00

ENGRAVINGS OF HOGARTH, William Hogarth. 101 of Hogarth's greatest works: *Rake's Progress, Harlot's Progress, Illustrations for Hudibras, Before and After, Beer Street and Gin Lane,* many more. Full commentary. 256pp. 11 x 13¾. 22479-1 Pa. $12.95

DAUMIER: 120 GREAT LITHOGRAPHS, Honore Daumier. Wide-ranging collection of lithographs by the greatest caricaturist of the 19th century. Concentrates on eternally popular series on lawyers, on married life, on liberated women, etc. Selection, introduction, and notes on plates by Charles F. Ramus. Total of 158pp. 9⅜ x 12¼. 23512-2 Pa. $6.00

DRAWINGS OF MUCHA, Alphonse Maria Mucha. Work reveals draftsman of highest caliber: studies for famous posters and paintings, renderings for book illustrations and ads, etc. 70 works, 9 in color; including 6 items not drawings. Introduction. List of illustrations. 72pp. 9⅜ x 12¼. (Available in U.S. only) 23672-2 Pa. $4.00

GIOVANNI BATTISTA PIRANESI: DRAWINGS IN THE PIERPONT MORGAN LIBRARY, Giovanni Battista Piranesi. For first time ever all of Morgan Library's collection, world's largest. 167 illustrations of rare Piranesi drawings—archeological, architectural, decorative and visionary. Essay, detailed list of drawings, chronology, captions. Edited by Felice Stampfle. 144pp. 9⅜ x 12¼. 23714-1 Pa. $7.50

NEW YORK ETCHINGS (1905-1949), John Sloan. All of important American artist's N.Y. life etchings. 67 works include some of his best art; also lively historical record—Greenwich Village, tenement scenes. Edited by Sloan's widow. Introduction and captions. 79pp. 8⅜ x 11¼. 23651-X Pa. $4.00

CHINESE PAINTING AND CALLIGRAPHY: A PICTORIAL SURVEY, Wan-go Weng. 69 fine examples from John M. Crawford's matchless private collection: landscapes, birds, flowers, human figures, etc., plus calligraphy. Every basic form included: hanging scrolls, handscrolls, album leaves, fans, etc. 109 illustrations. Introduction. Captions. 192pp. 8⅞ x 11¾. 23707-9 Pa. $7.95

DRAWINGS OF REMBRANDT, edited by Seymour Slive. Updated Lippmann, Hofstede de Groot edition, with definitive scholarly apparatus. All portraits, biblical sketches, landscapes, nudes, Oriental figures, classical studies, together with selection of work by followers. 550 illustrations. Total of 630pp. 9⅛ x 12¼. 21485-0, 21486-9 Pa., Two-vol. set $15.00

THE DISASTERS OF WAR, Francisco Goya. 83 etchings record horrors of Napoleonic wars in Spain and war in general. Reprint of 1st edition, plus 3 additional plates. Introduction by Philip Hofer. 97pp. 9⅜ x 8¼. 21872-4 Pa. $4.00

THE COMPLETE BOOK OF DOLL MAKING AND COLLECTING, Catherine Christopher. Instructions, patterns for dozens of dolls, from rag doll on up to elaborate, historically accurate figures. Mould faces, sew clothing, make doll houses, etc. Also collecting information. Many illustrations. 288pp. 6 x 9. 22066-4 Pa. $4.50

THE DAGUERREOTYPE IN AMERICA, Beaumont Newhall. Wonderful portraits, 1850's townscapes, landscapes; full text plus 104 photographs. The basic book. Enlarged 1976 edition. 272pp. 8¼ x 11¼.
23322-7 Pa. $7.95

CRAFTSMAN HOMES, Gustav Stickley. 296 architectural drawings, floor plans, and photographs illustrate 40 different kinds of "Mission-style" homes from *The Craftsman* (1901-16), voice of American style of simplicity and organic harmony. Thorough coverage of Craftsman idea in text and picture, now collector's item. 224pp. 8⅛ x 11. 23791-5 Pa. $6.00

PEWTER-WORKING: INSTRUCTIONS AND PROJECTS, Burl N. Osborn. & Gordon O. Wilber. Introduction to pewter-working for amateur craftsman. History and characteristics of pewter; tools, materials, step-by-step instructions. Photos, line drawings, diagrams. Total of 160pp. 7⅞ x 10¾. 23786-9 Pa. $3.50

THE GREAT CHICAGO FIRE, edited by David Lowe. 10 dramatic, eye-witness accounts of the 1871 disaster, including one of the aftermath and rebuilding, plus 70 contemporary photographs and illustrations of the ruins—courthouse, Palmer House, Great Central Depot, etc. Introduction by David Lowe. 87pp. 8¼ x 11. 23771-0 Pa. $4.00

SILHOUETTES: A PICTORIAL ARCHIVE OF VARIED ILLUSTRATIONS, edited by Carol Belanger Grafton. Over 600 silhouettes from the 18th to 20th centuries include profiles and full figures of men and women, children, birds and animals, groups and scenes, nature, ships, an alphabet. Dozens of uses for commercial artists and craftspeople. 144pp. 8⅜ x 11¼.
23781-8 Pa. $4.50

ANIMALS: 1,419 COPYRIGHT-FREE ILLUSTRATIONS OF MAMMALS, BIRDS, FISH, INSECTS, ETC., edited by Jim Harter. Clear wood engravings present, in extremely lifelike poses, over 1,000 species of animals. One of the most extensive copyright-free pictorial sourcebooks of its kind. Captions. Index. 284pp. 9 x 12. 23766-4 Pa. $8.95

INDIAN DESIGNS FROM ANCIENT ECUADOR, Frederick W. Shaffer. 282 original designs by pre-Columbian Indians of Ecuador (500-1500 A.D.). Designs include people, mammals, birds, reptiles, fish, plants, heads, geometric designs. Use as is or alter for advertising, textiles, leathercraft, etc. Introduction. 95pp. 8¾ x 11¼. 23764-8 Pa. $3.50

SZIGETI ON THE VIOLIN, Joseph Szigeti. Genial, loosely structured tour by premier violinist, featuring a pleasant mixture of reminiscenes, insights into great music and musicians, innumerable tips for practicing violinists. 385 musical passages. 256pp. 5⅝ x 8¼. 23763-X Pa. $4.00

TONE POEMS, SERIES II: TILL EULENSPIEGELS LUSTIGE STREICHE, ALSO SPRACH ZARATHUSTRA, AND EIN HELDEN-LEBEN, Richard Strauss. Three important orchestral works, including very popular *Till Eulenspiegel's Marry Pranks,* reproduced in full score from original editions. Study score. 315pp. 9⅜ x 12¼. (Available in U.S. only)
23755-9 Pa. $8.95

TONE POEMS, SERIES I: DON JUAN, TOD UND VERKLARUNG AND DON QUIXOTE, Richard Strauss. Three of the most often performed and recorded works in entire orchestral repertoire, reproduced in full score from original editions. Study score. 286pp. 9⅜ x 12¼. (Available in U.S. only)
23754-0 Pa. $7.50

11 LATE STRING QUARTETS, Franz Joseph Haydn. The form which Haydn defined and "brought to perfection." (*Grove's*). 11 string quartets in complete score, his last and his best. The first in a projected series of the complete Haydn string quartets. Reliable modern Eulenberg edition, otherwise difficult to obtain. 320pp. 8⅜ x 11¼. (Available in U.S. only)
23753-2 Pa. $7.50

FOURTH, FIFTH AND SIXTH SYMPHONIES IN FULL SCORE, Peter Ilyitch Tchaikovsky. Complete orchestral scores of Symphony No. 4 in F Minor, Op. 36; Symphony No. 5 in E Minor, Op. 64; Symphony No. 6 in B Minor, "Pathetique," Op. 74. Bretikopf & Hartel eds. Study score. 480pp. 9⅜ x 12¼.
23861-X Pa. $10.95

THE MARRIAGE OF FIGARO: COMPLETE SCORE, Wolfgang A. Mozart. Finest comic opera ever written. Full score, not to be confused with piano renderings. Peters edition. Study score. 448pp. 9⅜ x 12¼. (Available in U.S. only)
23751-6 Pa. $11.95

"IMAGE" ON THE ART AND EVOLUTION OF THE FILM, edited by Marshall Deutelbaum. Pioneering book brings together for first time 38 groundbreaking articles on early silent films from *Image* and 263 illustrations newly shot from rare prints in the collection of the International Museum of Photography. A landmark work. Index. 256pp. 8¼ x 11.
23777-X Pa. $8.95

AROUND-THE-WORLD COOKY BOOK, Lois Lintner Sumption and Marguerite Lintner Ashbrook. 373 cooky and frosting recipes from 28 countries (America, Austria, China, Russia, Italy, etc.) include Viennese kisses, rice wafers, London strips, lady fingers, hony, sugar spice, maple cookies, etc. Clear instructions. All tested. 38 drawings. 182pp. 5⅜ x 8.
23802-4 Pa. $2.50

THE ART NOUVEAU STYLE, edited by Roberta Waddell. 579 rare photographs, not available elsewhere, of works in jewelry, metalwork, glass, ceramics, textiles, architecture and furniture by 175 artists—Mucha, Seguy, Lalique, Tiffany, Gaudin, Hohlwein, Saarinen, and many others. 288pp. 8⅜ x 11¼.
23515-7 Pa. $6.95

THE AMERICAN SENATOR, Anthony Trollope. Little known, long unavailable Trollope novel on a grand scale. Here are humorous comment on American vs. English culture, and stunning portrayal of a heroine/villainess. Superb evocation of Victorian village life. 561pp. 5⅜ x 8½.
23801-6 Pa. $6.00

WAS IT MURDER? James Hilton. The author of *Lost Horizon* and *Goodbye, Mr. Chips* wrote one detective novel (under a pen-name) which was quickly forgotten and virtually lost, even at the height of Hilton's fame. This edition brings it back—a finely crafted public school puzzle resplendent with Hilton's stylish atmosphere. A thoroughly English thriller by the creator of Shangri-la. 252pp. 5⅜ x 8. (Available in U.S. only)
23774-5 Pa. $3.00

CENTRAL PARK: A PHOTOGRAPHIC GUIDE, Victor Laredo and Henry Hope Reed. 121 superb photographs show dramatic views of Central Park: Bethesda Fountain, Cleopatra's Needle, Sheep Meadow, the Blockhouse, plus people engaged in many park activities: ice skating, bike riding, etc. Captions by former Curator of Central Park, Henry Hope Reed, provide historical view, changes, etc. Also photos of N.Y. landmarks on park's periphery. 96pp. 8½ x 11.
23750-8 Pa. $4.50

NANTUCKET IN THE NINETEENTH CENTURY, Clay Lancaster. 180 rare photographs, stereographs, maps, drawings and floor plans recreate unique American island society. Authentic scenes of shipwreck, lighthouses, streets, homes are arranged in geographic sequence to provide walking-tour guide to old Nantucket existing today. Introduction, captions. 160pp. 8⅞ x 11¾.
23747-8 Pa. $6.95

STONE AND MAN: A PHOTOGRAPHIC EXPLORATION, Andreas Feininger. 106 photographs by *Life* photographer Feininger portray man's deep passion for stone through the ages. Stonehenge-like megaliths, fortified towns, sculpted marble and crumbling tenements show textures, beauties, fascination. 128pp. 9¼ x 10¾.
23756-7 Pa. $5.95

CIRCLES, A MATHEMATICAL VIEW, D. Pedoe. Fundamental aspects of college geometry, non-Euclidean geometry, and other branches of mathematics: representing circle by point. Poincare model, isoperimetric property, etc. Stimulating recreational reading. 66 figures. 96pp. 5⅝ x 8¼.
63698-4 Pa. $2.75

THE DISCOVERY OF NEPTUNE, Morton Grosser. Dramatic scientific history of the investigations leading up to the actual discovery of the eighth planet of our solar system. Lucid, well-researched book by well-known historian of science. 172pp. 5⅜ x 8½.
23726-5 Pa. $3.50

THE DEVIL'S DICTIONARY. Ambrose Bierce. Barbed, bitter, brilliant witticisms in the form of a dictionary. Best, most ferocious satire America has produced. 145pp. 5⅜ x 8½.
20487-1 Pa. $2.25

HISTORY OF BACTERIOLOGY, William Bulloch. The only comprehensive history of bacteriology from the beginnings through the 19th century. Special emphasis is given to biography-Leeuwenhoek, etc. Brief accounts of 350 bacteriologists form a separate section. No clearer, fuller study, suitable to scientists and general readers, has yet been written. 52 illustrations. 448pp. 5⅝ x 8¼.   23761-3 Pa. $6.50

THE COMPLETE NONSENSE OF EDWARD LEAR, Edward Lear. All nonsense limericks, zany alphabets, Owl and Pussycat, songs, nonsense botany, etc., illustrated by Lear. Total of 321pp. 5⅜ x 8½. (Available in U.S. only)   20167-8 Pa. $3.95

INGENIOUS MATHEMATICAL PROBLEMS AND METHODS, Louis A. Graham. Sophisticated material from Graham Dial, applied and pure; stresses solution methods. Logic, number theory, networks, inversions, etc. 237pp. 5⅜ x 8½.   20545-2 Pa. $4.50

BEST MATHEMATICAL PUZZLES OF SAM LOYD, edited by Martin Gardner. Bizarre, original, whimsical puzzles by America's greatest puzzler. From fabulously rare Cyclopedia, including famous 14-15 puzzles, the Horse of a Different Color, 115 more. Elementary math. 150 illustrations. 167pp. 5⅜ x 8½.   20498-7 Pa. $2.75

THE BASIS OF COMBINATION IN CHESS, J. du Mont. Easy-to-follow, instructive book on elements of combination play, with chapters on each piece and every powerful combination team—two knights, bishop and knight, rook and bishop, etc. 250 diagrams. 218pp. 5⅜ x 8½. (Available in U.S. only)   23644-7 Pa. $3.50

MODERN CHESS STRATEGY, Ludek Pachman. The use of the queen, the active king, exchanges, pawn play, the center, weak squares, etc. Section on rook alone worth price of the book. Stress on the moderns. Often considered the most important book on strategy. 314pp. 5⅜ x 8½.
                     20290-9 Pa. $4.50

LASKER'S MANUAL OF CHESS, Dr. Emanuel Lasker. Great world champion offers very thorough coverage of all aspects of chess. Combinations, position play, openings, end game, aesthetics of chess, philosophy of struggle, much more. Filled with analyzed games. 390pp. 5⅜ x 8½.
                     20640-8 Pa. $5.00

500 MASTER GAMES OF CHESS, S. Tartakower, J. du Mont. Vast collection of great chess games from 1798-1938, with much material nowhere else readily available. Fully annotated, arranged by opening for easier study. 664pp. 5⅜ x 8½.   23208-5 Pa. $7.50

A GUIDE TO CHESS ENDINGS, Dr. Max Euwe, David Hooper. One of the finest modern works on chess endings. Thorough analysis of the most frequently encountered endings by former world champion. 331 examples, each with diagram. 248pp. 5⅜ x 8½.   23332-4 Pa. $3.75

SECOND PIATIGORSKY CUP, edited by Isaac Kashdan. One of the greatest tournament books ever produced in the English language. All 90 games of the 1966 tournament, annotated by players, most annotated by both players. Features Petrosian, Spassky, Fischer, Larsen, six others. 228pp. 5⅜ x 8½.   23572-6 Pa. $3.50

ENCYCLOPEDIA OF CARD TRICKS, revised and edited by Jean Hugard. How to perform over 600 card tricks, devised by the world's greatest magicians: impromptu, spelling tricks, key cards, using special packs, much, much more. Additional chapter on card technique. 66 illustrations. 402pp. 5⅜ x 8½. (Available in U.S. only)   21252-1 Pa. $4.95

MAGIC: STAGE ILLUSIONS, SPECIAL EFFECTS AND TRICK PHO-TOGRAPHY, Albert A. Hopkins, Henry R. Evans. One of the great classics; fullest, most authorative explanation of vanishing lady, levitations, scores of other great stage effects. Also small magic, automata, stunts. 446 illus-trations. 556pp. 5⅜ x 8½.   23344-8 Pa. $6.95

THE SECRETS OF HOUDINI, J. C. Cannell. Classic study of Houdini's incredible magic, exposing closely-kept professional secrets and revealing, in general terms, the whole art of stage magic. 67 illustrations. 279pp. 5⅜ x 8½.   22913-0 Pa. $4.00

HOFFMANN'S MODERN MAGIC, Professor Hoffmann. One of the best, and best-known, magicians' manuals of the past century. Hundreds of tricks from card tricks and simple sleight of hand to elaborate illusions involving construction of complicated machinery. 332 illustrations. 563pp. 5⅜ x 8½.   23623-4 Pa. $6.00

MADAME PRUNIER'S FISH COOKERY BOOK, Mme. S. B. Prunier. More than 1000 recipes from world famous Prunier's of Paris and London, specially adapted here for American kitchen. Grilled tournedos with anchovy butter, Lobster a la Bordelaise, Prunier's prized desserts, more. Glossary. 340pp. 5⅜ x 8½. (Available in U.S. only)   22679-4 Pa. $3.00

FRENCH COUNTRY COOKING FOR AMERICANS, Louis Diat. 500 easy-to-make, authentic provincial recipes compiled by former head chef at New York's Fitz-Carlton Hotel: onion soup, lamb stew, potato pie, more. 309pp. 5⅜ x 8½.   23665-X Pa. $3.95

SAUCES, FRENCH AND FAMOUS, Louis Diat. Complete book gives over 200 specific recipes: bechamel, Bordelaise, hollandaise, Cumberland, apri-cot, etc. Author was one of this century's finest chefs, originator of vichyssoise and many other dishes. Index. 156pp. 5⅜ x 8.   23663-3 Pa. $2.75

TOLL HOUSE TRIED AND TRUE RECIPES, Ruth Graves Wakefield. Authentic recipes from the famous Mass. restaurant: popovers, veal and ham loaf, Toll House baked beans, chocolate cake crumb pudding, much more. Many helpful hints. Nearly 700 recipes. Index. 376pp. 5⅜ x 8½.   23560-2 Pa. $4.50

THE CURVES OF LIFE, Theodore A. Cook. Examination of shells, leaves, horns, human body, art, etc., in *"the* classic reference on how the golden ratio applies to spirals and helices in nature . . . . "—Martin Gardner. 426 illustrations. Total of 512pp. 5⅜ x 8½.   23701-X Pa. $5.95

AN ILLUSTRATED FLORA OF THE NORTHERN UNITED STATES AND CANADA, Nathaniel L. Britton, Addison Brown. Encyclopedic work covers 4666 species, ferns on up. Everything. Full botanical information, illustration for each. This earlier edition is preferred by many to more recent revisions. 1913 edition. Over 4000 illustrations, total of 2087pp. 6⅛ x 9¼.   22642-5, 22643-3, 22644-1 Pa., Three-vol. set $25.50

MANUAL OF THE GRASSES OF THE UNITED STATES, A. S. Hitchcock, U.S. Dept. of Agriculture. The basic study of American grasses, both indigenous and escapes, cultivated and wild. Over 1400 species. Full descriptions, information. Over 1100 maps, illustrations. Total of 1051pp. 5⅜ x 8½.   22717-0, 22718-9 Pa., Two-vol. set $15.00

THE CACTACEAE,, Nathaniel L. Britton, John N. Rose. Exhaustive, definitive. Every cactus in the world. Full botanical descriptions. Thorough statement of nomenclatures, habitat, detailed finding keys. The one book needed by every cactus enthusiast. Over 1275 illustrations. Total of 1080pp. 8 x 10¼.   21191-6, 21192-4 Clothbd., Two-vol. set $35.00

AMERICAN MEDICINAL PLANTS, Charles F. Millspaugh. Full descriptions, 180 plants covered: history; physical description; methods of preparation with all chemical constituents extracted; all claimed curative or adverse effects. 180 full-page plates. Classification table. 804pp. 6½ x 9¼.   23034-1 Pa. $12.95

A MODERN HERBAL, Margaret Grieve. Much the fullest, most exact, most useful compilation of herbal material. Gigantic alphabetical encyclopedia, from aconite to zedoary, gives botanical information, medical properties, folklore, economic uses, and much else. Indispensable to serious reader. 161 illustrations. 888pp. 6½ x 9¼. (Available in U.S. only)   22798-7, 22799-5 Pa., Two-vol. set $13.00

THE HERBAL or GENERAL HISTORY OF PLANTS, John Gerard. The 1633 edition revised and enlarged by Thomas Johnson. Containing almost 2850 plant descriptions and 2705 superb illustrations, Gerard's *Herbal* is a monumental work, the book all modern English herbals are derived from, the one herbal every serious enthusiast should have in its entirety. Original editions are worth perhaps $750. 1678pp. 8½ x 12¼.   23147-X Clothbd. $50.00

MANUAL OF THE TREES OF NORTH AMERICA, Charles S. Sargent. The basic survey of every native tree and tree-like shrub, 717 species in all. Extremely full descriptions, information on habitat, growth, locales, economics, etc. Necessary to every serious tree lover. Over 100 finding keys. 783 illustrations. Total of 986pp. 5⅜ x 8½.   20277-1, 20278-X Pa., Two-vol. set $11.00

YUCATAN BEFORE AND AFTER THE CONQUEST, Diego de Landa. First English translation of basic book in Maya studies, the only significant account of Yucatan written in the early post-Conquest era. Translated by distinguished Maya scholar William Gates. Appendices, introduction, 4 maps and over 120 illustrations added by translator. 162pp. 5⅜ x 8½.
23622-6 Pa. $3.00

THE MALAY ARCHIPELAGO, Alfred R. Wallace. Spirited travel account by one of founders of modern biology. Touches on zoology, botany, ethnography, geography, and geology. 62 illustrations, maps. 515pp. 5⅜ x 8½.
20187-2 Pa. $6.95

THE DISCOVERY OF THE TOMB OF TUTANKHAMEN, Howard Carter, A. C. Mace. Accompany Carter in the thrill of discovery, as ruined passage suddenly reveals unique, untouched, fabulously rich tomb. Fascinating account, with 106 illustrations. New introduction by J. M. White. Total of 382pp. 5⅜ x 8½. (Available in U.S. only)
23500-9 Pa. $4.00

THE WORLD'S GREATEST SPEECHES, edited by Lewis Copeland and Lawrence W. Lamm. Vast collection of 278 speeches from Greeks up to present. Powerful and effective models; unique look at history. Revised to 1970. Indices. 842pp. 5⅜ x 8½.
20468-5 Pa. $8.95

THE 100 GREATEST ADVERTISEMENTS, Julian Watkins. The priceless ingredient; His master's voice; 99 44/100% pure; over 100 others. How they were written, their impact, etc. Remarkable record. 130 illustrations. 233pp. 7⅞ x 10 3/5.
20540-1 Pa. $5.95

CRUICKSHANK PRINTS FOR HAND COLORING, George Cruickshank. 18 illustrations, one side of a page, on fine-quality paper suitable for watercolors. Caricatures of people in society (c. 1820) full of trenchant wit. Very large format. 32pp. 11 x 16.
23684-6 Pa. $5.00

THIRTY-TWO COLOR POSTCARDS OF TWENTIETH-CENTURY AMERICAN ART, Whitney Museum of American Art. Reproduced in full color in postcard form are 31 art works and one shot of the museum. Calder, Hopper, Rauschenberg, others. Detachable. 16pp. 8¼ x 11.
23629-3 Pa. $3.00

MUSIC OF THE SPHERES: THE MATERIAL UNIVERSE FROM ATOM TO QUASAR SIMPLY EXPLAINED, Guy Murchie. Planets, stars, geology, atoms, radiation, relativity, quantum theory, light, antimatter, similar topics. 319 figures. 664pp. 5⅜ x 8½.
21809-0, 21810-4 Pa., Two-vol. set $11.00

EINSTEIN'S THEORY OF RELATIVITY, Max Born. Finest semi-technical account; covers Einstein, Lorentz, Minkowski, and others, with much detail, much explanation of ideas and math not readily available elsewhere on this level. For student, non-specialist. 376pp. 5⅜ x 8½.
60769-0 Pa. $4.50

CATALOGUE OF DOVER BOOKS

THE EARLY WORK OF AUBREY BEARDSLEY, Aubrey Beardsley. 157 plates, 2 in color: *Manon Lescaut, Madame Bovary, Morte Darthur, Salome,* other. Introduction by H. Marillier. 182pp. 8⅛ x 11. 21816-3 Pa. $4.50

THE LATER WORK OF AUBREY BEARDSLEY, Aubrey Beardsley. Exotic masterpieces of full maturity: *Venus and Tannhauser, Lysistrata, Rape of the Lock, Volpone,* Savoy material, etc. 174 plates, 2 in color. 186pp. 8⅛ x 11. 21817-1 Pa. $5.95

THOMAS NAST'S CHRISTMAS DRAWINGS, Thomas Nast. Almost all Christmas drawings by creator of image of Santa Claus as we know it, and one of America's foremost illustrators and political cartoonists. 66 illustrations. 3 illustrations in color on covers. 96pp. 8⅜ x 11¼. 23660-9 Pa. $3.50

THE DORÉ ILLUSTRATIONS FOR DANTE'S DIVINE COMEDY, Gustave Doré. All 135 plates from Inferno, Purgatory, Paradise; fantastic tortures, infernal landscapes, celestial wonders. Each plate with appropriate (translated) verses. 141pp. 9 x 12. 23231-X Pa. $4.50

DORÉ'S ILLUSTRATIONS FOR RABELAIS, Gustave Doré. 252 striking illustrations of *Gargantua and Pantagruel* books by foremost 19th-century illustrator. Including 60 plates, 192 delightful smaller illustrations. 153pp. 9 x 12. 23656-0 Pa. $5.00

LONDON: A PILGRIMAGE, Gustave Doré, Blanchard Jerrold. Squalor, riches, misery, beauty of mid-Victorian metropolis; 55 wonderful plates, 125 other illustrations, full social, cultural text by Jerrold. 191pp. of text. 9⅜ x 12¼. 22306-X Pa. $7.00

THE RIME OF THE ANCIENT MARINER, Gustave Doré, S. T. Coleridge. Dore's finest work, 34 plates capture moods, subtleties of poem. Full text. Introduction by Millicent Rose. 77pp. 9¼ x 12. 22305-1 Pa. $3.50

THE DORE BIBLE ILLUSTRATIONS, Gustave Doré. All wonderful, detailed plates: Adam and Eve, Flood, Babylon, Life of Jesus, etc. Brief King James text with each plate. Introduction by Millicent Rose. 241 plates. 241pp. 9 x 12. 23004-X Pa. $6.00

THE COMPLETE ENGRAVINGS, ETCHINGS AND DRYPOINTS OF ALBRECHT DURER. "Knight, Death and Devil"; "Melencolia," and more—all Dürer's known works in all three media, including 6 works formerly attributed to him. 120 plates. 235pp. 8⅜ x 11¼. 22851-7 Pa. $6.50

MECHANICK EXERCISES ON THE WHOLE ART OF PRINTING, Joseph Moxon. First complete book (1683-4) ever written about typography, a compendium of everything known about printing at the latter part of 17th century. Reprint of 2nd (1962) Oxford Univ. Press edition. 74 illustrations. Total of 550pp. 6⅛ x 9¼. 23617-X Pa. $7.95

HOLLYWOOD GLAMOUR PORTRAITS, edited by John Kobal. 145 photos capture the stars from 1926-49, the high point in portrait photography. Gable, Harlow, Bogart, Bacall, Hedy Lamarr, Marlene Dietrich, Robert Montgomery, Marlon Brando, Veronica Lake; 94 stars in all. Full background on photographers, technical aspects, much more. Total of 160pp. 8⅜ x 11¼.   23352-9 Pa. $6.00

THE NEW YORK STAGE: FAMOUS PRODUCTIONS IN PHOTO-GRAPHS, edited by Stanley Appelbaum. 148 photographs from Museum of City of New York show 142 plays, 1883-1939. *Peter Pan, The Front Page, Dead End, Our Town,* O'Neill, hundreds of actors and actresses, etc. Full indexes. 154pp. 9½ x 10.   23241-7 Pa. $6.00

DIALOGUES CONCERNING TWO NEW SCIENCES, Galileo Galilei. Encompassing 30 years of experiment and thought, these dialogues deal with geometric demonstrations of fracture of solid bodies, cohesion, leverage, speed of light and sound, pendulums, falling bodies, accelerated motion, etc. 300pp. 5⅜ x 8½.   60099-8 Pa. $4.00

THE GREAT OPERA STARS IN HISTORIC PHOTOGRAPHS, edited by James Camner. 343 portraits from the 1850s to the 1940s: Tamburini, Mario, Caliapin, Jeritza, Melchior, Melba, Patti, Pinza, Schipa, Caruso, Farrar, Steber, Gobbi, and many more—270 performers in all. Index. 199pp. 8⅜ x 11¼.   23575-0 Pa. $7.50

J. S. BACH, Albert Schweitzer. Great full-length study of Bach, life, background to music, music, by foremost modern scholar. Ernest Newman translation. 650 musical examples. Total of 928pp. 5⅜ x 8½. (Available in U.S. only)   21631-4, 21632-2 Pa., Two-vol. set $11.00

COMPLETE PIANO SONATAS, Ludwig van Beethoven. All sonatas in the fine Schenker edition, with fingering, analytical material. One of best modern editions. Total of 615pp. 9 x 12. (Available in U.S. only)
                          23134-8, 23135-6 Pa., Two-vol. set $15.50

KEYBOARD MUSIC, J. S. Bach. Bach-Gesellschaft edition. For harpsichord, piano, other keyboard instruments. English Suites, French Suites, Six Partitas, Goldberg Variations, Two-Part Inventions, Three-Part Sinfonias. 312pp. 8⅛ x 11. (Available in U.S. only)   22360-4 Pa. $6.95

FOUR SYMPHONIES IN FULL SCORE, Franz Schubert. Schubert's four most popular symphonies: No. 4 in C Minor ("Tragic"); No. 5 in B-flat Major; No. 8 in B Minor ("Unfinished"); No. 9 in C Major ("Great"). Breitkopf & Hartel edition. Study score. 261pp. 9⅜ x 12¼.
                                                        23681-1 Pa. $6.50

THE AUTHENTIC GILBERT & SULLIVAN SONGBOOK, W. S. Gilbert, A. S. Sullivan. Largest selection available; 92 songs, uncut, original keys, in piano rendering approved by Sullivan. Favorites and lesser-known fine numbers. Edited with plot synopses by James Spero. 3 illustrations. 399pp. 9 x 12.   23482-7 Pa. $9.95

A MAYA GRAMMAR, Alfred M. Tozzer. Practical, useful English-language grammar by the Harvard anthropologist who was one of the three greatest American scholars in the area of Maya culture. Phonetics, grammatical processes, syntax, more. 301pp. 5⅜ x 8½.   23465-7 Pa. $4.00

THE JOURNAL OF HENRY D. THOREAU, edited by Bradford Torrey, F. H. Allen. Complete reprinting of 14 volumes, 1837-61, over two million words; the sourcebooks for *Walden*, etc. Definitive. All original sketches, plus 75 photographs. Introduction by Walter Harding. Total of 1804pp. 8½ x 12¼.   20312-3, 20313-1 Clothbd., Two-vol. set $70.00

CLASSIC GHOST STORIES, Charles Dickens and others. 18 wonderful stories you've wanted to reread: "The Monkey's Paw," "The House and the Brain," "The Upper Berth," "The Signalman," "Dracula's Guest," "The Tapestried Chamber," etc. Dickens, Scott, Mary Shelley, Stoker, etc. 330pp. 5⅜ x 8½.   20735-8 Pa. **$4.50**

SEVEN SCIENCE FICTION NOVELS, H. G. Wells. Full novels. *First Men in the Moon, Island of Dr. Moreau, War of the Worlds, Food of the Gods, Invisible Man, Time Machine, In the Days of the Comet.* A basic science-fiction library. 1015pp. 5⅜ x 8½. (Available in U.S. only)
                     20264-X Clothbd. $8.95

ARMADALE, Wilkie Collins. Third great mystery novel by the author of *The Woman in White* and *The Moonstone*. Ingeniously plotted narrative shows an exceptional command of character, incident and mood. Original magazine version with 40 illustrations. 597pp. 5⅜ x 8½.
                     23429-0 Pa. $6.00

MASTERS OF MYSTERY, H. Douglas Thomson. The first book in English (1931) devoted to history and aesthetics of detective story. Poe, Doyle, LeFanu, Dickens, many others, up to 1930. New introduction and notes by E. F. Bleiler. 288pp. 5⅜ x 8½. (Available in U.S. only)
                     23606-4 Pa. $4.00

FLATLAND, E. A. Abbott. Science-fiction classic explores life of 2-D being in 3-D world. Read also as introduction to thought about hyperspace. Introduction by Banesh Hoffmann. 16 illustrations. 103pp. 5⅜ x 8½.
                     20001-9 Pa. $2.00

THREE SUPERNATURAL NOVELS OF THE VICTORIAN PERIOD, edited, with an introduction, by E. F. Bleiler. Reprinted complete and unabridged, three great classics of the supernatural: *The Haunted Hotel* by Wilkie Collins, *The Haunted House at Latchford* by Mrs. J. H. Riddell, and *The Lost Stradivarious* by J. Meade Falkner. 325pp. 5⅜ x 8½.
                     22571-2 Pa. $4.00

AYESHA: THE RETURN OF "SHE," H. Rider Haggard. Virtuoso sequel featuring the great mythic creation, Ayesha, in an adventure that is fully as good as the first book, *She*. Original magazine version, with 47 original illustrations by Maurice Greiffenhagen. 189pp. 6½ x 9¼.
                     23649-8 Pa. $3.50

UNCLE SILAS, J. Sheridan LeFanu. Victorian Gothic mystery novel, considered by many best of period, even better than Collins or Dickens. Wonderful psychological terror. Introduction by Frederick Shroyer. 436pp. 5⅜ x 8½. 21715-9 Pa. $6.00

JURGEN, James Branch Cabell. The great erotic fantasy of the 1920's that delighted thousands, shocked thousands more. Full final text, Lane edition with 13 plates by Frank Pape. 346pp. 5⅜ x 8½. 23507-6 Pa. $4.50

THE CLAVERINGS, Anthony Trollope. Major novel, chronicling aspects of British Victorian society, personalities. Reprint of Cornhill serialization, 16 plates by M. Edwards; first reprint of full text. Introduction by Norman Donaldson. 412pp. 5⅜ x 8½. 23464-9 Pa. $5.00

KEPT IN THE DARK, Anthony Trollope. Unusual short novel about Victorian morality and abnormal psychology by the great English author. Probably the first American publication. Frontispiece by Sir John Millais. 92pp. 6½ x 9¼. 23609-9 Pa. $2.50

RALPH THE HEIR, Anthony Trollope. Forgotten tale of illegitimacy, inheritance. Master novel of Trollope's later years. Victorian country estates, clubs, Parliament, fox hunting, world of fully realized characters. Reprint of 1871 edition. 12 illustrations by F. A. Faser. 434pp. of text. 5⅜ x 8½. 23642-0 Pa. $5.00

YEKL and THE IMPORTED BRIDEGROOM AND OTHER STORIES OF THE NEW YORK GHETTO, Abraham Cahan. Film *Hester Street* based on *Yekl* (1896). Novel, other stories among first about Jewish immigrants of N.Y.'s East Side. Highly praised by W. D. Howells—Cahan "a new star of realism." New introduction by Bernard G. Richards. 240pp. 5⅜ x 8½. 22427-9 Pa. $3.50

THE HIGH PLACE, James Branch Cabell. Great fantasy writer's enchanting comedy of disenchantment set in 18th-century France. Considered by some critics to be even better than his famous *Jurgen*. 10 illustrations and numerous vignettes by noted fantasy artist Frank C. Pape. 320pp. 5⅜ x 8½. 23670-6 Pa. $4.00

ALICE'S ADVENTURES UNDER GROUND, Lewis Carroll. Facsimile of ms. Carroll gave Alice Liddell in 1864. Different in many ways from final Alice. Handlettered, illustrated by Carroll. Introduction by Martin Gardner. 128pp. 5⅜ x 8½. 21482-6 Pa. $2.50

FAVORITE ANDREW LANG FAIRY TALE BOOKS IN MANY COLORS, Andrew Lang. The four Lang favorites in a boxed set—the complete *Red, Green, Yellow* and *Blue* Fairy Books. 164 stories; 439 illustrations by Lancelot Speed, Henry Ford and G. P. Jacomb Hood. Total of about 1500pp. 5⅜ x 8½. 23407-X Boxed set, Pa. $15.95

CATALOGUE OF DOVER BOOKS

AMERICAN ANTIQUE FURNITURE, Edgar G. Miller, Jr. The basic coverage of all American furniture before 1840: chapters per item chronologically cover all types of furniture, with more than 2100 photos. Total of 1106pp. 7⅞ x 10¾. 21599-7, 21600-4 Pa., Two-vol. set $17.90

ILLUSTRATED GUIDE TO SHAKER FURNITURE, Robert Meader. Director, Shaker Museum, Old Chatham, presents up-to-date coverage of all furniture and appurtenances, with much on local styles not available elsewhere. 235 photos. 146pp. 9 x 12. 22819-3 Pa. $6.00

ORIENTAL RUGS, ANTIQUE AND MODERN, Walter A. Hawley. Persia, Turkey, Caucasus, Central Asia, China, other traditions. Best general survey of all aspects: styles and periods, manufacture, uses, symbols and their interpretation, and identification. 96 illustrations, 11 in color. 320pp. 6⅛ x 9¼. 22366-3 Pa. $6.95

CHINESE POTTERY AND PORCELAIN, R. L. Hobson. Detailed descriptions and analyses by former Keeper of the Department of Oriental Antiquities and Ethnography at the British Museum. Covers hundreds of pieces from primitive times to 1915. Still the standard text for most periods. 136 plates, 40 in full color. Total of 750pp. 5⅝ x 8½. 23253-0 Pa. $10.00

THE WARES OF THE MING DYNASTY, R. L. Hobson. Foremost scholar examines and illustrates many varieties of Ming (1368-1644). Famous blue and white, polychrome, lesser-known styles and shapes. 117 illustrations, 9 full color, of outstanding pieces. Total of 263pp. 6⅛ x 9¼. (Available in U.S. only) 23652-8 Pa. $6.00

Prices subject to change without notice.

Available at your book dealer or write for free catalogue to Dept. GI, Dover Publications, Inc., 180 Varick St., N.Y., N.Y. 10014. Dover publishes more than 175 books each year on science, elementary and advanced mathematics, biology, music, art, literary history, social sciences and other areas.